Algorithms for Intelligent Systems

This book series publishes research on the analysis and development of algorithms for intelligent systems with their applications to various real world problems. It covers research related to autonomous agents, multi-agent systems, behavioral modeling, reinforcement learning, game theory, mechanism design, machine learning, meta-heuristic search, optimization, planning and scheduling, artificial neural networks, evolutionary computation, swarm intelligence and other algorithms for intelligent systems.

The book series includes recent advancements, modification and applications of the artificial neural networks, evolutionary computation, swarm intelligence, artificial immune systems, fuzzy system, autonomous and multi agent systems, machine learning and other intelligent systems related areas. The material will be beneficial for the graduate students, post-graduate students as well as the researchers who want a broader view of advances in algorithms for intelligent systems. The contents will also be useful to the researchers from other fields who have no knowledge of the power of intelligent systems, e.g. the researchers in the field of bioinformatics, biochemists, mechanical and chemical engineers, economists, musicians and medical practitioners.

The series publishes monographs, edited volumes, advanced textbooks and selected proceedings.

More information about this series at http://www.springer.com/series/16171

Aboul Ella Hassanien · Ashraf Darwish ·
Sherine M. Abd El-Kader ·
Dabiah Ahmed Alboaneen
Editors

Enabling Machine Learning Applications in Data Science

Proceedings of Arab Conference for Emerging Technologies 2020

Springer

Editors
Aboul Ella Hassanien
Faculty of Computers and Artificial Intelligence
Cairo University
Giza, Egypt

Sherine M. Abd El-Kader
Electronics Research Institute
Cairo, Egypt

Ashraf Darwish
Faculty of Science
Helwan University
Cairo, Egypt

Dabiah Ahmed Alboaneen
College of Science and Humanities
Imam Abdulrahman Bin Faisal University
Jubail, Saudi Arabia

ISSN 2524-7565 ISSN 2524-7573 (electronic)
Algorithms for Intelligent Systems
ISBN 978-981-33-6128-7 ISBN 978-981-33-6129-4 (eBook)
https://doi.org/10.1007/978-981-33-6129-4

This Springer imprint is published by the registered company Springer Nature Singapore Pte Ltd.
The registered company address is: 152 Beach Road, #21-01/04 Gateway East, Singapore 189721, Singapore

Preface

Today, it is all about the connected smarter world based on the new emerging technologies and artificial intelligence (AI) for sustainable development. Sustainable development purposes have been designed for meeting the human development goals while simultaneously sustaining the ability of natural resources in order to maintain the natural resources. AI and emerging technologies are currently used with many areas of applications in most sectors of the states. They can help humankind in achieving sustainability through technology for a more sustainable future for all.

This volume constitutes the refereed proceedings of the 1st International Conference on Enabling Machine Learning Applications in Data Science: Innovation Vision and Approach, which took place in Cairo, Egypt, during July 19–21, 2020, and due to the COVID-19 pandemic, the conference had been held online. This international interdisciplinary conference is covering research and development in the field of intelligent systems and applications. In response to the call for papers for this conference, 87 research papers had been submitted to the conference. Among these papers, 28 papers were selected for inclusion in the conference proceeding. The papers were evaluated and ranked on the basis of their significance, novelty, and technical quality by at least two reviewers for each paper. The topics of these papers are covering research and development in the field of artificial intelligence and emerging technologies.

The first edition of this conference is organized by the Scientific Research Group in Egypt (SRGE). It had been held to provide an international forum that brings together those who are actively involved in the areas of interest, to report on up-to-the-date innovations and developments, to summarize the state of the art, and to exchange ideas and advances in all aspects of artificial intelligence. The conference proceeding has three major tracks. Track (I) presents the AI and machine learning applications in different applications and disciplines. Nowadays, deep learning technology is widely used in many areas of applications. Therefore, Track (II) presents deep learning applications. Track (III) covers the applications of blockchain technology with the social and smart networks applications. This proceeding aims at emphasizing the role and recent developments in the field of AI and emerging technologies and related technologies with a special focus on sustainable development purposed in the Arab world. This volume targets high-quality scientific research papers with applications, including theory, practical, prototypes, new ideas, case studies, and surveys, which cover all aspects of emerging technologies and AI for sustainable development goals.

All submissions were reviewed by three reviewers on average, with no distinction between papers submitted for all conference tracks. We are convinced that the quality and diversity of the topics covered will satisfy both the attendees and the readers of these conference proceedings. We express our sincere thanks to the plenary speakers and international program committee members for helping us to formulate a rich technical program. We would like to extend our sincere appreciation for the outstanding work contributed over many months by the organizing committee: local organization chair and publicity chair. We also wish to express our appreciation to the SRGE members for their assistance. We would like to emphasize that the success of this conference would not have been possible without the support of many committed volunteers who generously contributed their time, expertise, and resources toward making the conference an unqualified success. Finally, thanks to Springer team for their support in all stages of the production of the proceedings. We hope that you will enjoy the conference program.

Giza, Egypt — Aboul Ella Hassanien
Cairo, Egypt — Ashraf Darwish
Cairo, Egypt — Sherine M. Abd El-Kader
Jubail, Saudi Arabia — Dabiah Ahmed Alboaneen

Contents

Part II Deep Learning Applications

Part III Blockchain Technology, Social and Smart Networks Applications

About the Editors

Dr. Aboul Ella Hassanien is Founder and Head of the Egyptian Scientific Research Group (SRGE) and Professor of Information Technology at the Faculty of Computer and Information, Cairo University. Professor Hassanien is Ex-Dean of the faculty of computers and information, Beni Suef University. Professor Hassanien has more than 800 scientific research papers published in prestigious international journals and over 40 books covering such diverse topics as data mining, medical images, intelligent systems, social networks and smart environment. Professor Hassanien won several awards including the Best Researcher of the Youth Award of Astronomy and Geophysics of the National Research Institute, Academy of Scientific Research (Egypt, 1990). He was also granted a scientific excellence award in humanities from the University of Kuwait for the 2004 Award and received the superiority of scientific—University Award (Cairo University, 2013). Also he honored in Egypt as the best researcher in Cairo University in 2013. He also received the Islamic Educational, Scientific and Cultural Organization (ISESCO) prize on Technology (2014) and received the state Award for excellence in engineering sciences 2015. He was awarded the medal of Sciences and Arts of the first class by President of the Arab Republic of Egypt, 2017.

Prof. Ashraf Darwish has bachelor's and master's degrees in mathematics from Tanta University in Egypt. He received his Ph.D. in computer science from computer science department at Saint Petersburg State University in 2006 from Russian Federation. He has worked as Associate Professor and then Professor of computer science at the mathematics and computer science department, faculty of science, Helwan University in Cairo. Currently, he is Adjunct Professor of computer science at ConTech University, CA, USA. Prior to this, he was Assistant Professor in the same department. From 2017 to 2018, he worked as Acting Department Chair. In 2014, he received the prestigious research distinguished award in computer science and information technology. Professor Darwish research interests span both computer science and information technology. Much of his work has been focusing on artificial intelligence, mainly through the application of data mining, machine learning, deep learning and robotics in different areas of research.

Dr. Sherine M. Abd El-Kader received the M.Sc. degree from the Electronics and Communications Department, Faculty of Engineering, Cairo University, in 1998, and the Ph.D. degree from the Computers Department, Faculty of Engineering, Cairo University, in 2003. She is currently Acting Vice President of Electronics Research Institute (ERI). She is Head of the Computers and Systems Department, ERI, since 2018. She is also Head of the Technology, Innovation and Commercialization Office (TICO), since 2018. She is a member of the technical evaluation office and performance evaluation of the president of the Academy of Scientific Research and Technology (ASRT) since 2019. She supervised more than 20 M.Sc. and Ph.D. students. She has published more than 50 papers and 9 book chapters in computer networking area and Editor for two books in 5G and Precision Agriculture Technologies for Food Security and Sustainability. She is working in many computer networking hot topics such as: the IoT, 5G, cognitive radio, Wi-MAX, Wi-Fi, IP mobility, QoS, wireless sensors networks, ad hoc networking, real-time traffics, vehicular networks biomedical and localization algorithms. She is also Technical Reviewer for many international journals. She had the first best researcher award in Electronics Research Institute in 2020.

Dr. Dabiah Ahmed Alboaneen has received Ph.D. in Cloud Computing and Artificial Intelligence from Glasgow Caledonian University (GCU) in 2019 and M.Sc. in Advanced Computer Networking with distinction from the GCU in 2013. In 2006, she joined the department of computer science, University of Imam Abdulrahman Bin Faisal. Currently, Dr. Dabiah is Assistant Professor and Head of the computer science department at the College of Science and Humanities in Jubail.

Part I
Machine Learning and Intelligent Systems Applications

Chapter 1
Who Is Typing? Automatic Gender Recognition from Interactive Textual Chats Using Typing Behaviour

Abeer Buker and Alessandro Vinciarelli

1 Introduction

Gender plays a fundamental role in human–human interaction. The availability of nonverbal behavioural cues such as facial expressions, eye gaze, and gestures in face-to-face interaction allows one to easily recognise the gender of others. In addition, all cultures provide further cues to distinguish between genders, like different clothes, hairstyles, etc. In social psychology, such phenomenon has been extensively investigated in the case of face-to-face interaction. However, when it comes to computer-mediated technologies (e.g. live chats), there is an element of uncertainty regarding whether such mediums comprise nonverbal cues. In the case where this is true, it would be unclear whether such nonverbal cues convey information about the characteristics and identity of individuals. Therefore, the goal of this work is to investigate the interplay between gender and typing behaviour through the analysis of interactive textual chats.

The widespread use of the Internet has provided many ways for communicating information using various social media platforms like Whatsapp, Facebook, and Twitter. In such mediums that mostly depend on textual communication, the absence of physical and cultural cues makes the task of recognising the gender a challenging one. This has promoted different types of misuse in hyperspaces including terrorism, catfishing, and child predators. According to the 2018 Annual Report of Internet Crime Complaint Centre (IC3) [1], from 2014 to 2018, IC3 received a total of 1,509,679 complaints regarding different types of Internet crimes in various coun-

A. Buker (✉) · A. Vinciarelli
Glasgow University, Glasgow G12 8QQ, UK
e-mail: a.buker.1@research.gla.ac.uk; alessandro.vinciarelli@glasgow.ac.uk

A. Vinciarelli
e-mail: aabuker@iau.edu.sa

A. Buker
Imam Abdulrahman Bin Faisal University, Dammam 34212, Saudi Arabia

A. E. Hassanien et al. (eds.), *Enabling Machine Learning Applications in Data Science*, Algorithms for Intelligent Systems, https://doi.org/10.1007/978-981-33-6129-4_1

tries. The number of victims in crimes related to confidence fraud/romance, identity thefts, and child abuse is 18,493, 16,128, and 1394, respectively. Moreover, around 40,198 of the criminals used social media platforms to facilitate crimes.

The remaining parts of the paper proceed as follows: Sect. 2 surveys related work, Sect. 3 describes the data, Sect. 4 illustrates the approach, Sect. 5 reports on results and discusses the findings, and Sect. 6 draws some conclusions.

2 Background and Related Work

Identifying the gender of the author of a written piece of text has been a topic of interest for researchers since texts were only available in a written format. Recently, the proliferation of social media has made available a rich amount of information and self-disclosed data concerning the author's identity and social factors (e.g. gender) which can be used for research and experimental purposes. Gender recognition is typically treated as a binary classification problem where an anonymous text is assigned to one of the two following classes: female and male. The growing body of the literature in this area suggests a common approach where the features accounting for gender are extracted from the input data. Then, the resulting feature vector is fed into a machine learning algorithm which is being trained and tested to learn make decisions about a particular class.

Alsmearat et al. [2] conducted a gender recognition experiment on Arabic articles by comparing stylometric features and content-based features using different machine learning algorithms. They concluded that style-based features are more accurate in gender recognition than content-based ones. Cheng et al. [3] explored the gender recognition problem from a psycho-linguistic point of view to investigate whether men and women use language differently. The experimental results indicated that certain linguistic cues, such as function words, are significant gender discriminators. Bayot and Gonçalves [4] applied deep learning methods for gender recognition using PAN 2018 dataset, which consists of Twitter data in three languages. Their approach mainly depends on word embedding and long short-term memory units (LTSM) to generate feature representations suitable for classification at run time. Such a technique is the new trend in author profiling tasks. However, the obtained accuracy compared to traditional machine learning approaches was low. The researchers suggested that the use of stylometric features may enhance the performance of the model. Interestingly, this was the approach adopted by Bsir and Zrigui [5] who used a combination of stylometric features with word embedding to predict gender from Facebook and Twitter. They obtained 62% and 79% accuracy for Facebook and Twitter corpus, respectively. Their results confirmed the assumption made by other researchers that a combination of stylometric features, and word embedding may optimise the performance of the deep learning models for gender recognition tasks.

Although extensive research has been carried out to understand humans' behaviour and automatically inferring their gender from various forms of electronic discourses, they were nearly entirely concerned with the text itself. Written text transmits more than just words and language. It has been suggested that as individuals are typing, they unconsciously produce nonverbal behavioural cues which can manifest in their typing behaviour [6], e.g. the speed of typing is linked to the author emotional state [7]. Consequently, nonverbal cues are perceived as an authentic, truthful, and candour form of nonverbal behaviours [8] which leaks information related to the actual state of an individual instead of what that individual wants to appear like [9]. Typing behaviour (also known as keystroke dynamics) is the line of research concerning how people are typing instead of what are they typing. It studies the unique timing pattern of the typing process unfolding real-time, fine-grained information on the author of the texts [10], leveraging the nonverbal behavioural aspect of the discourse. Typing behaviour is a branch of behavioural biometrics that studies technologies for written verification and identification. For such a reason, gender and age were the most researched characteristics because these two are the first information one might be interested to know about an unknown individual [11].

Tsimperidis et al. [12] used keystroke duration to determine the feasibility of predicting writers' gender from their typing behaviour. One limitation of this study is the fact that the researchers used a fixed-text approach where participants used a predefined text to create their profile, which might have contributed to the promising results they have produced (in the region of 70% accuracy). Following their previous study, Tsimperidis et al. [11] carried out another experiment for gender recognition from keystrokes using free-text approach. In their recent work, they categorised features in two sets, temporal and non-temporal features. The temporal features include the time a key was pressed and the interval between two consecutive keys, while the non-temporal ones include typing speed, frequency of errors, and pressure applied to keys. This study is considered one of the prominent ones in keystroke dynamics as it yielded the highest accuracy in the field (95%) using the proposed feature set for gender recognition. One drawback of this work is the large number of features used for classification (350 features) resulting in long training time and a high computational cost. Overall, except for [11], a common limitation across the research in typing behaviour to date is that most of them have reported accuracies that are not satisfactory, suggesting that typing behaviour studies are still in their early phases and more work is anticipated to be carried out in the future.

This work is one of the first studies that adopt typing behaviour for gender recognition. From an application point of view, the use of typing behaviour for user's authentication has the advantage of providing continuous verification of the user's identity even after the user has successfully logged to the system [13]. Furthermore, the non-intrusive typing behaviour methods are relatively inexpensive; they do not rely on special hardware apart from the keyboard.

3 The Data

The experiments were performed over a collection of dyadic chats where conversations are synchronous (i.e. such as in WhatsApp where interlocutors exchange messages simultaneously) and involved unacquainted individuals. The corpus contains 30 conversations for a total of 60 participants.

Out of the 60 participants, 35 are females, and 25 are males. Participants were randomly grouped into pairs (63% same-gender pairs and 37% mixed-gender pairs). The topic of the conversations was based on the Winter Survival Task (WST), a frequently used scenario in social psychology experiments. During the experiments, participants were required to consider 12 items and to decide the ones that are most likely to increase their chances of surviving a plane crash in Northern Canada [14]. The main benefit of adopting the WST is that most people do not have much experience regarding what has to be done in such a situation, and hence, the outcome of the discussion is not based on the actual knowledge of the problem but on social and psychological dynamics.

The data was collected using a key-logging platform where every key pressed has been time-stamped. Therefore, a chat can be considered as a sequence of keys that have been pressed by participants at different times. Overall, the data compromises a total of 191,410 keys, 119,639 authored by females, and 71,771 authored by males. The upper chart of Fig. 1 shows the distribution of keys across participants. During the chats, when participants write something, the "Enter" key is pressed to send the written text to their interlocutor. The chats can then be segmented into chunks where each chunk is a sequence of keys delimited by enter keys. Thus, a chat can be considered as a set of chunks $C = \{c_1, c_2, \ldots, c_l\}$ where each chunk is represented by $c_i = \{(k_j, t_j), j = 1, 2, \ldots, m_i\}$ where k_j is the jth key pressed, t_j is the timestamp of that key and m_i is the total number of keys in chunk i. One reason underlying such a choice is that chunks are considered to be semantically meaningful units. In total, there are 3177 chunks, 1128 authored by males and 2049 authored by females, and they are the analysis unit of the experiments. The lower chart of Fig. 1 shows the chunk distribution across the participants.

4 The Approach

Two main steps underlie the proposed approach to the experiments of this work. The first one is feature extraction which involves converting each chunk into a feature vector. The second step is classification where each chunk is classified in terms of gender, followed by aggregation on the chunk level to acquire the gender of a given participant. The following sections describe in greater detail these two steps.

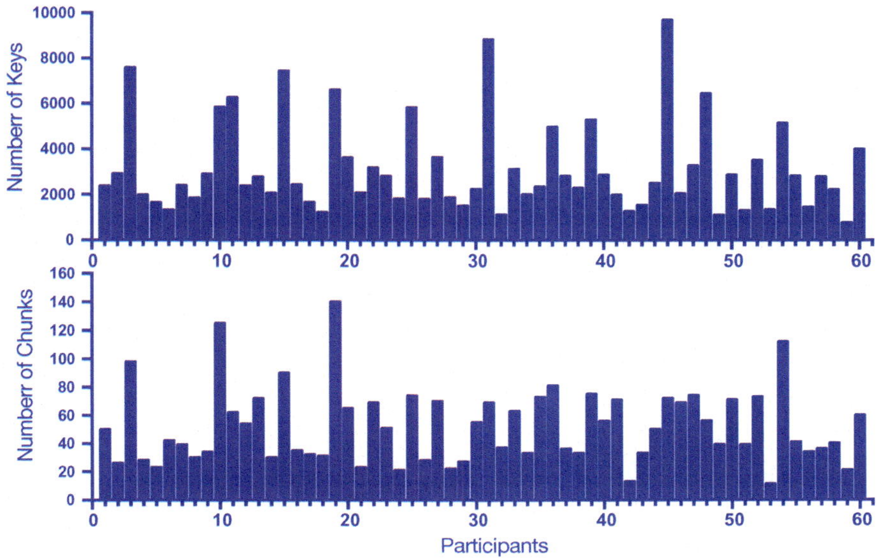

Fig. 1 Distribution of chunks and keys over participants. The upper chart shows the number of keys per participant. The lower chart shows the number of chunks per participant

4.1 Feature Extraction

The feature set includes features that were proposed in the extensive psychological literature aimed at the analysis of chats (see Table 2). In addition to these features, the experiments consider timing behaviour that can be accessed only in the data at disposition because they have been captured with a key-logging platform (every key has been time-stamped). The proposed feature set does not take into consideration what the participant has typed (the content of the conversation), but how the participant has typed it (the typing behaviour). Two main reasons underlying such a choice. The first is to preserve the participants' privacy. The second one is that nonverbal cues are more likely to account for honest information, i.e. to reflect the actual attitude of the participants in relation to social and psychological phenomena [9].

The features have been designed to capture four dimensions. The first, called "Social Presence", aims at assessing the tendency to include more or less content (in terms of the amount of characters) in a chunk. This is important because the tendency to fragment a message into chunks of different length accounts for the way the participants manage their social presence [15]. In other words, the shorter the chunks are, and the more frequently the enter key is pressed, the more a participant is attempting to show that they are socially present.

The second category called "Cognitive Load" which measures the need of thinking about what to write. The notion behind cognitive load is that the performance of a task is linked to the capacity of the working memory to perform cognitive processing and the cognitive demand of the task [16]. The literature has established a link between typing behaviour and cognitive load [17], for instance, once a task demand increases, typing speed decreases [18], more error might be made, and hence, the use of backspace for deletion increases [17].

The third category is "Affect" accounts for the emotions an individual might experience during typing. Typing speed and median latency capture the implicit expression of effect because they are known to interplay with cognitive load, thus leading people to type slower or faster [19]. On the contrary, the density of using emoticons, exclamation marks, and uppercase tokens are known to measure the explicit expression of emotions in written texts [20–23].

The last category which is the "Style" aims at quantifying the writer style as they are inspired from the field of authorship analysis. The literature suggests that the use of non-alphabetical characters, punctuation, uppercase letters, and question marks may help in identifying the writer of a text [3, 24–26].

Using the dataset at disposition, every extracted chunk was converted into a feature vector of dimension D = 15. Once the features were extracted, a chat can then be thought of as a sequence of feature vectors $\mathbf{x} = (\mathbf{x_1}, \ldots, \mathbf{x_i}, \ldots, \mathbf{x_c})$ where $\mathbf{x_i}$ has been extracted from the ith chunk and $c = 3177$ is the total number of chunks.

4.2 Classification

The goal of the classification is to assign the features extracted from each chunk to one of the two following classes: female and male. Such a task was performed using random forests (RFs). The primary challenge in the data at hand is that classes are imbalanced (the number of female participants is larger), and most conventional machine learning algorithms assume balanced data distributions among classes [27]. The primary advantage of random forests is that they allow one to adjust the weights of the two classes by assigning a higher weight to the minority class (i.e. higher misclassification cost), hence reducing the classifier bias towards the majority class. The training process was repeated 10 times, and results presented in Table 1 are the average performance over the ten repetitions. However, due to the limited variance, the ten repetitions were considered to be sufficient, and no further training was needed.

Once the chunks were classified in terms of gender, the results obtained were aggregated at the chunk level to acquire the gender of a given participant. That is because the gender of a particular participant does not change across the chunks they authored. Therefore, participants are assigned the class that has been assigned to the majority of their chunks. Meaning that, if most of the chunks authored by participant *S* have been classified as male, then *S* will be assigned to class male. The performances are reported in terms of chunk level and participan level according to the majority vote heuristic.

Table 1 Classification results

Level	Baseline (%)	Accuracy (%)	Precision (%)	Recall (%)
Chunk	54.5	85.0 ± 0.2	86.1	91.9
participant	51.4	98.8 ± 1.2	97.2	100

The table reports the average performances for gender recognition with ten repetitions and randomly shuffled data. The performance evaluation measures are the accuracy with their respective standard deviations across ten repetitions, precision, and recall. The participant-level classification is performed by applying the majority vote to the individual chunks of a given participant. The table also presents the accuracy of the random classifier according to the prior probability

In the corpus, the prior probability of female and male participants is $p_f = 58.3\%$ and $p_m = 41.7\%$, respectively. Therefore, the baseline approach adopted for comparison is subject to the following rule:

$$p_c = {p^2}_f + {p^2}_m \tag{1}$$

The resulting baseline accuracy is 54.5% which is the class that corresponds to the highest a priori probability. The developed classification model will then be considered to work better than chance if its accuracy is higher than 54.5% to a statistically significant extent.

5 Results and Discussion

The experiments aimed at two primary goals. The first is to automatically infer the gender of a chat participant, and the second is to identify gender markers in typing behaviour. The rest of this section presents and discusses the results of the experiments.

5.1 Classification Results

Table 1 reports the results of the classification for gender recognition task from typing behaviour in terms of accuracy, precision, and recall. The difference is statistically significant on the participant level and chunk level compared to the baseline (see Sect. 4.2). The results indicate that the developed classifier performed better than chance in terms of gender recognition task. However, it is interesting to note that the participant-level performance is better than the chunk-level one which can be explained by the fact that even if the majority vote at the participant-level leads to the correct gender, that does not necessarily mean that all the chunks written by that particular participant were correctly classified.

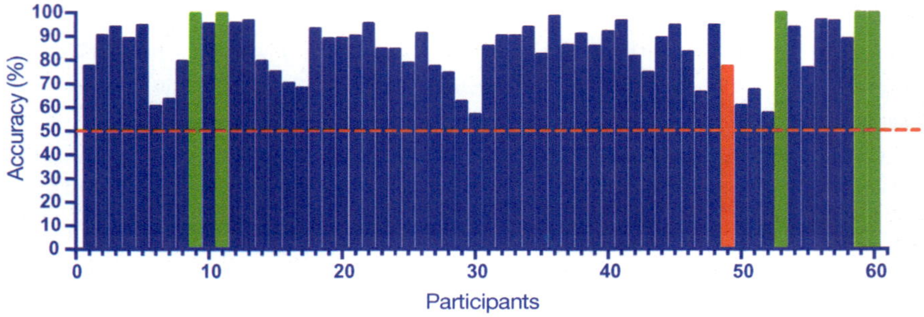

Fig. 2 Classification results. The graph shows for each participant the percentage of the classified chunks that lead to the decision. The red bar identifies the participant who was misclassified, whereas the green bars indicate the cases where all the chunks belonging to a certain participant were correctly classified

Figure 2 elaborates on that concept further showing for each participant how confident the classifier was in making the decision. For the vast majority of the participants, the chunk classification was correct in most cases (over the borderline assuming that the borderline is 50%). In fact, for some participants, the classifier was able to classify all the chunks written by that participant correctly. The only misclassified participant is participant (49) which was classified as a female, whereas the real gender of that participant is a male.

5.2 *Identification of Gender Makers*

One of the key advantages of random forests is its ability to evaluate the importance of the features on a giving classification task using Gini importance, i.e. measures how effectively a feature can split the dataset into subsets of lower impurity. In the case of binary classification problems, the maximum Gini impurity is 0.5, and as it can be seen from Fig. 3, this corresponds to a reduction in impurity by 26%.

However, the literature suggests that the use and interpretation of Gini importance require caution as such a metric is biased in favour of features with many possible split points [28]. For such a reason, Table 2 presents for each of the features whether there are statistically significant differences in typing behaviour between females and males. T-test and X^2 were used depending on the type of the feature being assessed (i.e. continuous or count). A difference is considered statistically significant when p-value < 0.05. Whenever the T-test results in a statistically significant output, an adjustment was carried out based on the suggestion made by [29]. The top 1% of outliers were removed from the data before carrying out the tests to avoid any influences these outliers might have. False discovery rate(FDR) correction was performed to limit the possibility of some of the results being statistically significant by chance.

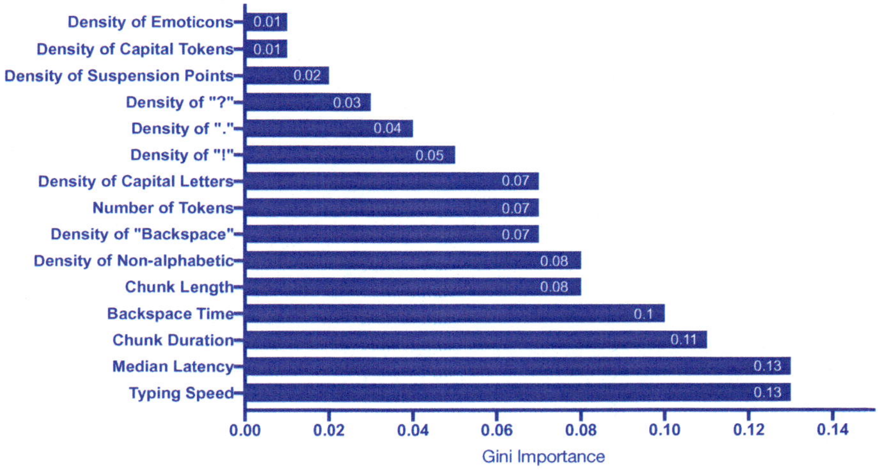

Fig. 3 Feature importance. The length of the bars is proportional to the average Gini importance of the features

Table 2 Feature analysis results with T-test adjustment and FDR correction

#	Feature	Category	Female	Male	References
F01	Length of chunks	Social presence	(↓)	(↑)	[15, 19, 20]
F02	Duration of each chunk	Social presence	(↓)	(↑)	[15, 19, 20]
F03	Number of tokens	Social presence	–	–	[15, 19]
F04	Density of “Backspaces”	Cognitive load	(↓)	(↑)	[30]
F05	Backspace time	Cognitive load	(↓)	(↑)	[30]
F06	Typing speed	Cognitive load, affect	(↑)	(↓)	[18]
F07	Median of latency time	Cognitive load, affect	(↓)	(↑)	[30]
F08	Density of “!”	Affect	–	–	[20]
F09	Density of emoticon	Affect	–	–	[20]
F10	Density of uppercase tokens	Affect	–	–	[21–23]
F11	Density of capital letters	Style	–	–	[20]
F12	Density of “?”	Style	–	–	[20]
F13	Density of “.”	Style	–	–	[20]
F14	Density of suspension points	Style	(↑)	(↓)	[31]
F15	Density of non-alphabetic characters	Style	–	–	[24]

The Table shows, for every feature, whether there is a statistically significant difference between males and females according to the performed statistical test. The top (↑) and down (↓) arrows indicate whether the feature is higher for males or females. The dash sign (–) indicates that there is no statistically significant difference between males and females

In the following, an assessment of the importance of each feature in identifying the gender of the chat participants is provided according to the features categories. The analysis takes into considerations both Gini importance and the results of the feature analysis presented in Table 2. The outcome of the analysis will identify the features that mostly account for gender, which will be considered as gender markers.

- **Social Presence**: The Gini importance is high for all features related to social presence. However, Table 2 shows that male participants spend more time writing their chunks (duration of each chunk), and as a result, they produce longer chunks than females (length of chunks). One possible explanation is that male participants less worry about their social presence, meaning that they tend to press the enter key less frequently and, as a consequence, spend more time producing their chunks. This is in line with the academic research on psychological gender differences which has shown that males use communications as a tool to exert their dominance [32]. Conversely, females feel the need to project their social presence, and hence, they produce shorter but more frequent chunks that can indicate that they are socially present.
- **Cognitive Load**: The Gini importance is high for all cognitive load features, and furthermore, Table 2 shows that for all these features, there are statistical differences in typing behaviour between males and females. For instance, male participants tend to use backspace for deletion more frequently (density of backspace), consequently, spend more time in deletion (backspace time), which leads to a style that is less spontaneous and more formal, i.e. less like a face-to-face conversation and more like a written exchange. On the contrary, women writing style is close to a spoken language. This is in line with the empathising/systemising theory of the gender difference, showing that male participants tend to be higher in systemising skills [33].
- **Affect**: Typing speed and median latency are the two features with the highest Gini importance across all features, and they are designed to capture the typing speed of a giving participant. Beside cognitive load, these two features also measure affect. Research suggests that typing speed is influenced by arousing emotions as high arousal leads to a higher keyboard typing speed [34]. Consequently, females in this experiments might have felt under pressure to complete the task of the experiment (Winter Survival Task) as quickly as possible bounded by the time constrains which in turn have resulted in the higher arousal which has lead them to type faster.
- **Style**: The last feature that has shown a significant difference between females and males participants in terms of typing behaviour is (density of suspension points) a feature related to style. As stated earlier, women's writing is more of a spoken language, and women's spoken language is often characterised by speed [35]. This phenomenon can be reflected in the use of ellipses marks to reduce typing constructs, hence speeding the conversation. However, the corresponding Gini importance of the feature is relatively low, suggesting that it has a limited impact in conveying the gender of the chat participants.

6 Conclusion

This paper has presented experiments aimed at automatically recognising the gender of people involved in interactive live chats. The experiments were performed using a set of 30 chats (a total of 60 participants) with the Winter Survival Task being the topic of the conversations. The extracted features do not take into consideration what the participant has typed but how the participant has typed it. The finding suggests that features related to social presence, cognitive load and affect are considered to be gender markers. Thus, gender tends to leave physical, machine detectable traces in typing behaviour. In addition, the experiments show that there is a difference in typing behaviour between males and females which is possible to predict with an accuracy up to 98.8%. The present experiments have gone some way towards enhancing our understanding of the interplay between gender and typing behaviour and are of interest to the market for technologies supporting live chats. Future work will target other social and psychological phenomena underlying the data used in this work (e.g. predicting the personality traits and conflict handling style of the chat participants).

References

1. Federal Bureau of Investigation (2018) Internet crime report. Technical report, Internet Crime Complaint Center
2. Alsmearat K, Al-Ayyoub M, Al-Shalabi R, Kanaan G (2017) Author gender identification from Arabic text. J Inf Secur Appl 35:85–95. https://doi.org/10.1016/j.jisa.2017.06.003
3. Cheng N, Chandramouli R, Subbalakshmi KP (2011) Author gender identification from text. Digit Invest 8:78–88. https://doi.org/10.1016/j.diin.2011.04.002
4. Bayot RK, Gonçalves T (2018) Multilingual author profiling using LSTMs: notebook for PAN at CLEF 2018. In: CEUR workshop proceedings
5. Bsir B, Zrigui M (2018) Enhancing deep learning gender identification with gated recurrent units architecture in social text. Comput Sist 22:757–766. https://doi.org/10.13053/cys-22-3-3036
6. Plank B (2018) Predicting authorship and author traits from Keystroke dynamics. In: Proceedings of the second workshop on computational modeling of people's opinions, personality, and emotions in social media. Association for Computational Linguistics, Stroudsburg, PA, USA, pp 98–104
7. Trojahn M, Arndt F, Weinmann M, Ortmeier F (2013) Emotion recognition through Keystroke dynamics on touchscreen keyboards. In: ICEIS 2013 - Proceedings of the 15th international conference on enterprise information systems, pp 31–37
8. Burgoon JK (2016) Nonverbal communication. Routledge, Abingdon
9. Vinciarelli A, Pantic M, Bourlard H (2009) Social signal processing: survey of an emerging domain. Image Vis Comput 27:1743–1759. https://doi.org/10.1016/j.imavis.2008.11.007
10. Fairhurst M, Da Costa-Abreu M (2011) Using keystroke dynamics for gender identification in social network environment. In: 4th international conference on imaging for crime detection and prevention 2011 (ICDP 2011). IET, pp P27–P27
11. Tsimperidis I, Arampatzis A, Karakos A (2018) Keystroke dynamics features for gender recognition. Digit Invest 24:4–10. https://doi.org/10.1016/j.diin.2018.01.018

12. Tsimperidis I, Katos V, Clarke N (2015) Language-independent gender identification through Keystroke analysis. Inf Comput Secur 23:286–301. https://doi.org/10.1108/ICS-05-2014-0032
13. Alsultan A, Warwick K (2013) Keystroke dynamics authentication: a survey of free-text methods. Int J Comput Sci 10:1–10
14. Joshi MP, Davis EB, Kathuria R, Weidner CK (2005) Experiential learning process: exploring teaching and learning of strategic management framework through the winter survival exercise. J Manag Educ 29:672–695. https://doi.org/10.1177/1052562904271198
15. Weinel M, Bannert M, Zumbach J et al (2011) A closer look on social presence as a causing factor in computer-mediated collaboration. Comput Human Behav 27:513–521. https://doi.org/10.1016/j.chb.2010.09.020
16. Sweller J (1988) Cognitive load during problem solving: effects on learning. Cogn Sci 12:257–285. https://doi.org/10.1207/s15516709cog1202_4
17. Conijn R, Roeser J, van Zaanen M (2019) Understanding the Keystroke log: the effect of writing task on keystroke features. Read Writ 32:2353–2374. https://doi.org/10.1007/s11145-019-09953-8
18. Lim YM, Ayesh A, Stacey M (2015) Using mouse and keyboard dynamics to detect cognitive stress during mental arithmetic. In: Arai K, Kapoor S, Bhatia R (eds) Intelligent systems in science and information 2014. Springer International Publishing, Cham, pp 335–350
19. Buker AAN, Roffo G, Vinciarelli A (2019) Type like a man! Inferring gender from Keystroke dynamics in live-chats. IEEE Intell Syst 34:53–59. https://doi.org/10.1109/MIS.2019.2948514
20. Brooks M, Aragon CR, Kuksenok K, et al (2013) Statistical affect detection in collaborative chat. In: Proceedings of the 2013 conference on computer supported cooperative work, p 317
21. Alm CO, Roth D, Sproat R (2005) Emotions from text: machine learning for text-based emotion prediction. In: Proceedings of the conference on human language technology and empirical methods in natural language processing. Association for Computational Linguistics, USA, pp 579–586
22. Yadollahi A, Shahraki AG, Zaiane OR (2017) Current state of text sentiment analysis from opinion to emotion mining. ACM Comput Surv 50 https://doi.org/10.1145/3057270
23. Riordan MA, Kreuz RJ (2010) Cues in computer-mediated communication: a corpus analysis. Comput Human Behav 26:1806–1817. https://doi.org/10.1016/j.chb.2010.07.008
24. Cristani M, Roffo G, Segalin C, et al (2012) Conversationally-inspired stylometric features for authorship attribution in instant messaging. In: Proceedings of the 20th ACM international conference multimedia - MM '12 1121. https://doi.org/10.1145/2393347.2396398
25. Kucukyilmaz T, Cambazoglu BB, Aykanat C, Can F (2008) Chat mining: Predicting user and message attributes in computer-mediated communication. Inf Process Manag 44:1448–1466. https://doi.org/10.1016/j.ipm.2007.12.009
26. Zheng R, Li J, Chen H, Huang Z (2006) A framework for authorship identification of online messages: writing-style features and classification techniques. J Am Soc Inf Sci Technol 57:378–393. https://doi.org/10.1002/asi.20316
27. Wu Q, Ye Y, Zhang H et al (2014) ForesTexter: an efficient random forest algorithm for imbalanced text categorisation. Knowl-Based Syst 67:105–116. https://doi.org/10.1016/j.knosys.2014.06.004
28. Nembrini S, König IR, Wright MN (2018) The revival of the Gini importance? Bioinformatics 34:3711–3718. https://doi.org/10.1093/bioinformatics/bty373
29. Hedges LV (2007) Correcting a significance test for clustering. J Educ Behav Stat 32:151–179. https://doi.org/10.3102/1076998606298040
30. Vizer LM (2009) Detecting cognitive and physical stress through typing behavior. In: CHI '09 extended abstracts on human factors in computing systems. Association for Computing Machinery, New York, NY, USA, pp 3113–3116
31. Epp C, Lippold M, Mandryk RL (2011) Identifying emotional states using Keystroke dynamics. In: Proceedings of the SIGCHI conference on human factors in computing systems. Association for Computing Machinery, New York, NY, USA, pp 715–724

32. Merchant K (2012) How men and women differ: gender differences in communication styles, influence tactics, and leadership styles. C Sr Theses 0–62
33. Baron-Cohen S (2003) The essential difference: men, women and the extreme male brain. Penguin
34. Kumari V (2004) Personality predicts brain responses to cognitive demands. J Neurosci 24:10636–10641. https://doi.org/10.1523/JNEUROSCI.3206-04.2004
35. Volkova M (2015) The peculiarities of using and perceiving ellipsis in men's and women's speech. In: Proceedings of the 3rd Patras international conference of graduate students in linguistics, pp 233–241

Chapter 2
An Efficient Framework to Build Up Heart Sounds and Murmurs Datasets Used for Automatic Cardiovascular Diseases Classifications

Sami Alrabie, Mrhrez Boulares, and Ahmed Barnawi

1 Introduction

Human heart is a very important organ in the body. Heart delivers blood to all portions of the body. Heart diseases known as Cardiovascular Diseases (CVDs) are considered as the main reason responsible for the death. CVDs are caused death for 33% wholly around the world, according to the surveyed conducted belonged to World Health Organization (WHO) [1]. The mechanical activity of hearts produces four different sounds called first normal sound S_1, second normal sound S_2, third abnormal sound S_3, fourth abnormal sound S_4 or murmur. S_3, S_4 heart sounds and murmur are signed for abnormalities of the heart. Stethoscope device is used for auscultating and recording heart sound or murmur. Murmur is the sound of blood flow turbulence in the heart. Murmurs occur within systole or diastole. Each murmur signs for specific disease and can be heard from specific area on the chest also has a unique feature. S_3, S_4 heart sounds and murmurs are difficult for hearing and determining the specific diseases. Physicians and Cardiologists depend on heart sounds and take a decision for two type of patients. Therefore, there is a need to automate the classification of heart diseases from the heart sounds and murmurs. The automation will be helping experts to avoid human medical errors associated with lack of experience.

The automation of the heart diseases classification using machine learning and deep learning techniques has been widely investigated. This topic of research is facing issue which is the scarcity of public accessible datasets. The available public datasets do not multiclass labelled heart sounds and murmurs of heart diseases in addition to that the number of recorded samples are relatively limited. Those datasets consist of two or three classes, namely normal class, abnormal class and murmur with no specific murmur type. Public datasets as shown in Table 1 include normal

S. Alrabie (✉) · M. Boulares · A. Barnawi
King Abdulaziz University, Jeddah 21589, Saudi Arabia
e-mail: sawadalrabie@stu.kau.edu.sa

A. E. Hassanien et al. (eds.), *Enabling Machine Learning Applications in Data Science*, Algorithms for Intelligent Systems, https://doi.org/10.1007/978-981-33-6129-4_2

Table 1 Heart sounds public data sets

Data set	References	Total of recordings	Sound categories
PASCAL A	[2]	176	Normal beats, Murmurs, Extra systoles
PASCAL B	[2]	656	Normal, Murmurs additional systoles and heart sounds, Artifacts
PhysioNet CinC	[3]	3126	Normal sounds, Unclear, Abnormal sounds
E-general medical	[4]	64	Normal sounds, varied diseases
Heart sounds, Murmurs Library	[5]	23	Normal, Abnormal

heart sound, abnormal heart sounds and (murmur: extra systeiolc). To the best of our knowledge, there is no available public dataset that has been built in proper standard methodology that includes a volume of samples of several classes of the most common heart diseases from their known areas on the chest as per medical practices.

This article aims to propose a framework for building heart sounds and murmurs dataset hat fulfillment of researchers are conducting classifications using machine learning and deep learning techniques.

The reminder of this article is organized as follows. Section 2 presents related work. Section 3 describes the proposed framework. Finally, Sect. 4 concludes the conclusion and future work.

2 Related Work

In [6], a model was proposed to detect heart sounds diseases from heart sounds. Time–frequency transformation was used to convert heart sounds to spectrogram images. Convolutional Neural Networks (CNNs) models were used to extract the features. The results of the model were outstanding compared to other models. The PASCAL heart sound challenge dataset A and B were used which have small number of samples. For data set A, it has three classes and for data set B, it has four classes as shown in Table 1.

Dominguez-Morales et al. in [7] proposed a tool to help hearing heart sounds process. Audio has been proceeded using neuromorphic auditory sensor and Neuromorphic Auditory VISualizer (NAVIS) Tool has been used for generating sonogram, images. CNNs models were used for classification. Modified AlexNet model has showed the best performance among other models. Heart sounds samples were taken from the PhysioNet/CinC Challenge data set. Two classes have been classified as normal and abnormal.

In [8], an algorithm was improved for classifying heart sounds using several features. Mel Frequency Cepstral Coefficient (MFCCs) was combined with Discrete Wavelets Transform (DWT) features. Support Vector Machine (SVM), Deep Neural Network (DNN) and K-Nearest Neighbors (KNNs) were used for classification. The authors have collected the data set from different sources. Data set consists of 5 classes, one normal and four classes abnormal (murmurs) namely, murmur of aortic stenosis, murmur of mitral Stenosis, murmur of mitral regurgitation, and murmur of mitral valve prolapse. So, five classes have been classified.

Wua et al. [9], proposed an automatic classification model for Phonocardiogram recording of the heart sounds and murmurs. Deep learning and ensemble learning have been used along with Savitzky–Golay filter. Results of the model was outperforming. They have been classified a Binary classifier and PhysioNet 2016 data set was used.

In [10], a deep learning model was developed for heartbeat sounds classification. The proposed model is based on Recurrent Neural Network (RNN) that includes Long Short Term Memory (LSTM), Dropout, Dense and Softmax layer. The performance of the model is outstanding compared to others models. The PASCAL heart sound challenge B was used which contains three classes of heart sounds and murmurs, normal, abnormal and Extra-systole.

Form the above studies, we can conclude that the majority of related works relies on the public available datasets which have limited of classes and recording of samples.

3 Proposed Framework

3.1 Dataset Logical Structure

The area where murmur is best heard is a significant factor in determining the cause of the disease. There are four main areas are used for auscultation heart sounds and murmurs. Aortic area, Pulmonic, Tricuspid area and Mitral area (known as Apex) (see Fig. 1). The first normal sound S_1 and the second normal S_2 can be heard at all areas. The third abnormal Heart sound S_3, known as ventricular gallop occurs in early diastole and can be beat heard at Mitral area (Apex) with the patient in the left lateral decubitus. The fourth abnormal Heart sound S_3, known as atrial gallop occurs in late diastole and can be beat heard at Mitral area (Apex) with the patient in the left lateral decubitus. The S_3 and S_4 heart sounds are signs for many heart diseases and are so difficult to determine the disease by hearing the sound [11]. Patients with S_3 or S_4 sound always are sent to more inspection tools such as Echocardiogram (Table 2).

The murmur is an additional sound generated as result of turbulent blood flow in the heart and blood vessels. Murmurs are produced by heart diseases as a result of defect valves or structure of heart itself. Murmurs are expressed by many factors that

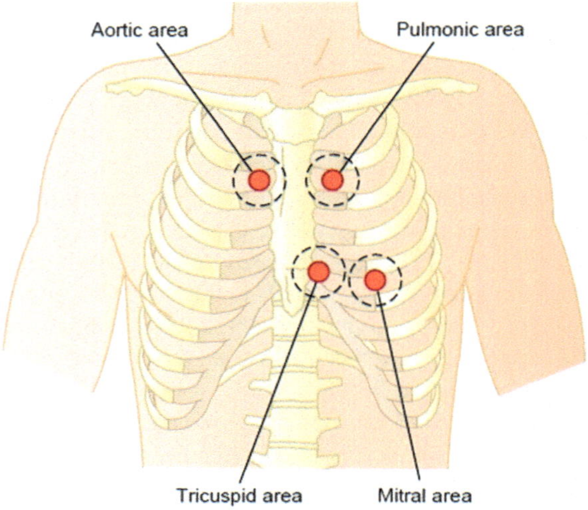

Fig. 1 Areas of auscultation to heart sounds and murmurs [11]

Table 2 Heart sounds occurrence timing and auscultation areas [11]

Heart sound	Occurrence timing	Auscultation area
S_1	The start of systole	At all areas
S_2	The beginning of diastole	At all areas
S_3	Early diastole	Best heard at Mitral area
S_4	Latte diastole	Best heard Mitral area

are timing in the cardiac cycle, intensity, shape, pitch, area, radiation and reaction to dynamic maneuvers. The location of murmurs is an important factor to determine the diseases [12]. The cardiologists take all these characteristics in the account to determine the cause of the murmur (the disease). Of course, the Cardiologists are the main factor in collecting the data sets. So, we will consider the location of murmur in the proposed framework and the Cardiologists will diagnose the diseases. There are common murmurs of heart diseases that we will consider. As mentioned above, there are four main areas are used for listening to heart sounds and murmurs then the diseases will be determined by the Cardiologists. At Aortic area, murmur of aortic stenosis, murmur of aortic regurgitation and innocent flow murmur can be best heard. At Pulmonic area, murmur pulmonic stenosis, murmur of pulmonary regurgitation and murmur of atrial septal defect can be best heard. Tricuspid area, murmur tricuspid stenosis, murmur of tricuspid regurgitation, murmur of ventricular septal defect and murmur of Hypertrophic obstructive cardiomyopathy (HOCM) can be best heard. Murmurs can be best heard at other areas but also heard at Tricuspid area, murmur of aortic regurgitation, murmur pulmonic stenosis and murmur of pulmonary regurgitation. At Mitral area (Apex), murmur of mitral stenosis and

Table 3 Areas and related heart sounds and murmurs

Area	Murmur	Murmur best heard somewhere else but also heard at this area
Aortic	S_1 OR S_2	–
	Aortic stenosis	
	Aortic regurgitation	
	Innocent flow murmur	
	S_1 OR S_2	
Pulmonic	Pulmonic stenosis	
	Pulmonary regurgitation	
	Atrial septal defect	
Tricuspid	S_1 OR S_2	Aortic regurgitation
	Tricuspid stenosis	
	Tricuspid regurgitation	Pulmonic stenosis
	Ventricular septal defect	Pulmonary regurgitation
Mitral area (Apex)	S_1 OR S_2	Aortic stenosis
	S_3	Aortic regurgitation
	S_4	Tricuspid stenosis
	Mitral stenosis	
	Mitral regurgitation	
		Tricuspid regurgitation
		Ventricular septal defect

murmur of mitral regurgitation. Murmurs can be best heard at other areas but also heard at Mitral area (Apex), murmur of aortic stenosis, murmur of aortic regurgitation, murmur tricuspid stenosis, murmur of tricuspid regurgitation and murmur of ventricular septal defect (see Table 3) [13, 14].

3.2 Mobile Application and Database Design

Mobile application has been created for collecting heart sounds and murmurs data sets. The mobile application was designed with a good User Interface and it is easy for use. The mobile application consists of many pages, the first page is designed to add new patient, then there is a page contains a picture of chest, which has buttons on each area (see Figs. 2 and 3). When clicks on each area, a page will be appeared, which consists of recording/replying buttons, heart sounds drop menu, diagnoses drop menu, and add note option. The mobile application works through steps; first the Cardiologist will choose "add new patient" then the page with the figure of four main areas appears. He will be required to press on Aortic area (see Fig. 4), he

Fig. 2 Main page of the mobile application

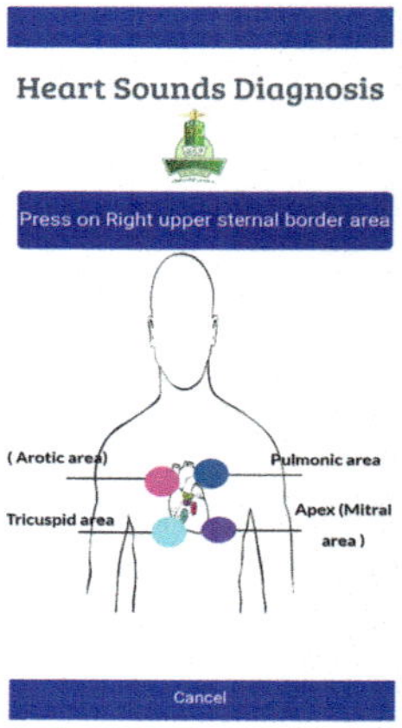

Fig. 3 Page of auscultation areas

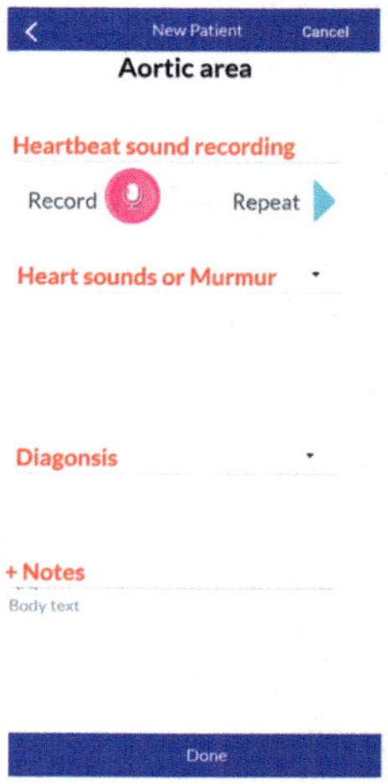

Fig. 4 Page of Aortic area

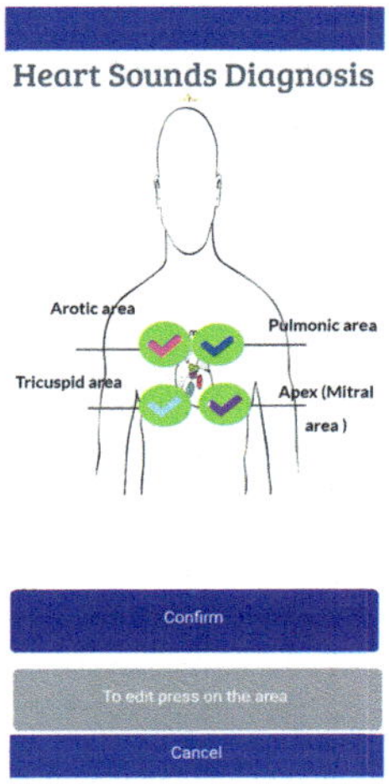

Fig. 5 Page of completing recordings and diagnosis from all areas

will press on recording button to record the sound or murmur, then he will press on heart sound drop menu to select the heart sounds or murmur. If he chooses S_1 or S_2 in diagnose drop menu will just Normal, if he chooses murmur, in diagnose drop menu the related diseases will be appeared according to this area as mentioned in Table 3, then he will choose the disease based on the sound or the murmur that have been recording, then he will press on done button to go back to the figure of chest. He will press on the rest of areas to do the same steps in the Aortic area. Finally, he will confirm all what has been done or edit (see Fig. 5). Then, the first page to add new patients will be appeared. The files of sounds and murmurs labeled with related diseases will be saved as wav format. Folder for each patient will be created and each folder has four sounds files named with area plus the disease. Same steps for others area will be done.

As there are murmurs of heart diseases best heard in specific areas in the chest and also can be heard in somewhere else (others areas) (see Figs. 6 and 7). This gives advantage to have more samples of murmurs and related diagnoses. To avoid the ambiguity in labeling when we add in drop menus the murmurs of best heard in specific areas in the chest and also can be heard in somewhere else (others areas). This problem has been solved by doing workflow logic. For example, if the Cardiologist diagnoses in the Aortic area S_1 or S_2 sound and labeled normal, so, aortic stenosis and aortic regurgitation will not appear in the Tricuspid area and the Mitral area. Another example, if the Cardiologist diagnoses in the Aortic area aortic stenosis murmur and labeled aortic stenosis disease. Hence, aortic regurgitation disease will not appear in the Tricuspid area nor in the Mitral area. Another example, if the Cardiologist diagnoses in the Aortic area functional murmur and labeled functional, then aortic stenosis and aortic regurgitation will not appear in the Tricuspid area and the Mitral area and so on. Moreover, if patient has two diseases from two different areas and those two disease can be heard from third area, they will not be appeared in the third area to avoid the ambiguity. For example, if patient has been diagnosed with aortic

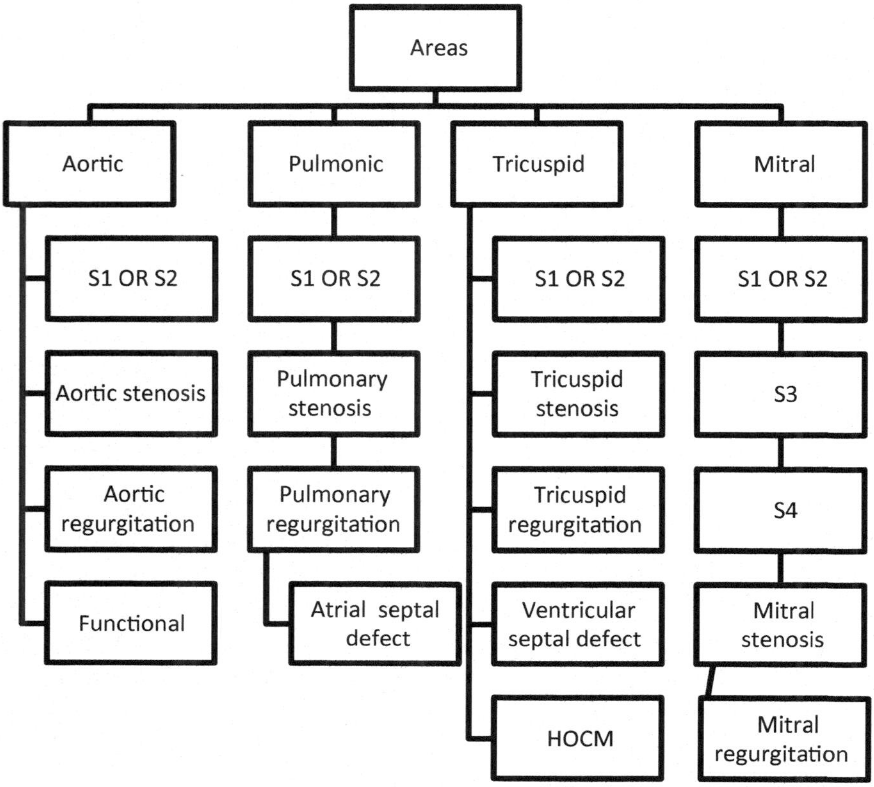

Fig. 6 Murmurs of heart diseases best heard in this areas

stenosis in the Aortic area and tricuspid stenosis in the Tricuspid area. Hence, they will not be appeared in the Mitral area.

3.3 Composition Stethoscope

The electronic Stethoscope is a very expensive. When make trying to make e electronic Stethoscope connect with the mobile application to record the heart sounds or the murmur, the license is needed to use the Application Programming Interface (API) of the electronic Stethoscope. The license is very costly. Therefore, we have composited a new Stethoscope, using normal Stethoscope and small sensitive microphone (see Fig. 8). We cut the tube of the normal Stethoscope from headphones side then we plugged the microphone inside the slot of the tube and covered by glue, then

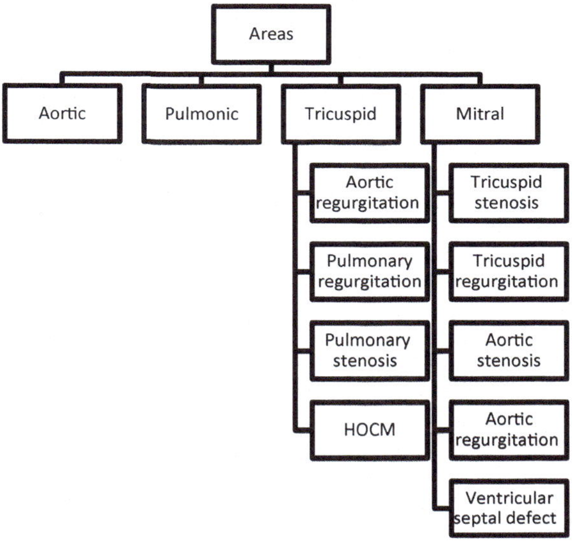

Fig. 7 Murmurs of heart diseases can be heard somewhere else and also can be heard in these areas

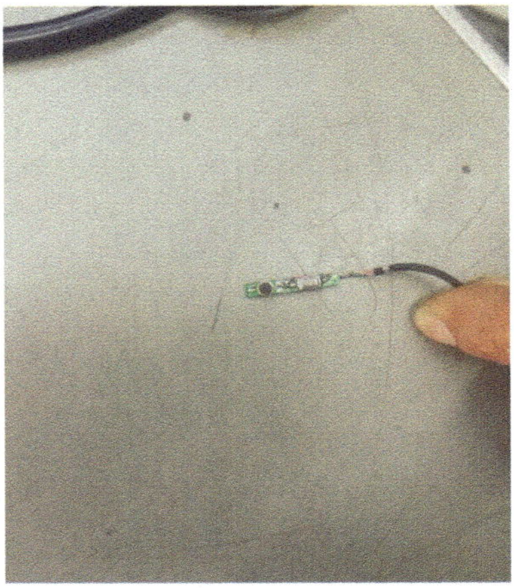

Fig. 8 A sensitive small microphone

we connected the cable of the microphone in the Tablet (see Fig. 9). The composition Stethoscope is very cheap and can be used for collecting data with the mobile applications.

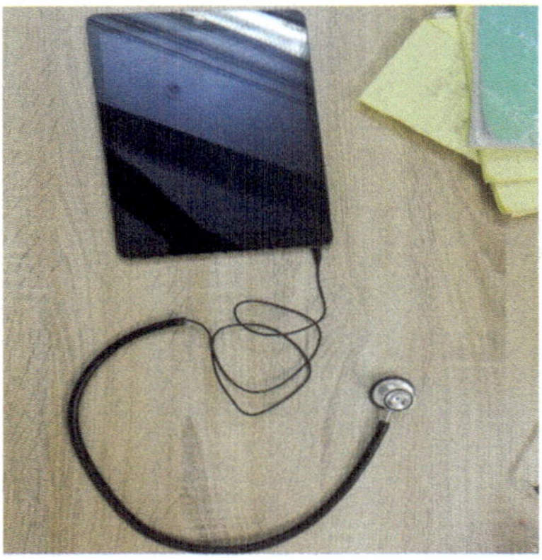

Fig. 9 The composition Stethoscope connected to iPad

4 Conclusion

In this work, we presented an efficient framework to create a data set, which is used for automatic Cardiovascular diseases classifications. The framework will contribute to create an organized data sets with multiclass of heart diseases, an easy method to collect huge of samples of heart sounds and murmurs labeled with related heart diseases and with inexpensive tools. The sounds and murmurs of heart have been mapped to their locations. Mobile application and database has been designed. Composition Stethoscope has been also developed.

For future work, we will apply the proposed framework to create huge an accessible public data set that will help researchers to use it in machine learning and deep learning research.

References

1. Mortality due Cardiovascular diseases in world. World Health Organization. https://www.medicalnewstoday.com/articles/282929.php
2. Bentley P et al (2011) PASCAL Classifying heart sounds challenge, https://www.peterjbentley.com/heartchallenge/. Accessed 20 Apr 2018
3. National Institute of General Medical Sciences and the National Institute of Biomedical Imaging and Bioengineering. PhysioNet/CinC challenge (2016). NIGMS & NIBIB. (2016). https://www.physionet.org/physiobank/ database/challenge/2016/. Accessed 20 Apr 2018
4. eGeneral Medical Inc. USA. eGeneralMedical.com. https://www.egeneralmedical.com/listohearmur.html. Accessed 20 Apr 2018
5. Michigan Medicine. Heart Sound & Murmur Library University of Michigan Health System. Ann Arbor. MI. https://www.med.umich.edu/lrc/psb_open/html/repo/primer_heartsound/primer_heartsound.html. Accessed 20 Apr 2018
6. Demir F, Şengür A, Bajaj V et al (2019) Towards the classification of heart sounds based on convolutional deep neural network. Health Inf Sci Syst 7:16

7. Dominguez-Morales JP, Jimenez-Fernandez AF, Dominguez-Morales MJ, Jimenez-Moreno G (2017) Deep neural networks for the recognition and classification of heart murmurs using neuromorphic auditory sensors. IEEE Trans Biomed Circuits Syst 12(1):24–34
8. Son GY, Kwon S (2018) Classification of heart sound signal using multiple features. Appl Sci 8(12):2344
9. Wu JMT, Tsai MH, Huang YZ, Islam SH, Hassan MM, Alelaiwi A, Fortino G (2019) Applying an ensemble convolutional neural network with Savitzky-Golay filter to construct a phonocardiogram prediction model. Appl Soft Comput 78:29–40
10. Raza A, Mehmood A, Ullah S, Ahmad M, Choi GS, On BW (2019) Heartbeat sound signal classification using deep learning. Sensors 19(21):4819
11. Arthur C, Guyton J, Sounds N (2020) Normal heart sounds. [online] BrainKart. Available at: https://www.brainkart.com/article/Normal-Heart-Sounds_19347/. Accessed 23 July 2020
12. Healio.com (2020) Learn the heart|Learntheheart.Com. [online] Available at: https://www.healio.com/cardiology/learn-the-heart. Accessed 23 July 2020
13. American College of Physicians. MKSAP: medical knowledge self-assessment program VIII.American College of Physicians, Philadelphia, PA 19881989
14. Slideshare.net. 2020. Heart Sounds And Murmur. [online] Available at: https://www.slideshare.net/vitrag24/heart-sounds-and-murmur. Accessed 23 July 2020

Chapter 3
Facial Recognition and Emotional Expressions Over Video Conferencing Based on Web Real Time Communication and Artificial Intelligence

Sally Ahmed Mosad Eltenahy

1 Introduction

Reactions and impressions of audiences are vital in any people meeting. Since the coronavirus pandemic, physical people meetings are replaced with online meetings globally to avoiding crowding. It is widely used for various objectives—for example but not limited to—education, business meeting, social events, … etc. In each situation, there is a need to know each audience's impression whether is sad, happy, natural, or surprised in real time. In Education, the instructor needs to know if each student gives attention or not. In Business, The Sales or Presales need to know if each customer is satisfied or not. The employer or supervisor needs to measure the proper performance according to the customers' and employees' facial emotions. This research paper's core is an application of videoconference with a feature of face recognition and emotions. This application is based on web real-time communication (WebRTC) and machine learning (ML) technologies. WebRTC technology is built-in the latest browsers' versions, so there is no need to install plugins or downloads. It is used to transmit and receive video and audio in real time. It is utilized to set up videoconferences in real-time, offer to join the others to the call and share data. Although, WebRTC hides the complicated tasks in browsers' components. WebRTC provides several JavaScript APIs to enable developers to create their own applications. There are a lot of open source platforms and applications which are developed based on WebRTC and can be customized according to our needs. In other hand, Machine Learning (ML) libraries such as Tensorflow.js enables us to use some models to detect facial expressions, gender and age. The proposal application is an integration between a videoconference application and some machine learning models that can detect and then store facial expressions like sad, happy, surprise, or nature of subscribers and detect their age and gender, and store that in the database as a reference.

S. A. M. Eltenahy (✉)
Faculty of Engineering, Mansoura University, Mansoura, Egypt

A. E. Hassanien et al. (eds.), *Enabling Machine Learning Applications in Data Science*, Algorithms for Intelligent Systems, https://doi.org/10.1007/978-981-33-6129-4_3

This paper is indexed as follows. Section 2 defines WebRTC Project. Section 3 introduces machine learning and tensorflow.js. Section 4 implementation of the application and evaluation. Finally, Sect. 5 the conclusion and future work.

2 WebRTC

In May 2011, Google released the WebRTC project as an OpenSource project built-in browser to support real-time communication. In July 2014, Google hangout had been released as an application based on WebRTC. Recently, WebRTC is supported in most recent common browsers' versions. Also, it is supported by Apple, Google, Microsoft and Mozilla, amongst others [1]. WebRTC is an integration of protocols, standards and APIs. There is no need to install plugins or downloads to use WebRTC. It is already built-in modern browsers. It has several JavaScript APIs that enable any developer to handle commands to access the hardware component like camera and microphone, share desktop, and exchange data. To handle requests across network , WebRTC needs four certain types of servers. It needs a NAT traversal server, a hosted server, a Signaling server and a media server. Thanks to those facilities, Developers became able to build applications based on WebRTC to transfer and receive data, audio and video. Also many OpenSource platforms and applications based on WebRTC are available for developers to customize them according to their requirements. In this study, Illustrate the architecture of WebRTC and the methodology of using WebRTC.

2.1 WebRTC Client-Side

WebRTC is a collection of protocols and standards. It is supported in modern browsers. WebRTC is built-in the browser and consists of two main layers. The first layer is being improved and developed by browser makers. It contains WebRTC C++ API and three main components (Voice Engine, Video Engine and Transport) that are dealing with the hardware of the client's device to handle the streaming video, voice and data. The second layer, which is called the web APIs layer, enables developers to use JavaScript APIs to develop their own applications. The following explains the client-side components and the web APIs layer.

WebRTC C++ API (PeerConnection). Enables browser makers to easily implement the Web API proposal [2].

Voice Engine. A framework for accessing the sound card to encode and decode the audio to the network. It is a collection of standards pointed in Table 1.

Video Engine. A framework for handling the camera streaming to the network or from the network to the screen. It is a collection of standards pointed in Table 2.

Table 1 WebRTC voice engine standards

Protocol/standard	Explanation
iSAC Codec	Wideband or super wideband, 16 kHz or 32 kHz respectively, speech and audio codec [3]
iLBC Codec	Narrowband, Sampling frequency 8 kHz/16 bit (160 samples for 20 ms frames, 240 samples for 30 ms frames) [4]
NetEQ for Voice	A dynamic jitter buffer and error concealment algorithm used for concealing the negative effects of network jitter and packet loss [2]. Keeps latency as low as possible while maintaining the highest voice quality [2]
Acoustic Echo Canceler	Improves voice quality, in real-time, by removing the acoustic echo resulting
Noise Reduction (NR)	Removes background noise

Table 2 WebRTC video engine's standards

Protocol/standard	Explanation
VP8 Codec	Compressed video data format, together with a discussion of the decoding procedure for the format [5]
Video Jitter Buffer	Collects and stores incoming media packets and decides when to pass them along to the decoder and playback [6]
Image enhancements	Adjusts the video picture

Transport. Allows calls across various types of network in real time as peer to peer connection. It is depend on standards pointed in Table 3.

Web API layer. JavaScript **APIs which developers need** to create applications based on WebRTC **such as**:

RTCPeerConnection. Responsible for peer to peer connection, is containing STUN/TURN server information and handles the streaming of voice and video from remote users.

RTCDataChannel. Responsible for shared files in the browser and takes into consideration file size and chunk size.

MediaRecorder. Responsible for recording media-audio and video.

Table 3 WebRTC transport's standards

Protocol/standard	Explanation
SRTP	Provides confidentiality, message authentication, and replay protection to the RTP traffic and to the control traffic for RTP, the Real-time Transport Control Protocol (RTCP) [7]
Multiplexing	Method by which multiple analog or digital signals are combined into one signal over a shared medium [8]
STUN + TURN + ICE	Allows establish call over various types of network

getUserMedia(). Responsible for getting permission from the user to use the camera, shared screen or a microphone.

2.2 WebRTC Server-Side

WebRTC needs four types of servers to can handle request over several types of networks. It needs hosted server, signaling server, STUN/TURN servers and Media server. The following explains those main types of servers.

Hosted Server. The server which hosted the application files and database.

Signaling Server. Passes signaling messages such as Session control messages, Error messages, Media metadata, Key data and Network data between the clients. The following explains the usage of those messages.

Session control messages. Establish or end the connection.

Media metadata. Exchange information of codec, bandwidth and media type.

Key data. Exchanges information of security.

Network data. Exchanges information about the public IP and port.

STUN/TURN Server. To keep clients/browsers connection as peer to peer connection behind NAT/Firewall in which each client/browser has a public IP address, STUN server and its extension TURN server are required.

STUN server. Session Traversal Utilities for NAT used to answer the question "what is my public IP address?" and then share the answer with the other user in the session, so he can try and use that address to send media directly [9].

TURN server. Relay NAT between peers, it costs more in bandwidth and is used when it is not possible p2p connection directly.

Media Server. Provides Group calling, Broadcast and live streaming, Recording, Server-side machine learning, Gateway to other networks/protocols and Cloud rendering (gaming or 3D).

3 Machine Learning

Machine Learning is an application of Artificial intelligent [10]. It trains a prediction model using the labeled input data or observations. To develop the models without any need to understand the underlying algorithms, we use machine learning frameworks or libraries. In the following, some examples of ML models and ML frameworks and libraries.

3.1 *Machine Learning Models Examples*

Face Detection Models. Returns the probability score with a bounding box referring to all human faces' possibility in an image and the location coordinates of those faces. It is useful for monitoring or detecting human activity. The following gives Face Detection Models examples.

SSD Mobilenet V1. Compute each face's coordinates in an image and re-turns the bounding boxes of each face together with its probability. It is quantized near 5.4 MB.

Tiny Face Detector. Low performance in detecting small faces. But it is faster, smaller, and lighter resources consuming than SSD Mobilenet V1. It is quantized to about 190 KB.

MTCNN. Ability to detect face bounding box sizes in a wide range. It is quantized only 2 MB.

68 Point Face Landmark Detection Model. very lightweight and fast, detecting accurately 68 face landmark points. it is trained on a dataset of ~35 k face images labelled with 68 face landmark points. It is quantized 350 kb in default edition and 80 kb in the tiny.

Face Expression Recognition Model. Lightweight, fast and provides reasonable accuracy. It is trained on a dataset of a variety of public images available as well as images scraped from the web. Quantized roughly 310 kb.

3.2 *Machine Learning Libraries Examples*

OpenCV. Relevant to image processing and management category. It is used in computer vision applications. It is written in C++ and binding in Python, Java, and MATLAB/OCTAVE.

TensorFlow.js. a library for machine learning in JavaScript [11]. Tensorflow.js has several models such as image classification, pose estimate, Object detection, text toxicity detection, speech command recognition, Body segmentation, face landmark detection, simple face detection, and more.

4 Implementation

Figure 1, showing the approximate topology of the proposed Application. The idea of the proposed application is about integration between one of OpenSource platforms used for videoconferencing which is called OpenVidu and one of the face recognition APIs which is called Face-api.js. The following describes the purposed application's topology and methodology.

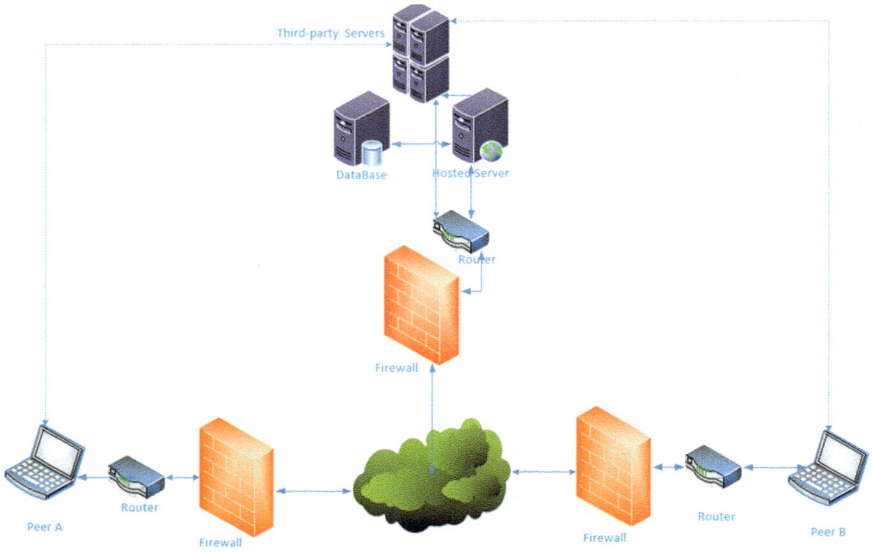

Fig. 1 The approximate topology of the proposed application

4.1 Network Topology

Third-Party Servers. To keep a live connection and quality of services between peers across different types of networks, there is a need for certain types of servers that WebRTC depends on. The following explains those kinds of servers used as third-party.

OpenVidu Server. It is acting as a signaling server. It handles the server-side stuff [12]. It receives the operations from OpenVidu Browser, which will be defined later, and do whatever is necessary to establish and manage your video-calls [12].

Kurento Media Server (KMS). It is acting as a media server. It is responsible for including group communications, transcoding, recording, mixing, broadcasting and routing of audiovisual flows.[13].

Coturn. It is acting as a NAT traversal server. It is an opensource STUN/TURN server. It is responsible for relaying between peers' browsers. It supports several protocols like UDP, TCP, TLS, DTLS and SCTP.

Redis. It is integrated with CoTurn to be used as a cash database. It is used to store the user with the key or password for authentication to coturn server, if it is required.

Nginx. It is an HTTP and reverses proxy server, a mail proxy server, and a generic TCP/UDP proxy server [14].

Hosted Server. In which, the Videoconference application with its features is hosted. The client can access it directly via URL. It is connected to the database server to store the data of the facial expressions detections. It consists of the following.

OpenVidu Browser. used as a library implemented by javascript and typescript. It allows-videocalls, sending and receiving video and audio, joining calls. The All actions are managed by OpenVidu Browser.

OpenVidu videoconference application. It is a web application implemented on top of OpenVidu browser and using its command. In this proposal, openvidu-call-react application is used. Which is available as an opensource application.

Face-api.js. a JavaScript face recognition API for the browser and nodejs implemented on top of tensorflow.js core (tensorflow/tfjs-core) [15]. Face-api.js uses the ML models like Face Detection model, Face Landmark Detection model, Face Recognition model, Face Expression Recognition model and Age Estimation and Gender Recognition.

Database Server. In which, storing the records of facial recognition and emotions over the videoconference per time. In this proposal, Mongodb is used.

Other Components in Network Topology. Firewall, routers and others are the expected components across real network topology.

4.2 Deployment

Concerning the Deploying of OpenVidu on-premises [16], The on-premises machine has to be ubuntu 16.04 or higher with at least 4 GB RAM. The docker and docker-compose have to be installed. Then, Install OpenVidu-server-KMS:2.15.0 container and also install Coturn and Redis. Follow the instructions which are available in the official website page of OpenVidu to deploy OpenVidu environment on-premises (version 2.15.0). besides that, establish a separate machine/server as a hosted server. The hosted server contains the videoconference application which is openvidu-call-react [17] integrated with Faceapi.js. The openvidu-call-react has to be configured to be linked to the OpenVidu server's Public IP or URL with the password. Note, some ports are needed to be open as showing in (Table 4).

Table 4 Port configurations

Port	Explanation
22 TCP	To connect using SSH to admin OpenVidu [16]
80 TCP	For SSL certificate verification
443 TCP	OpenVidu server and application are published by default in standard https port [16]
3478 TCP + UDP	Used by TURN server to resolve clients Ips [16]
40,000–57,000 TCP + UDP	Used by Kurento Media Server to establish media connections [16]
57,001–65,535 TCP + UDP	Used by TURN server to establish relayed media connections [16]

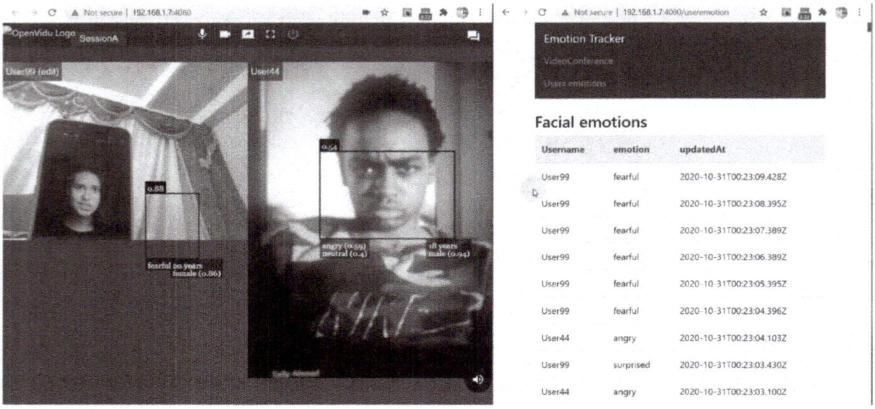

Fig. 2 Emotional tracker interface

In openvidu-call-react's repository. Download the models of face recognition and Install the required TensorFlow and face-api.js dependencies as following:

```
>npm i @tensorflow-models/coco-ssd @tensorflow/tfjs-converter @tensorflow/tfjs-core face-api.js
```

In openvidu-call-react's stream component file. Import the face-api to the stream component and integrate it to each streaming video of each client.

4.3 Results and Observations

Figure 2, Displaying the result of integration OpenVidu videoconference application with Face-api.js api and storing the results in mongodb in real-time.

5 Conclusion and Future Work

It is kind to have emotions and reactions detection feature in our online meeting for all members at a certain time for several purposes. The proposal application does that through the proposal application which is integration between Open-Source videoconference framework based on WebRTC and machine learning framework. The proposal application's utilization is optimized as it is divided utilization consuming into each client-side. The future work of this proposal is to make it with statistics, graphs and notifications for the publisher of the online meeting to know if everything is normal or he/she has to make a decision or handling something if there is anything unnormal.

References

1. Team GW Real-time communication for the web. Available from: https://webrtc.org/
2. WebRTC Architecture 13,07,2020; Available from: https://webrtc.github.io/webrtc-org/architecture
3. RTP Payload Format for the iSAC Codec draft-ietf-avt-rtp-isac-04 2013
4. Internet Low Bitrate Codec. 25 January 2020; Available from: https://en.wikipedia.org/wiki/Internet_Low_Bitrate_Codec
5. VP8 Data Format and Decoding Guide. 2011; Available from: https://tools.ietf.org/html/rfc6386
6. webrtcglossary. Jitter Buffer. 2017; Available from: https://webrtcglossary.com/jitter-buffer/
7. The Secure Real-time Transport Protocol (SRTP). 2004; Available from: https://tools.ietf.org/html/rfc3711
8. Multiplexing. 1 August 2020; Available from: https://en.wikipedia.org/wiki/Multiplexing
9. Levent-Levi T (2020) WebRTC Server: What is it exactly? Available from: https://bloggeek.me/webrtc-server/
10. What is Machine Learning? A definition. 6 May 2020; Available from: https://expertsystem.com/machine-learning-definition/
11. TensorFlow For JavaScript. Available from: https://www.tensorflow.org/js
12. What is OpenVidu? 2020
13. About Kurento and WebRTC. 2018; Available from: https://doc-kurento.readthedocs.io/en/6.14.0/user/about.html
14. nginx. Available from: https://nginx.org/en/.
15. face-api.js. 2020; Available from: https://github.com/justadudewhohacks/face-api.js/
16. Deploying OpenVidu on premises. 2020 [cited 2020 Augest]; Available from: https://docs.openvidu.io/en/2.15.0/deployment/deploying-on-premises
17. openvidu-call-react 2020; Available from: https://docs.openvidu.io/en/2.15.0/demos/openvidu-call-react/

Chapter 4
Recent Advances in Intelligent Imaging Systems for Early Prediction of Colorectal Cancer: A Perspective

Debapriya Banik, Debotosh Bhattacharjee, and Mita Nasipuri

1 Introduction

Colorectal cancer (CRC) is a challenging health problem worldwide. It is the third most frequently occurring cancers in men (1,026,215 cases) and second among women (823,303 cases) [1]. It accounts for nearly 9.2% of all cancer-related deaths. This dramatic increase in incidence and mortality rate is due to the rise in urbanization, improper diet, and lifestyle, which generally prevail in the developed countries. Moreover, it is predicted to rise approximately by 80% in 2035, with an incidence of 114,986 new cases and mortality of 87,502 [2]. However, the incidence rate of CRC is reported low in India. CRC ranks 4th among men with an incidence rate of 6.4% and ranks 5th among women with an incidence rate of 3.4% [3, 4]. CRC generally arises from the protrusion in the colon surface, which is known as a polyp [5]. Although most polyps are benign in its initial stage, they can turn into cancer eventually. CRC is highly curable if polyps are detected in its early stages of development. Therefore, the early diagnosis of CRC with proper treatment can save more lives. Colonoscopy is considered as a gold standard for the diagnosis of CRC [6]. It allows visual screening of the entire colon and rectal surface during the exploration and biopsy to decide the grade of abnormality and further resection of polyp via polypectomy based on the biopsy report [7]. However, colonoscopy is an operator-depended procedure where there is a high probability of polyp miss-rates due to human factors such as workload, lethargy, and insufficient attentiveness [8]. The average miss-rates of polyp are approximately 4–12%. The miss-detected polyps could lead to late diagnosis of CRC with a survival rate of less than 10%. Recently, wireless capsule endoscopy (WCE)

D. Banik (✉) · D. Bhattacharjee · M. Nasipuri
Department of Computer Science and Engineering, Jadavpur University, Kolkata 700032, India

D. Bhattacharjee
e-mail: debotosh@ieee.org

A. E. Hassanien et al. (eds.), *Enabling Machine Learning Applications in Data Science*, Algorithms for Intelligent Systems, https://doi.org/10.1007/978-981-33-6129-4_4

has also emerged as a promising technique for colon exploration, where a capsule is ingested, and more than 50,000 images are captured for manual analysis, which is also a time-consuming and a tiresome procedure [9]. Due to the high incidence rate and mortality rate of CRC worldwide, several computer-aided diagnosis (CAD) strategies have been exploited over the last few years to decrease the polyp miss-rates. However, the detection of the polyp by CAD techniques is still a challenging problem due to the wide variation of polyp in terms of shape, size, scale color, and texture.

This chapter provides different insights into colorectal cancer. Additionally, different sophisticated diagnostic techniques for screening of CRC are also elaborately discussed. The necessity and challenges in early prediction of CRC are also highlighted in this chapter. This chapter is an attempt to explore different computer-aided intelligent imaging techniques for polyp detection that can be feasible and efficient in clinical practice. The chapter is organized as follows: A brief outline including incidence and mortality rate, symptoms, and risk factors is elaborately discussed in Sect. 2. In Sect. 3, we have presented different diagnostic modalities for the early prediction of CRC. We have explored the need for early prediction of CRC in Sect. 4. In Sect. 5, we have highlighted different challenges in the early prediction of CRC by intelligent imaging techniques. We have elaborately explored different computer-aided intelligent techniques proposed by different research groups across the globe for early detection of CRC in Sect. 6. Some significant remarks from the related literature and future direction are discussed in Sect. 7. Finally, concluding remarks for this chapter are reported in Sect. 8.

2 Epidemiology of Colorectal Cancer

Cancer is a disease that is characterized by an uncontrolled division of abnormal cells and its survival, and when this abnormal growth arises in colon or rectum, it is termed as colorectal cancer (CRC) [10]. The end part of the gastrointestinal (GI) system comprises the colon and rectum (colorectum), which together constitute the large intestine [11]. The beginning part of the large intestine is the colon, which is divided into four parts [12]: The first part is the ascending colon, which begins from the cecum and travels upwards to the right side of the abdomen. The next part is the transverse colon, which travels from the right side to the left side of the body, which is followed by a descending colon, which descends on the left side of the body. The final part of the colon is the sigmoid colon, which ends up at the rectum. The rectum is the end part of the large intestine. The ascending and the transverse colon are collectively termed as proximal colon, whereas the descending and sigmoid colon are collectively termed as the distal colon. An overview of the large intestine is depicted in Fig. 1.

CRC usually arises from abnormal growth of glandular tissue called as a "polyp" that develops in the inner lining of the colon or rectum. Clinicians generally group polyps based on various parameters. Based on size [13], they may be grouped as

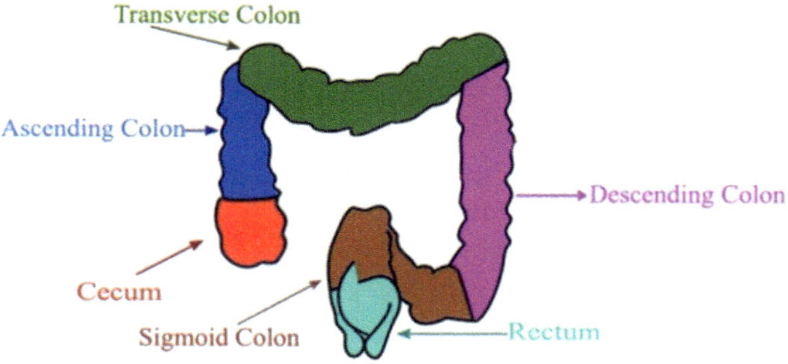

Fig. 1 Anatomy of large intestine

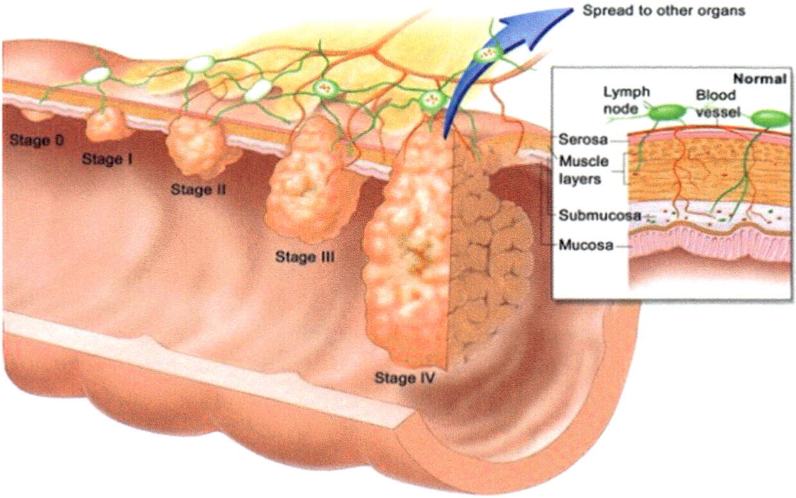

Fig. 2 Stages of growth of CRC [17]

(i).Diminutive (≤5 mm) (ii) Small (6–9 mm) (iii) Large (>10 mm). Based on shape, they may be grouped according to Paris Criteria [14] as (a) Ip (protruded, pedunculated) (b) Is (protruded, sessile) (c) IIa (superficial, elevated) (d) IIb (flat) (e) IIc (superficial, shallow, depressed) (f) III (excavated). Furthermore, based on histological analysis, they may be categorized into hyperplastic or adenomatous [15]. Hyperplastic polyps are benign in nature, and they have no risk of developing into cancer, whereas adenomatous polyps are premalignant and more likely to become malignant if left unchecked and not removed via polypectomy. Once cancer is formed in the inner lining of the colorectum, it can slowly invade the wall of the colorectum and thereby penetrate the blood or lymph vessels. Different stages of growth of a polyp are shown in Fig. 2. Cancer cells can travel through blood vessels to other vital organs and tissues in the body, which is generally termed as metastasis [16].

Depending on the extent of invasion of the abnormal cancerous growth, CRC may be described based on different stages [18].

- Stage 0: Cancer is in the inner lining of the colon and has not begun to invade the wall of the colorectum, which is generally termed as in-situ.
- Stage 1: Cancer has grown into the muscle layer but not spread through the wall or invaded nearby tissues.
- Stage 2: Cancers invaded through the wall of the colorectum and has grown into nearby structures.
- Stage 3.Cancers spread to nearby tissues and invade nearby lymph nodes.
- Stage 4.Cancers spread to other body parts and invade vital organs like liver, lung, or bones.

CRC's survival rate depends on the stage it is detected on, going from rates higher than 95% in early stages to rates lower than 35% in latter ones [19].

2.1 Current Trends in Incidence and Mortality of Colorectal Cancer

Colorectal cancer (CRC) is one of the most frequently occurring gastrointestinal (GI) related cancers. In 2018, there were over 1.8 million new cases of CRC and is one of the most prevalent cancers affecting both genders is CRC [20]. Globally, CRC is the third most common cause of cancer among men (10.0% of the total cancers) and the second most common cause in women (9.4% of the total cases) [21]. Figure 3 shows the incidence and mortality rate of CRC among all genders as per the Globocan'2018 report [22], whose incidence rate only behind lung and breast cancer, and the mortality rate is only behind lung cancer. Moreover, it is predicted to rise approximately by 80% in 2035, with an incidence of 114,986 new cases and mortality of 87,502 [23].

The incidence rate of CRC is higher in developed countries in comparison with the under-developed countries. However, the mortality rate is much higher in under-developed countries, which are mainly due to poor health care service and lack of awareness. The incidence and prevalence rate of CRC is much higher in western countries in contrast to Asian countries like India. However, in India, CRC ranked fourth among men with an incidence rate of 6.4% and fifth among women with an incidence rate of 3.4% [24]. Unfortunately, the low incidence rate of CRC in India also has a low 5-year survival rate. From Fig. 4, we can visualize the incidence rate of CRC varies across different geographical locations. Australia, Canada, New Zealand, Spain, France, Poland are the major countries among the highest age standard incidence rate (ASR) [25].

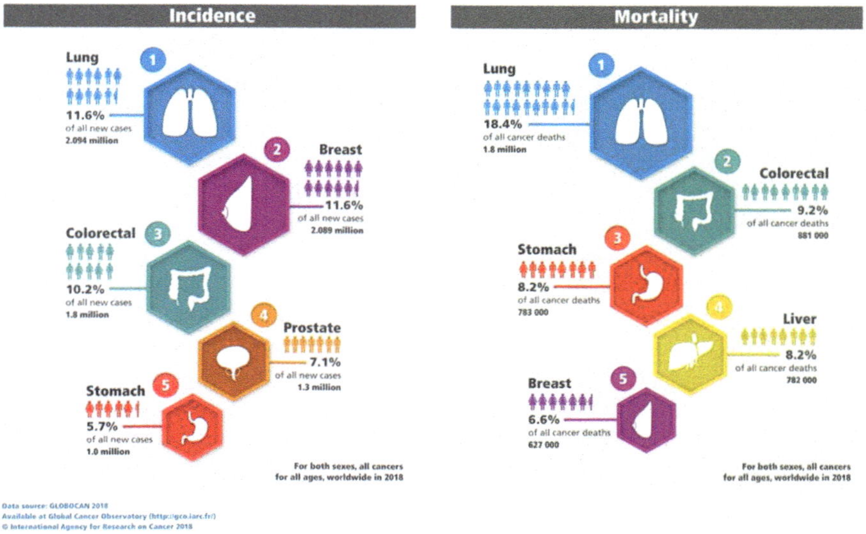

Fig. 3 Incidence and mortality rate of CRC

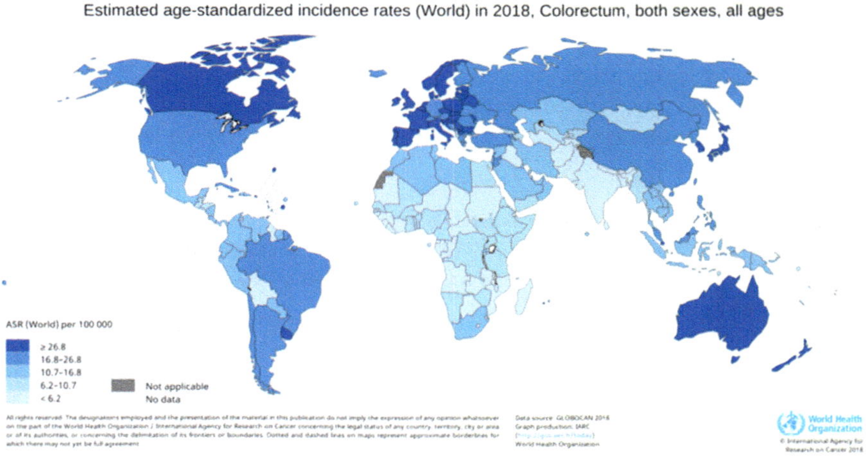

Fig. 4 Age standard incidence rate (ASR) of CRC across the globe

2.2 *Etiology of Colorectal Cancer*

Colorectal cancer is a significant health problem that appears from multiple tumorigenic pathways. Several risk factors are associated with the incidence of CRC. As discussed in the previous section, CRC is sporadic, and the incidence rate is higher in western countries, especially countries with adaptable lifestyle, unhealthy diet, physical inactivity, and obesity [26]. Some of the risk factors cannot be avoided, but some can be controlled so the disease can be prevented. Some of the relevant risk factors that are responsible for the cause of the disease are listed below [27].

- **Age**: CRC can be diagnosed at any age irrespective of gender. But the risk gets higher in older age, specifically after 50.
- **Personal history of benign polyps or adenoma**: If a person is diagnosed with a specific type of polyp earlier, then there is a high risk that the person will be prone to CRC in the future.
- **Inflammatory bowel conditions**: Person with a chronic inflammatory disease such as ulcerative colitis or Crohn's disease has a higher risk of CRC.
- **Inherited syndrome**: From first degree relatives (blood relative) of a person, certain genes are inherited, which may be a cause of CRC. These syndromes include familial adenomatous polyposis (FAP) and hereditary nonpolyposis colorectal cancer (HNPCC), also known as Lynch syndrome.
- **Ethnic background**: Individuals with the American-African race have a high risk of CRC.
- **Unhealthy diet**: Individuals associated with low fiber and high-fat diet and leading a sedentary lifestyle with no physical activity have a high risk of CRC.
- **Obesity**: Persons with unrestricted weight gain have a high risk of CRC.
- **Diabetes**: Diabetic patients having resistance to insulin have an increased risk of CRC.
- **Addiction to Smoking & Alcohol**: Individuals who are highly addicted to smoking and highly alcoholic are highly prone to CRC.
- **Radiation therapy**: Individuals diagnosed with other particular types of cancer, and if radiation therapy is directed toward the abdomen, then there is a high risk of CRC.

CRC does not have early warning signs and specific symptoms in its early stages. However, CRC is highly curable if detected in its early stages. So, it is recommended to consult experts if there is a change in bowel habits, blood in the stool or rectal bleeding, persistent abdominal pain.

3 Medical Diagnostic Procedures for Screening of Colorectal Cancer

Accurate and early screening is key to an early diagnosis of a disease. The survival rate of CRC is more than 95% if diagnosed in its early stage [19]. In this era of modern medical advancement, there is a development of sophisticated medical diagnosis modalities and techniques for early and timely detection and removal of polyp before it turns into CRC. The diagnosis tests and screening plays a vital role for doctors to plan the course of treatment. A concise overview of various tests and screening procedures which are commonly used for the diagnosis of CRC is given below [28].

1. Colonoscopy
   It is a minimally invasive screening modality that is widely used for the detection of CRC. It is considered as a "gold standard" for CRC screening. During colonoscopy examination, a long, flexible tube with a tiny video camera at the tip of the tube is inserted into the rectum to examine abnormality condition in the entire colon. However, the method might not detect polyp of smaller size or flat polyps and polyps hidden in the intestinal folds. As the method is completely manual, so it requires expert clinicians to explore the complete colon and acute concentration during the intervention as there are folds and angulations in the colon. Abnormal tissue samples for biopsy analysis or polyps can be removed during the exploration [29]. Figure 5 shows some samples of polyps captured during colonoscopy.
2. CT Colonoscopy
   It is a virtual colonoscopy examination of the colorectum. In this scheme, a CT-scan produces cross-sectional images where the colon can be visualized in the form of a 3D model. It helps to detect abnormality in the colorectum. To perceive a clear view of the colon, air is blown inside the colon via a catheter placed in the rectum. However, tiny polyps might not be detected by this examination, and if polyps are detected, colonoscopy is follow up for its removal and biopsy. Figure 6 shows different sample images from a CT colonoscopy examination [30].

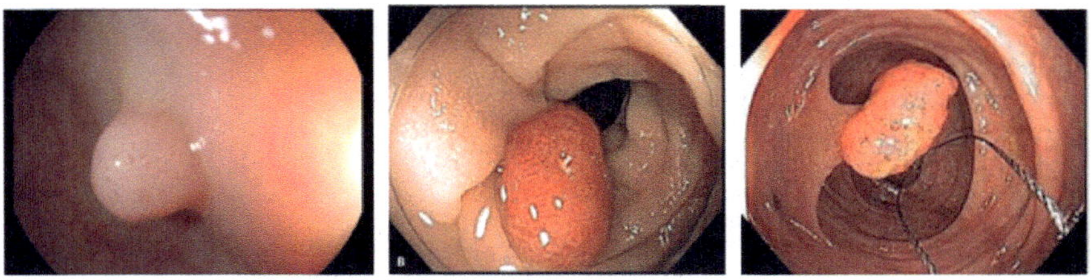

Fig. 5 Polyps captured during colonoscopy

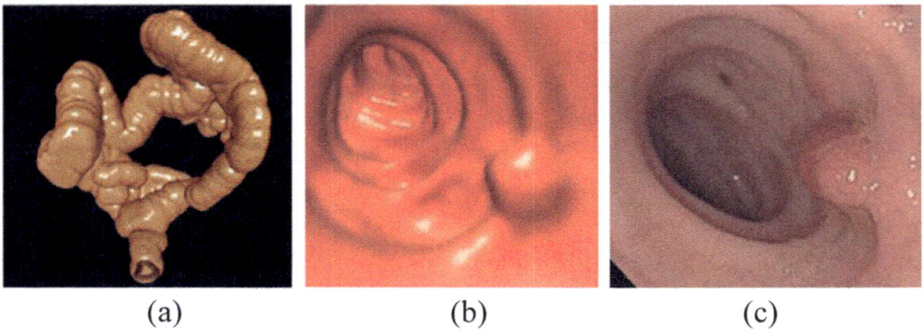

Fig. 6 **a** 3D model of Colon **b** Virtual CTC image **c** Colonosopy Image

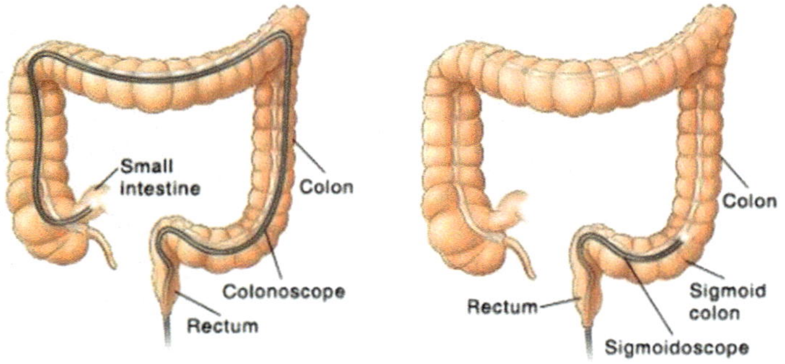

Fig. 7 Pictorial representation of a colonoscopy and sigmoidoscopy

3. Flexible Sigmoidoscopy
   In this examination, a thin, flexible tube with a tiny video camera at the tip of the tube is inserted into the rectum. This procedure explores the rectum and the lower part of the colon (sigmoid colon) in search of abnormality. However, the upper colon is not explored, and abnormality in the rest of the colon cannot be detected. Figure 7 shows the basic difference between colonoscopy and sigmoidoscopy in terms of its length of exploration [31].
4. Colon Capsule Endoscopy
   It is a relatively new non-invasive diagnostic procedure that enables physicians to explore the entire colon and identify abnormal growth [9]. In this scheme, a patient ingests a capsule containing two-minute cameras on either end. As the capsule travels the GI tract, it captures images and wirelessly transmits it to a recorder fitted outside the patient body. The camera captures more than 50,000 images during exploration, which is very much time consuming for a physician to detect polyp.
5. Endorectal Ultrasound
   In this procedure, a specialized transducer is inserted into the rectum, which helps clinicians to check how far the polyp invaded the wall of the rectum. Figure 8 shows a USG of a rectal tumor denoted by the white arrow [32].
6. Magnetic Resonance Imaging
   In this procedure, clinicians try to explore how far cancer has spread to local regions or nearby organs and subsequently check for any co-morbid condition to plan the treatment accordingly. Figure 9 shows an MRI image of a rectal polyp invaded the rectal wall (shown in red arrow) and proceeding toward the lymph nodes [33].
7. Fecal immunochemical test (FIT) or Fecal occult blood test (FOBT)
   These tests are generally pathological lab test where stool sample is tested for the presence of blood. However, there is a chance of false positive where it can predict the presence of abnormality when none is present.

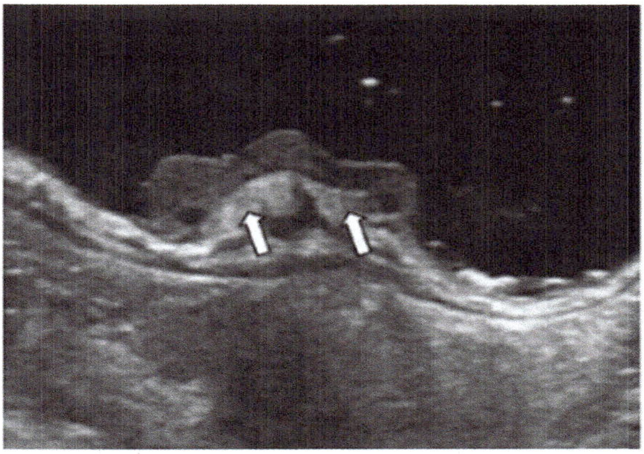

Fig. 8 Endorectal USG of a polyp in rectum

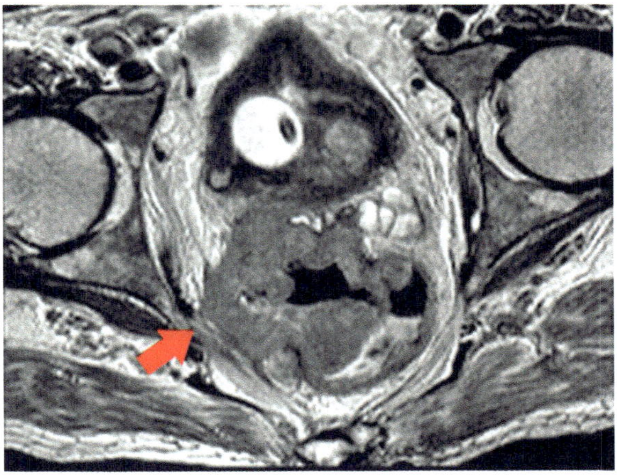

Fig. 9 Sample MRI image of a rectal polyp

8. Stool DNA test
   It is a non-invasive lab procedure, also known as the Cologuard test, where a stool sample is screened for suspicious DNA changes in the cells. It also considers signs of blood in the stool.

There is no specific screening guideline for CRC. So medical experts suggest a suitable test based on patient condition and staging, which may follow up several other diagnostic tests [34]. However, if an abnormality is detected in the colorectum by any of the diagnostic techniques apart from a colonoscopy, then a colonoscopy must be employed to remove the abnormal lesion. So, till date, colonoscopy is considered as the first choice for diagnostic examination. It may be noted that earlier the cancer is diagnosed, the easier and better is the treatment.

4 Necessity for Early Prediction of Colorectal Cancer

The early prognosis of the disease is closely related to the efficient decision-making process. Due to the high incidence rate of CRC over the past few decades, there is a huge demand for new technologies in the medical field for early diagnosis of CRC. The primary goal for early diagnosis of CRC is to detect the polyp when it is in the initial stage, and if not detected early, it may evolve to become malignant and spread to other vital organs [16]. When CRC is diagnosed in its early stage, the 5-year relative survival rate is 90% [35]. But unfortunately, only 4 out of 10 CRC is predicted in its early stage [36]. The survival rate descends to 15% when it is detected in the later advanced stage [37]. As per the statistics provided by the American Cancer Society's (ACS) publication, Cancer Facts, and Figures 2019 [38], it has been reported that for Stage 0, the 5-year survival rate is 90%. However, 39% of patients are fortunate to get diagnosed in this stage. If the cancer is spread to regional lymph nodes, regional organs/tissues, the rate declines to 71%. Furthermore, the 5-year survival rate drops to 14% if cancer spreads to distant organs in the body. Hence, there is a necessity for the early screening of colorectum.

5 Challenges in Early Detection of Colorectal Cancer

Early detection of polyps plays a crucial role in improving the survival rate of CRC patients. Medical imaging modalities, specifically colonoscopy, which is considered as a "gold standard" for the screening of CRC, has been widely employed for early detection and monitoring of the disease. Colonoscopy is an operator-dependent procedure, and detection of polyps completely lies in the dexterous ability of colonoscopists, so human factors such as lethargy, insufficient concentration, and back to back procedures may lead to miss-detection of polyps; moreover, the colonoscope cannot detect small or flat polyps which when diagnosed in later stage turns to be an adenoma [8]. However, early detection of polyp using computer-aided methods is very challenging due to high intraclass variation of polyps in terms of shape, size, color, texture, scale, and position and low interclass variation between polyps and artifacts like specular highlights, fecal content, and air bubbles [39]. Hence, several research groups across the world have reported different advanced and robust computer-aided intelligent techniques to reduce the polyp miss-detection rate, which may be clinically useful and assist clinicians in detecting CRC in its early stage of growth.

6 Analysis of Intelligent Medical Imaging Techniques for Early Detection of Colorectal Cancer

In this section, we explore different intelligent medical imaging techniques proposed for the early detection of polyp over the last few years. We can broadly categorize polyp detection techniques based on two approaches. The first approach is based on machine learning (ML) techniques where relevant domain-specific handcrafted features are extracted and trained using classifiers such as SVM, ANN for accurate detection of the region of interest. Handcrafted features-based approaches require strong domain knowledge to capture the optical primitives such as curves, edges, lines, color, texture by exploiting basic image processing methods. Recently, the second approach has achieved remarkable attention in the field of medical image analysis, which is based on deep learning (DL) techniques. The primary advantage of the deep networks is that it does not require feature engineering; rather, it automatically extracts the features while learning. In Fig. 10, we have depicted the generic steps for early detection of colorectal cancer using intelligent medical imaging techniques from colonoscopy frames.

6.1 Collection of CRC Frames

The collection of valid medical images with proper annotation by a medical expert is a primary step for any research in the biomedical domain. The acquisition of quality images during colonoscopy examination is a very big challenge. The images are captured in an uncontrolled environment and hence suffer from insufficient illumination conditions [40]. Moreover, there is a high conflict between positive and negative samples. There are different publicly available benchmark datasets, namely CVC-Clinic DB [41, 42], ETIS-LARIB [43], ASU-Mayo Clinic DB [44], and CVC-Colon DB [45]. CVC-Clinic DB consists of 612 SD frames of size 388 × 284 comprising 31 different variations of polyps from 34 video sequences. Etis Larib DB consists of 196 HD frames of size 1225 × 996 comprising 44 different polyps from 34 video sequences. ASU-Mayo Clinic DB consists of SD and HD frames of size 712 × 480 and 1920 × 1080, comprising long and short 20 video sequences containing approximately 400 polyp frames. CVC-colon DB consists of SD frames

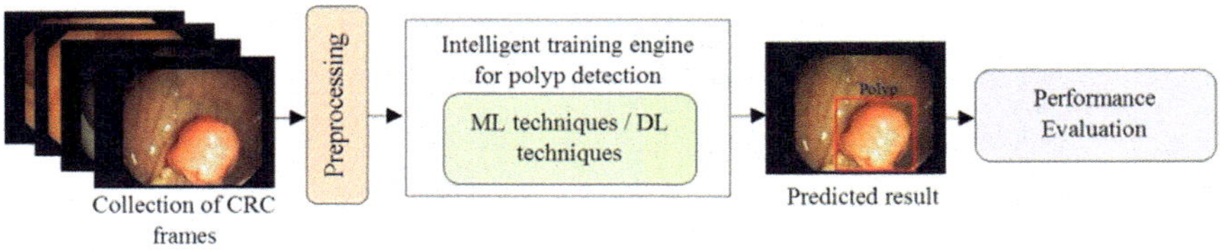

Fig. 10 Generic steps for early detection of CRC using intelligent medical imaging techniques

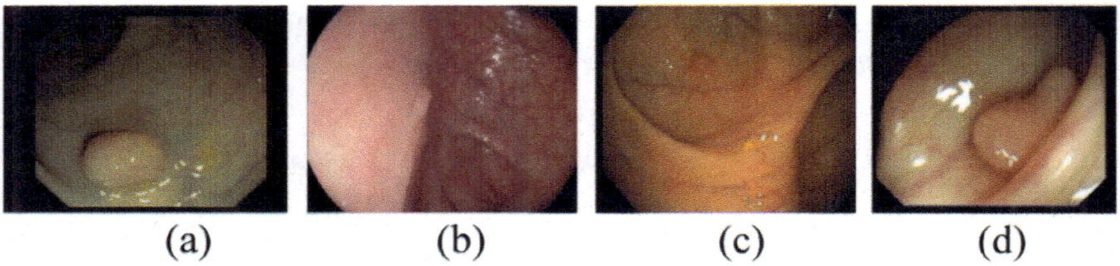

Fig. 11 Different databases and corresponding ground-truth **a** CVC-Clinic DB **b** Etis-Larib **c** Asu-Mayo Clinic DB **d** CVC-Colon DB

of size 574 × 500 comprising 300 polyp frames. All the images from the datasets are annotated by experts which are used as ground-truth for the purpose of performance evaluation of the developed techniques. Figure 11 shows some sample colonoscopy frames from each of the databases.

6.2 Preprocessing

Colonoscopy frames are generally more complex to understand due to their improper acquisition and artifacts. Usually, artifacts like specular highlights, over-exposed regions, and motion blur arise due to scene illumination conditions [46]. Figure 12 shows different artifacts in different colonoscopy frames. Furthermore, polyps may be falsely detected due to blood vessels and endoluminal folds. These artifacts vary from frame to frame and impede further processing of the frame. However, most of the techniques cited in the literature are focused on the removal of specular highlights as it is the most common issue in the majority of the frames. Jorge Bernal et al. [47] have proposed a preprocessing step to tackle the impact of specular highlights, blood vessels, and black mask in the colonoscopy frame. They have detected the specular highlights by assuming that the intensity inside the specular regions in relatively higher, which is followed by inpainting the detected specular regions via a diffusion method. Similarly, the black mask outside the image frame is also inpainted

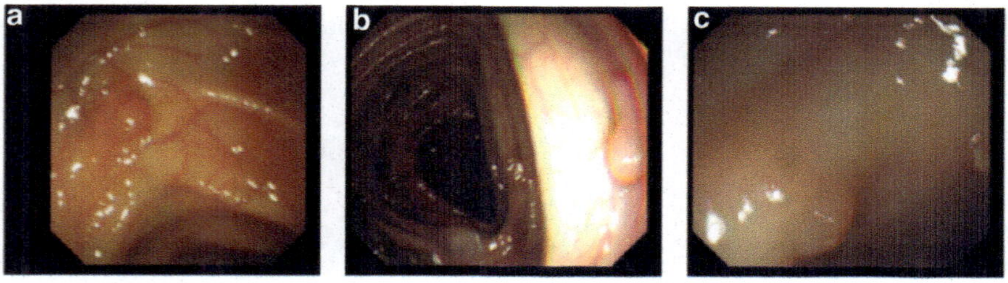

Fig. 12 Different scene illumination challenges **a** Presence of specular highlights **b** Presence of over-exposed regions **c** Impact of motion blur

in a similar way, such as the boundary information is retained. Blood vessels are partially tackled by exploring different color spaces and choosing the particular space which mitigates the blood vessels. Another interesting work for specular highlights removal is proposed by Isabel Funke et al. [48], which is based on deep learning, namely, generative adversarial network (GAN). In their work, CNN is employed to localize and remove the specular highlights. To train the CNN, GAN is being used, which introduces an adversary to judge the performance of the CNN. Recently, Alain Sanchez-Gonzalez et al. [49] proposed another method for the removal of specular highlights and black edges. The images are transformed into HSV color space, and the black edges are detected using a channel V thresholding. The detected black edge is replaced by the average pixel value of the neighborhood, and the image is reconstructed without a black edge. For the removal of specular highlights, initially, the specular regions are detected via a threshold, and a binary mask is being generated. The mask calculates the position of the specular highlights, and specular pixels are reconstructed.

6.3 Intelligent Training Engine for Polyp Detection

With the advancement of flexible medical imaging equipment, the development of computer-aided intelligent algorithms for assessment and early prediction of disease severity is taking on an increasingly critical role in the health industry. This will serve as a supportive diagnostic tool for the physicians to early predict the presence of abnormalities and hereby quantification of the disease severity. Different ML and DL techniques have been proposed by different groups of researchers across the globe. Yuan et al. [50] proposed a bottom-up and top-down saliency approach for the automatic detection of polyps. Initially, they have segmented the image using super-pixels, and sparse auto-encoder (SAE) is employed to learn discriminative features. Then, a bottom-up saliency map is constructed by fusion of contrast and object center-based saliency method, which is followed by a top-down saliency map that uses a strong classifier to classify the samples obtained from the bottom-up saliency. Finally, the two saliency maps are integrated to detect the polyps. In 2016, Tajbakhsh et al. [44] proposed an automated hybrid polyp detection method based on shape and context information. An edge map is being extracted using Canny's method, followed by estimating gradient orientation for each edge pixel. 1D DCT is being applied for each non-overlapping oriented patch, all the patches are concatenated, and a feature vector is formed, which is fed into a classifier for edge classification. Finally, polyps are localized based on the refined edge maps. An interesting model to localize polyps based on polyp appearance is proposed by Bernal et al. [41]. They define the polyp boundaries based on the valley information. The valley information is integrated via the window of radial sectors with other boundary parameters associated with polyps shape to generate Window Median Depth of Valleys Accumulation (WM-DOVA) energy maps, which determine the likelihood of the presence of a polyp. The approaches described in [41, 44, 50] are based on ML techniques, which will have to gather domain-specific handcrafted features based on color, shape, texture from the input images. But the chosen features might not detect the polyp efficiently.

In contrast, DL techniques drastically improved the former state-of-the-art and appeared to be more promising. Duran-Lopez et al. [51] proposed a deep learning algorithm to detect the presence of polyps in the frame. Initially, a fully convolutional neural network (FCNN) extracts features and predicts the region of interest(ROI). The saturation of accuracy due to the depth of the network is handled by ResNet50. Once the ROI is predicted, it is passed to faster regional convolutional neural network(faster R-CNN) detector to detect the polyp regions. Before the detection algorithm, some preprocessing steps are employed to remove the artifacts such as black edges and resize the image followed by augmentation. Another interesting scheme for polyp detection is proposed by Shin [52]. In their work, they have used a generative adversarial network(GAN) for generating synthetic polyp images and thereby increasing the number of training samples. For detection, they have employed a pre-trained CNN, namely Inception-Resnet, which is used for the F-RCNN detection method. F-RCNN employs a region proposal network (RPN) to propose the candidate object regions (ROI). A polyp detection scheme based on deeply learned hierarchical features is proposed by Park et al. [53]. Initially, a Canny edge detector is used to crop the boundaries lines, which is followed by the extraction of small patches. The patches are fed into CNN in different scales, which learns scale-invariant features. Finally, a post-processing step is involved in detecting the polyp candidate region. Mohammed et al. [54] proposed a deep network, namely Y-Net, for polyp detection, which consists of two encoder networks and a decoder network. One of the encoders uses a pre-trained VGG19 network trained on ImageNet, and the other encoder uses the same pre-trained network except for the fully connected layer. Both of the encoders are concatenated and fed into a decoder network with upsampling layers and a convolutional layer. Another novel polyp detection approach was proposed by Park et al. [55]. Initially, frames are preprocessed using a hidden Markov model combined with Shannon entropy and average—filtered value. Different patches at different angles are fed into CNN for multi-level feature extraction. Finally, in the fully connected layer of CNN, a conditional random field is used for intra and inter-spatial and temporal relationships in frames. Zhang et al. [56] proposed a transfer learning strategy for the detection and classification of polyps. In this scheme, low-level features are learned from two publicly available non-medical databases. The CNN architecture chosen for detection and classification is CaffeNet and an SVM with different kernels which are employed in the fully connected layer for detection and classification. The detected polyps are fed into another similar network to classify polyps into adenoma, hyperplasia, and non-polyp. A 3-D integration deep learning framework is proposed by Yu et al. [39]. In this scheme, a 3D CNN is employed for the detection of polyps as it can learn spatio-temporal features with more discrimination capabilities between intraclass and interclass variations. A 3D-offline FCN is initially employed to learn spatio-temporal features, which are followed by an online 3D-FCN to effectively remove the FPs generated in offline learning. Finally, the two networks are fused for the detection of polyps.

In the last few research works, it has been seen that the hybrid methodologies boost the detection performance in comparison with the individual ML and DL approaches. Recently in 2018, Pogorelov et al. [37] have proposed a hybrid technique for polyp detection and localization from medical images. They have employed frame-wise and block-wise approaches. Frame-wise detection is done by handcrafted global features (GF-D), fine-tuning, or retraining of existing DL architectures. GF-D handcrafted features are extracted using the LIRE framework and fed into the logistic model tree (LMT) for classification. Fine-tuning or retraining is done on 3 DL architectures. Localization of the detected frame is done by GAN. The frame-wise detection algorithms using the DL architectures are used block-wise localization using a window size of 128 × 128 with a partial overlap of 66%. Another hybrid method is proposed by Billah et al. [57]. Initially, the frames are preprocessed by removing the black borders. Then, patches of size 227 × 227 are extracted for feature extraction. Textural features are extracted by 3-level wavelet decomposition, and statistical features are also extracted using the co-occurrence matrix. A CNN with multiple layers is also employed to extract CNN features. Finally, the handcrafted features and DL features are combined and fed into an SVM classifier to detect the polyp. In Table 2, we have summarized different intelligent imaging techniques for early detection of colorectal cancer.

6.4 Performance Evaluation

Performance evaluation is very crucial to judge the clinical applicability of the developed intelligent medical imaging techniques. Some of the classical performance evaluation methods are described in terms of the confusion matrix [58], as shown in Fig. 13. Each of the attributes in the matrix is defined as follows: True positive (TP) means the system accurately detects the polyp in the predicted output, false positive (FP) means the system falsely detect the polyp in the predicted output but the absence of polyp in the ground-truth, true negative (TN) means the system correctly does not detect the presence of polyp in the predicated output, and lastly, false negative (FN) means the system does not indicate the presence the polyp in the predicated output although there is a polyp in the ground-truth.

Additionally, based on the aforementioned metrics, precision, recall, accuracy, specificity, sensitivity, F1, and F2 score are also evaluated to justify the efficacy of the developed technique [8]. Each of the measures is evaluated, as shown in Tables 1 and 2.

Fig. 13 Confusion matrix

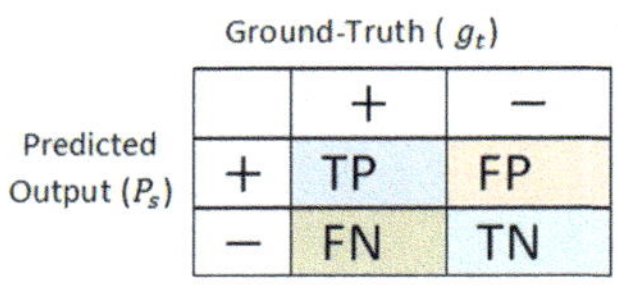

Table 1 Aggregation metrics

Metric	Calculation
Precision (Pre)	$\text{Pre} = \frac{\text{TP}}{\text{TP} + \text{FP}}$
Recall (Rec)	$\text{Rec} = \frac{\text{TP}}{\text{TP} + \text{FN}}$
Specificity (Spec)	$\text{Spec} = \frac{\text{TN}}{\text{FP} + \text{FN}}$
Accuracy (Acc)	$\text{Acc} = \frac{\text{TP} + \text{TN}}{\text{TP} + \text{TN} + \text{FP} + \text{FN}}$
F1-score (F1)	$F1 = \frac{2 \times \text{Pre} \times \text{Rec}}{\text{Pre} + \text{Rec}}$
F2-score (F2)	$F2 = \frac{5 \times \text{Pre} \times \text{Rec}}{4 \times \text{Pre} + \text{Rec}}$

7 Discussion and Future Directions

Computer-aided diagnosis (CAD) using advanced intelligent medical imaging techniques for the detection of polyps has shown promising intervention in providing decision support during colonoscopy examination. Intelligent imaging techniques viz. machine learning (ML) and deep learning (DL) algorithms proposed by various researchers, as discussed in the previous section, have shown significant technological developments in this domain. However, in the last few years, deep learning algorithms have gained tremendous success in the field of medical imaging due to the availability of advanced workstations powered with GPUs [59]. Moreover, deep learning algorithms overcome the burden of task-specific feature engineering as in machine learning algorithms. As discussed, polyps have wide variations in terms of shape, size, color, and texture from frame to frame. ML approaches proposed by different research groups for detection and localization of polyps cannot efficiently detect or localize the polyp, especially in different illumination conditions and for complex heterogeneous shapes as done by DL approaches. However, many DL approaches proposed by researchers cannot detect or localize flat polyps and also polyps, where the polyp region has the same intensity level as the non-polyp region. One major challenge in training the deep CNN models is the availability of large training annotated datasets. To cope with the challenge, researchers have adopted an augmentation scheme and transfer learning scheme to enhance the datasets and learn the low-level features. The acquisition process of a polyp is generally done in an uncontrolled environment, so images are susceptible to various artifacts like specular highlights, endoluminal folds, blood vessels, and hard mimics, which hinders polyp detection task and such artifacts, may be falsely detected as polyps. In few studies, researchers have proposed methodologies to remove those artifacts, specifically specular highlights. It is worth mentioning that there is a need for a balanced and annotated datasets for training the intelligent imaging techniques for polyp detection and its performance evaluation. Different researchers have proposed transfer learning schemes for the detection of polyps with pre-trained models trained on non-medical datasets. But transfer learning approach on medical data from the same minimally

Table 2 Summary of different intelligent imaging techniques for early detection of CRC

Approach	Author	Method	Database (images)	Evaluation metric
ML	[50]	Integrate bottom-up and top-down saliency map	CVC-Clinic DB (612)	Pre: 0.72, Rec: 0.83, F1: 0.75
ML	[44]	Context and shape-based approach	CVC-Colon DB (300) and Asu-Mayo DB (19,400)	CVC-Colon DB Sens: 88.0%, Asu-Mayo DB Sens: 48%
ML	[41]	Polyp appearance model based on WM-DOVA saliency maps	CVC-Colon DB (300) and CVC-Clinic DB(612)	CVC-Colon DB Acc: 72.33%, CVC-Clinic DB Acc:70.26%
DL	[51]	Faster Regional CNN to detect the polyp in a frame	CVC-clinic video DB (11,954)	TP: 3533, FP: 866, TN:1659,FN:1154,Pre:8031%,Rec:75.37%,Acc:71.99%,Spec:65.70%,F1:77.76 F2:76.30
DL	[52]	Pre-trained using Inception-ResNet and finally detection by F-RCNN	Training: CVC-clinic DB (372), Testing: CVC-clinicVideo DB (18 videos)	TP:6760, FP: 2981, FN:3265,TN:962, Pre:69.4%,Rec:67.4%
DL	[53]	CNN with scale-invariant patches to localize polyp followed by post-processing	Training: CVC-Clinic DB(550), Validation: CVC-Clinic DB (62)	TP:48,FP:25,FN:10,Pre:0.65,Rec:0.82,F1:0.73

(continued)

Table 2 (continued)

Approach	Author	Method	Database (images)	Evaluation metric
DL	[54]	Y-net with two encoders pre-trained with VGG19 network and a decoder network	Training: Asu-mayo clinic DB (4278), Testing: Asu-mayo clinic DB (4300)	TP:3582,FP:513, FN:662,Pre:87.4%,Rec:85.9%, F1:85.9%,F2:85%
DL	[55]	Enhancement of the quality of frames followed by extraction of patches at different angles which are fed into CNN and finally CRF for localization	Private Dataset	AUC: 0.86, Sens: 86%,Spec: 85%
DL	[56]	Transfer learning with low-level features from non-medical databases to target CaffeNet with SVM	Non-medical DB: Places205 and ILSVRC Medical DB: PWH (215)	Acc: 98%, Rec: 97.6%,Pre:99.4%,F1:0.98,AUC:1
DL	[39]	An offline and online 3D FCN	Asu-Mayo Clinic DB	TP:3062,FP:414, FN:1251,Pre:88.1%,Rec:71%,F1:78.6%,F2:73.9%
DL + ML	[37]	Detection and localization using handcrafted features and fine-tuning and retraining of DL architectures	Training: CVC (356),CVC(612), CVC(968), Nerthus.Testing: CVC(12 k), Kvasir	Detection: Spec: 94%, Acc: 90.9% Localization: Spec: 98.4%, Acc: 94.6%

(continued)

Table 2 (continued)

Approach	Author	Method	Database (images)	Evaluation metric
DL + ML	[57]	Extract color wavelet features, statistical features, and CNN features and classify using SVM	Private DB (100 videos)	Sens: 98.79%, Spec: 98.52%, Acc: 98.65%

invasive modality can be a direction for future research. Artifacts due to scene conditions have a high impact on the performance of the polyp detection task, so a strong and robust preprocessing step needs to be paid more attention.

8 Conclusion

Early diagnosis of CRC is highly curable and hence can save lives. In this chapter, we have highlighted recent advances in intelligent CAD systems for early diagnosis of CRC. However, cooperation between researchers and clinicians is very much essential in developing advanced and sophisticated computer-aided techniques as they understand their own challenges and limitations in their respective areas of research. The contributions shown by different researchers in the early detection of polyps could be a potential aid for clinicians. Intelligent systems can effectively improve the diagnosis system and grading process as well as enable quantitative studies of the mechanisms underlying disease onset and progression and plan further course of treatment.

Acknowledgements The first author is grateful to the Council of Scientific and Industrial Research (CSIR) for providing Senior Research Fellowship (SRF) under the SRF-Direct fellowship program (ACK No.143416/2K17/1, File No.09/096(0922)2K18 EMR-I). The authors are thankful for the Indo-Austrian joint project grant No. INT/AUSTRIA/BMWF/P-25/2018 funded by the DST, GOI, and the SPARC project (ID: 231) funded by MHRD, GOI.

References

1. The International Agency for Research on Cancer (IARC) Report W (2018) Latest global cancer data: Cancer burden rises to 18.1 million new cases and 9.6 million cancer deaths in 2018. Int Agency Res Cancer 13–15
2. Meyer B, Are C (2018) Current status and future directions in colorectal cancer. Indian J Surg Oncol 9:440–441
3. Cancer Statistics India, https://cancerindia.org.in/cancer-statistics/.last Accessed 10 Oct 2019
4. Sharma D, Singh G (2017) Clinico-pathological profile of colorectal cancer in first two decades of life: a retrospective analysis from tertiary health center. Indian J Cancer 54:397
5. Kerr J, Day P, Broadstock M, Weir R, Bidwell S (2007) Systematic review of the effectiveness of population screening for colorectal cancer
6. Hassan C, Quintero E, Dumonceau J-M, Regula J, Brandão C, Chaussade S, Dekker E, Dinis-Ribeiro M, Ferlitsch M, Gimeno-García A (2013) Post-polypectomy colonoscopy surveillance: European society of gastrointestinal endoscopy (ESGE) guideline. Endoscopy. 45:842–864
7. Hoff G, Sauar J, Hofstad B, Vatn MH (1996) The Norwegian guidelines for surveillance after polypectomy: 10-year intervals. Scand J Gastroenterol 31:834–836
8. Bernal J, Tajkbaksh N, Sanchez FJ, Matuszewski BJ, Chen H, Yu L, Angermann Q, Romain O, Rustad B, Balasingham I, Pogorelov K, Choi S, Debard Q, Maier-Hein L, Speidel S, Stoyanov D, Brandao P, Cordova H, Sanchez-Montes C, Gurudu SR, Fernandez-Esparrach G, Dray X, Liang J, Histace A (2017) Comparative validation of polyp detection methods in video

colonoscopy: results from the MICCAI 2015 endoscopic vision challenge. IEEE Trans Med Imaging 36:1231–1249
9. Iddan G, Meron G, Glukhovsky A, Swain P (2000) Wireless capsule endoscopy. Nature 405:417
10. What Is Colorectal Cancer? https://www.cancer.org/cancer/colon-rectal-cancer/about/what-is-colorectal-cancer.html. Last Accessed 18 Aug 2019
11. Hounnou G, Destrieux C, Desme J, Bertrand P, Velut S (2002) Anatomical study of the length of the human intestine. Surg Radiol Anat 24:290–294
12. Picture of the Colon, https://www.webmd.com/digestive-disorders/picture-of-the-colon#1.
13. Summers RM (2010) Polyp size measurement at CT colonography: what do we know and what do we need to know? Radiology 255:707–720
14. Axon A, Diebold MD, Fujino M, Fujita R, Genta RM, Gonvers JJ, Guelrud M, Inoue H, Jung M, Kashida H (2005) Update on the Paris classification of superficial neoplastic lesions in the digestive tract. Endoscopy 37:570–578
15. Polyps A (2014) Early detection of colorectal cancer (CRC) and adenomatous polyps clinical decision support tool. Gastroenterology 147:925–926
16. How cancer starts,grows and spreads, https://www.cancer.ca/en/cancer-information/cancer-101/what-is-cancer/how-cancer-starts-grows-and-spreads/?region=on. Last Accessed 02 Oct 2019
17. Advances in Colorectal Cancer, https://www.nih.gov/research-training/advances-colorectal-cancer-research. last Accessed 08 May 2019
18. Colon Cancer, https://www.fascrs.org/patients/disease-condition/colon-cancer-expanded-version. Last Accessed 18 Sept 2019
19. Brenner H, Jansen L, Ulrich A, Chang-Claude J, Hoffmeister M (2016) Survival of patients with symptom-and screening-detected colorectal cancer. Oncotarget 7:44695
20. American Institute of Cancer Research, https://www.wcrf.org/dietandcancer/cancer-trends/colorectal-cancer-statistics. Last Accessed 10 July 2019
21. Bray F, Ferlay J, Soerjomataram I, Siegel RL, Torre LA, Jemal A (2018) Global cancer statistics 2018: GLOBOCAN estimates of incidence and mortality worldwide for 36 cancers in 185 countries. CA Cancer J Clin 68:394–424
22. New Global Cancer Data, https://www.uicc.org/news/new-global-cancer-data-globocan-2018. Last Accessed 10 Aug 2019
23. Meyer B, Are C (2017) Current status and future directions in colorectal cancer. Indian J Surg Oncol 8:455–456. https://doi.org/10.1007/s13193-017-0717-3
24. Globocan 2018: India factsheet, https://cancerindia.org.in/globocan-2018-india-factsheet/. Last accessed 10 July 2019
25. Cancer Today, https://gco.iarc.fr/today/home. Last Accessed 10 July 2019
26. Deng Y (2017) Rectal cancer in asian vs. western countries: why the variation in incidence? Curr Treat Options Oncol 18:64
27. CRC risk factors, https://www.mayoclinic.org/diseases-conditions/colon-cancer/symptoms-causes/syc-20353669. Last Accessed 03 May 2019
28. CRC screening, https://www.cancer.org/cancer/colon-rectal-cancer.html. Last Accessed 05 May 2019
29. Min M, Su S, He W, Bi Y, Ma Z, Liu Y (2019) Computer-aided diagnosis of colorectal polyps using linked color imaging colonoscopy to predict histology. Sci Rep 9:2881
30. CT virtual Colonoscopy, https://www.sdimaging.com/ct-virtual-colonoscopy/. Last Accessed 01 Aug 2019
31. Flexible sigmoidoscopy, https://www.mountnittany.org/articles/healthsheets/7398. Last Accessed 01 Aug 2019
32. Radiology key, https://radiologykey.com/colon-rectum-and-anus/. Last Accessed 05 Jan 2019
33. MRI, https://www.massgeneral.org/imaging/news/radrounds/october_2011/. Last Accessed 06 Jan 2019
34. Geiger TM, Ricciardi R (2009) Screening options and recommendations for colorectal cancer. Clin Colon Rectal Surg 22:209–217

35. Society AC: Colorectal Cancer Facts & Figures 2017–2019 ; 1–40. Available from: https://www.cancer.org/content/dam/cancer-org/research/cancer-facts-and-statistics/colorectal-cancer-facts-and-figures/colorectal-cancer-facts-and-figures-2017-2019.pdf. Centers Dis. Control Prev. Behav. Risk Factor Surveill. Syst. 2014. Public use data file. Color. Cancer Screen. (2017).
36. Solving the Mystery of Why Colorectal Cancer Is on the Rise in Young Adults, https://www.ascopost.com/issues/june-25-2019/solving-the-mystery-of-why-colorectal-cancer-is-on-the-rise-in-young-adults/. Last Accessed 10 July 2019
37. Pogorelov K, Ostroukhova O, Jeppsson M, Espeland H, Griwodz C, De Lange T, Johansen D, Riegler M, Halvorsen P (2018) deep learning and hand-crafted feature based approaches for polyp detection in medical videos. In: Proceedings of IEEE Symposium on Computer Medical System, 381–386. https://doi.org/10.1109/CBMS.2018.00073
38. Screening CC (2019) American cancer society. In: Colorectal cancer facts & figures 2017–2019
39. Yu L, Chen H, Dou Q, Qin J, Heng PA (2017) Integrating online and offline three-dimensional deep learning for automated polyp detection in colonoscopy videos. IEEE J Biomed Heal Informatics 21:65–75
40. Wang P, Xiao X, Glissen Brown JR, Berzin TM, Tu M, Xiong F, Hu X, Liu P, Song Y, Zhang D, Yang X, Li L, He J, Yi X, Liu J, Liu X (2018) Development and validation of a deep-learning algorithm for the detection of polyps during colonoscopy. Nat Biomed Eng 2:741–748
41. Bernal J, Sánchez FJ, Fernández-Esparrach G, Gil D, Rodríguez C, Vilariño F (2015) WM-DOVA maps for accurate polyp highlighting in colonoscopy: validation vs. saliency maps from physicians. Comput Med Imaging Graph 43:99–111
42. Fernández-Esparrach G, Bernal J, López-Cerón M, Córdova H, Sánchez-Montes C, de Miguel CR, Sánchez FJ (2016) Exploring the clinical potential of an automatic colonic polyp detection method based on the creation of energy maps. Endoscopy. 48:837–842
43. Silva J, Histace A, Romain O, Dray X, Granado B (2014) Toward embedded detection of polyps in wce images for early diagnosis of colorectal cancer. Int J Comput Assist Radiol Surg 9:283–293
44. Tajbakhsh N, Gurudu SR, Liang J (2016) Automated polyp detection in colonoscopy videos using shape and context information. IEEE Trans Med Imaging 35:630–644
45. Bernal J, Sánchez J, Vilariño F (2012) Towards automatic polyp detection with a polyp appearance model. Pattern Recognit 45:3166–3182
46. Sánchez-González A, Soto BG-Z (2017) Colonoscopy image pre-processing for the development of computer-aided diagnostic tools. In: Surgical Robotics. IntechOpen
47. Bernal J, Sanchez J. Vilarino F (2013) Impact of image preprocessing methods on polyp localization in colonoscopy frames. In: Proceedings of annual international conference on IEEE engineering in medicine and biology society EMBS, 7350–7354. https://doi.org/10.1109/EMBC.2013.6611256
48. Funke I, Bodenstedt S, Riediger C, Weitz J, Speidel S (2018) Generative adversarial networks for specular highlight removal in endoscopic images. 1057604, 3
49. Sánchez-gonzález A, García-zapirain B, Sierra-sosa D, Elmaghraby A (2018) Automatized colon polyp segmentation via contour region analysis. Comput Biol Med 100:152–164
50. Yuan Y, Li D, Meng MQH (2018) Automatic polyp detection via a novel unified bottom-up and top-down saliency approach. IEEE J Biomed Heal Informatics 22:1250–1260
51. Duran-Lopez L, Luna-Perejon F, Amaya-Rodriguez I, Civit-Masot J, Civit-Balcells A, Vicente-Diaz S, Linares-Barranco A (2019) Polyp detection in gastrointestinal images using faster regional convolutional neural network. VISIGRAPP 2019 Proc 14th Int J Conf Comput Vision, Imaging Comput Graph Theory Appl 4:626–631
52. Shin Y, Qadir HA, Balasingham I (2018) Abnormal colon polyp image synthesis using conditional adversarial networks for improved detection performance. IEEE Access 6:56007–56017
53. Park S, Lee M, Kwak N (2015) Polyp detection in colonoscopy videos using deeply-learned hierarchical features. Seoul Natl. Univ. 1–4
54. Mohammed A, Yildirim S, Farup I, Pedersen M, Hovde Ø (2018) Y-Net: a deep convolutional neural network for polyp detection, 1–11

55. Park SY, Sargent D (2016) Colonoscopic polyp detection using convolutional neural networks. Med Imaging 2016 Comput Diagnosis 9785:978528
56. Zhang R, Zheng Y, Mak TWC, Yu R, Wong SH, Lau JYW, Poon CCY (2017) Automatic detection and classification of colorectal polyps by transferring low-level CNN features from nonmedical domain. IEEE J Biomed Heal Informatics 21:41–47
57. Billah M, Waheed S, Rahman MM (2017) An automatic gastrointestinal polyp detection system in video endoscopy using fusion of color wavelet and convolutional neural network features. Int J Biomed Imaging 2017:1–9
58. Prabha DS, Kumar JS (2016) Performance evaluation of image segmentation using objective methods. Indian J Sci Technol 9:1–8
59. Ahmad OF, Soares AS, Mazomenos E, Brandao P, Vega R, Seward E, Stoyanov D, Chand M, Lovat LB (2019) Artificial intelligence and computer-aided diagnosis in colonoscopy: current evidence and future directions. Lancet Gastroenterol Hepatol 4:71–80

Chapter 5
Optimal Path Schemes of Mobile Anchor Nodes Aided Localization for Different Applications in IoT

Enas M. Ahmed, Anar A. Hady, and Sherine M. Abd El-Kader

1 Introduction

A wireless sensor network (WSN) is a group of small sensor nodes that form a network in association to each other for monitoring and collecting data from a selected area [1]. WSNs are used in many applications such as event monitoring, military applications, and health-care applications [2]. In many of these applications, the required mission will not be achieved without knowing the positions of nodes gathering data from their surroundings. For instance, in the case of industry cities, identifying the position of a sensor node that monitors a certain pollution ratio is a must. Also, locations of sensor nodes are very important in many other applications like target tracking [3]. The classical method of finding the locations of nodes is global positioning system (GPS). However, it comes at a very high cost, and it consumes a lot of power which is a crucial issue in sensor networks as they rely mainly on batteries, and thus, GPS is not suitable for many applications.

Defining the location of sensor nodes is called localization; it is the action of determining the locations of the unknown sensor nodes in the network using known locations of other sensor nodes called anchor nodes. Localization in WSNs has been addressed widely in the literature, but there are still a lot of challenges facing proposed methods.

Localization techniques are classified into two main classes, range based and range free techniques [4]. Range-based techniques depend on the characteristics of

E. M. Ahmed (✉)
Benha Faculty Of Engineering, Benha University, Benha, Egypt
e-mail: enas.mohamed@bhit.bu.edu.eg

A. A. Hady · S. M. Abd El-Kader
Electronics Research Institute, Cario, Egypt
e-mail: anar_abdelhady@eri.sci.eg

S. M. Abd El-Kader
e-mail: Sherine@eri.sci.eg

A. E. Hassanien et al. (eds.), *Enabling Machine Learning Applications in Data Science*,
Algorithms for Intelligent Systems, https://doi.org/10.1007/978-981-33-6129-4_5

the signal as received signal strength (RSSI), time of arrival (TOA), time difference of arrival (TDOA), angle of arrival (AOA) and thus are affected by the surroundings. On the other hand, range free techniques are based on connectivity information and the relationship between the unknown nodes and anchor nodes. There are numerous range free techniques proposed in the literature as DV-Hop, APIT, and MDS.

As localization studies increased, a different classification appeared depending on mobility of unknown nodes and anchor nodes. In this paper, we explain this classification extensively and the different types of path models for mobile anchors. We show the advantages and disadvantages of each and the most suitable applications they could be used for. The rest of this paper is sectioned as follows; Sect. 2 explains the localization model based on mobility in WSNs. Section 3 introduces mobile anchor paths of mobile anchors static unknown nodes class. In Sect. 4, a comparative analysis is given. Section 5 includes the conclusions.

2 Mobility Based Localization Model in WSN

The localization model is organized in accordance with mobility into four classes as follows: static anchor nodes static unknown nodes, static anchor nodes mobile anchor nodes, mobile anchor nodes static unknown nodes, and mobile anchor nodes mobile unknown nodes. The fundamental class is the static anchor nodes static unknown nodes, and in this class, the unknown nodes and anchor nodes are fixed. The second class is mobile unknown nodes static anchor nodes [6–8], and in this class, the unknown nodes are mobile, but the anchor nodes are fixed. This class is suitable for applications where the unknown sensor is tied to a movable body as animals and the static anchor nodes give each unknown node its updated position at each motion. The third class is static unknown nodes and mobile anchor nodes, and the mobile node is responsible for providing the unknown nodes with the location information by traveling through the network and helping other unknown nodes to get their location. This class is divided into random mobility and programmed mobility. In the next section, this class is analyzed extensively. The fourth class is mobile unknown nodes and mobile anchor nodes [9] used in applications as robotics. There are many surveys in the literature that discuss the types of localization algorithm based on mobile anchor node. Authors in [10] present some selective localization techniques based on mobile anchor nodes. Authors in [11] present a classification of mobile wireless sensor network; they distinguish between classic wireless sensor network and mobile wireless sensor network and show the benefits of mobility. In [12], the authors categorized localization techniques relying on the mobility into the four categories adopted in this paper as mentioned earlier.

3 Static Unknown Nodes, Mobile Anchor Nodes Class

The principle of this class is to use mobile anchor nodes to get positions of other unknown sensor nodes using the characteristic of the signal such as TOA, RSSI, AOA or using its connectivity information such as the number of hops between unknown nodes and mobile anchor nodes. The mobile anchor node travels through the network, stops at stop points, and sends beacon messages to the static unknown nodes in its communication range, and by these messages and the information which included in the messages, the unknown nodes can get their locations. The main purpose of this class is to reduce the number of anchors by replacing the fixed anchor nodes with one mobile anchor node and thus reducing cost and power. Nonetheless, using a mobile anchor node faces a lot of challenges with respect to choosing the best path model which achieves good localization accuracy, short and consumes less power. In the following subsections, different types of paths are introduced, and a comparison between them is done.

3.1 Mobile Anchor Paths

This research area focuses on creating a best path model for mobile anchor node aided localization, so we need to distinguish between different types of path models. We categorized path models into three categories: random path model, static path model, and dynamic path model [13–15]. The major difference between them is the movement direction decision. Each type of path model will be presented and examples of it in the subsequent parts.

3.1.1 Random Path Model

In this type of path model, the movement direction and the movement velocity through traverse in the network are random. The mobile anchor node moves in a random motion as shown in Fig. 1 [5]. The path length of each movement is chosen randomly [16]. There is no promise in that model that all unknown nodes receive a sufficient number of messages from the mobile anchor node. Also, it does not resolve the collinearity problem. We use this type of path model where there is no need for high localization accuracy and no need for low power consumption.

An example for random path model is presented in [17], and the authors design a new model that relies on realized DV-hop localization technique in wireless sensor networks based on mobile anchor node not static anchor node. They execute three mobile anchor modules: unplanned waypoint, unplanned direction and unplanned stop points which are explained in the next subsections. The mobile anchor nodes go through the network randomly and contact each other during the break time of

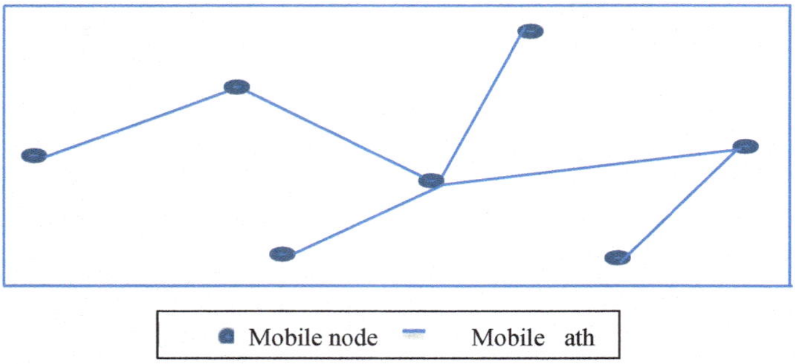

Fig. 1 Mobile anchor random path

their motion. The management of localization plan is the same for each localization model as follows as shown in Fig. 2:

1. Primary state: In this state, the unknown nodes and mobile anchor nodes are spread out in the area. Each of them has a particular ID. The first location of an anchor node i is p_i. In this step, the DV-hop localization technique is used by anchors to get locations of the unknown nodes for the first time.
2. Transfer state: There is t parameter, which is used to numerate the transfer time of mobile anchors. The mobile anchor moves in a predestined model. The value of t is reduced by 1 after each transfer step.
3. Position determination state: The mobile anchor stops for a period of time, and then, the unknown nodes can know their positions with the aid of DV-hop localization technique based on mobile anchor nodes.
4. Refresh state: Unknown nodes restore their calculated positions after each position determination, so they can calculate exact location information.

Localization Depending on Unplanned Waypoint Model

The idea of this model [18] relies on choosing the waypoint randomly in the simulation and does not take into consideration any detectable rules. Unplanned waypoint model involves five steps:

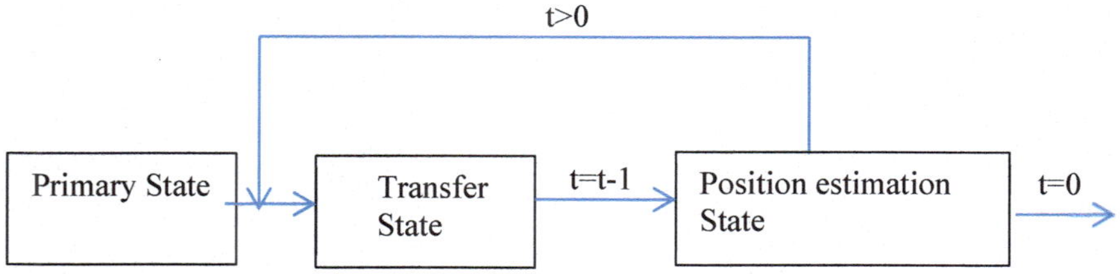

Fig. 2 Plan of localization model

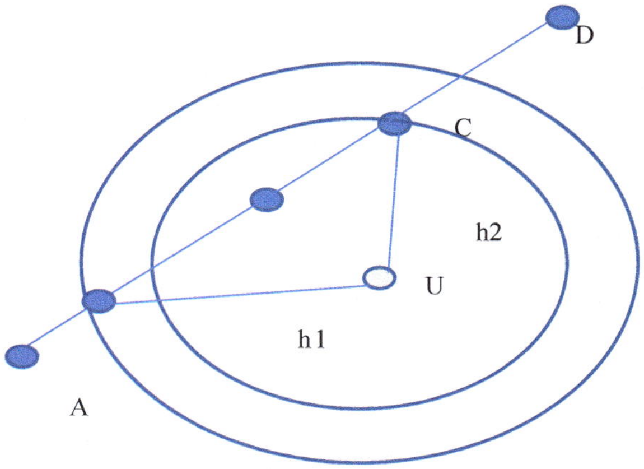

Fig. 3 Contact between *U* (unknown node) and *A* (mobile node) in noisy environment

1. Choosing a new stop point randomly.
2. Setting a speed which lies in the range from v_l to v_h, where v_l and v_h are the low and high velocities.
3. Moving to the determined stop point.
4. Stay in the determined stop point for a certain time.
5. Go to the first step and reiterate the steps.

An example of unplanned waypoint model is presented in [19], and the authors propose an algorithm named Yu which is based on a flight beacon which helps the unknown nodes to detect their locations. The authors present a different vision of the traditional terminology of receiver scope in the case of noisy wireless surrounding. The error possibility of getting messages reduces with the rise of signal to noise ratio. So, there are many centered circular shapes that can exist whose midpoint is the position of the unknown nodes as shown in Fig. 3. If the unknown node gets messages without any fault, a circular shape can exist. The center of all circular shapes is the location of unknown nodes. This method achieves high localization without using any characteristic information such as angle of arrival.

Localization Depended on Unplanned Direction Model

This model relies on choosing an unplanned direction by mobile anchor nodes and moving toward the boundary of the simulation area in that direction, then modifying the velocity and direction. This model is proposed to solve the problems of unplanned mobility model. In this model, the mobile anchor node selects an unplanned direction and travels to the boundary of the simulation area in that direction, as shown in Fig. 4. When it reaches its goal, it stops for a time duration and then chooses a new direction and velocity and repeats these steps again.

An example of unplanned direction model is presented in [20], and the authors propose a localization method relying on arrival and departure (ADO) of mobile

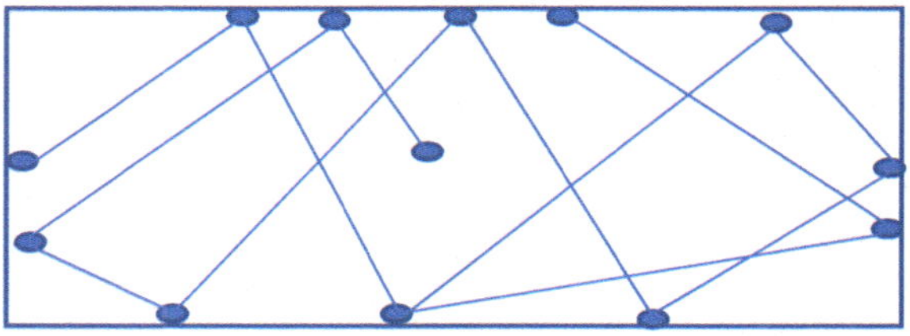

Fig. 4 Example of unplanned direction model

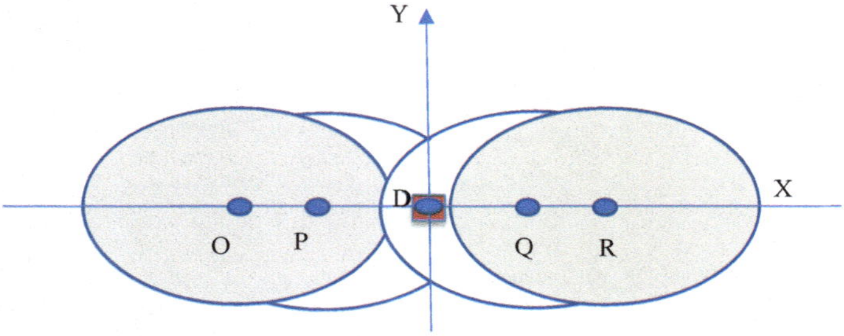

Fig. 5 Localization using landing and departure parameters of the mobile node

anchor nodes. They analyze three mobility models, namely the scattered-straight line move arrangement, intensive-straight line move arrangement, and the unplanned move arrangement. In the unplanned move arrangement, the mobile node traverses through an established straight line, then after a limited distance, it adjusts a new different distance. ADO uses landing and departure parameters to determine the position of unknown nodes as shown in Fig. 5. The mobile node moves from south to west in the line of *x*-axis. The unknown node *D* accepts the mobile node message at the beginning of the simulation time when the mobile anchor node reaches the departure point *D*. When the mobile anchor node moves forward again, the unknown node *F* accepts the mobile message. The mobile node reaches at the landing point.

Localization Depended on Unplanned Movement Model

The unplanned movement model is an uncomplicated model relying on unplanned velocity and directions. The mobile anchor transfers from the recent position to the new position by selecting the velocity randomly from the normal velocity distribution between [0, 2π]. Each transfer in the unplanned movement model happens in either fixed time or fixed distance. Then, after an interval, a new different velocity and direction are computed. The unplanned movement model has no memory, and the recent velocity and direction are not based on the last velocity and direction. An example of unplanned movement model is presented in [21], and the authors propose a distributed online localization method (DOL) based on unplanned transferring mobile anchor node. In this method, the unknown nodes get their locations

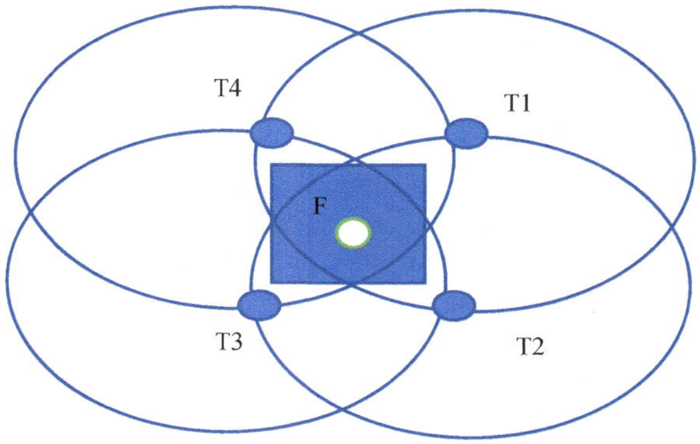

Fig. 6 Center of the crossing area *F*

by using rough-grained area free localization method. Each unknown node accepts messages from the mobile anchor node which lies in its scope. Then, after saving many messages from mobile anchor nodes at different locations, it uses a rectangular box which is bounded inside the overlapping zone. This step is very important, to assure that the real position of the unknown node is inside the rectangular box and to get an accurate the computation of the position of the unknown node. After that, the unknown node uses the center of the crossing area of many anchor messages as its position as shown in Fig. 6.

3.1.2 Static Path Model

The second type of path models is the static path model. The main thing that distinguishes between static path model and the others is the nature of the non-changeable movement path. When it faces any obstacles, it turns around the obstacle and completes its trajectory except in some cases. The static path model solves the problems of random path models; the problem of mobile anchor node messages which do not reach all unknown nodes in the network area. So, it achieves a high localization accuracy. Nonetheless, design path trajectory needs to be attentive to some details such as the number of mobile anchors stop points, the path length, and the shape of the path trajectory which have to avoid the problem of collinearity. There are many examples of static path model such as Scan and Hilbert path model [22], LMAT path model [23], *Z*-Curve [3], and *H*-Curve path model [16] as shown in Fig. 7. For example, in LMAT path model, the path model has similar edge triangle trajectory form. The mobile anchor node travels through the network and stops at each point of each vertex of the triangle this path model scheme solves the problem of collinearity.

In *Z*-Curve path model in Fig. 8, the path model forms *Z*-shape curves. The network area is divided into four sub-squares, and the mobile anchor node connects each center of each sub-square. The advantage of this model is that it guarantees that the area is full covered by *Z*-curve path models and solves the problem of collinearity.

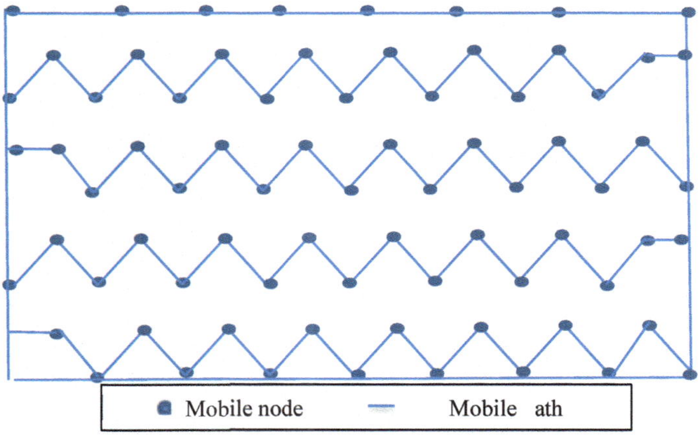

Fig. 7 LMAT path model

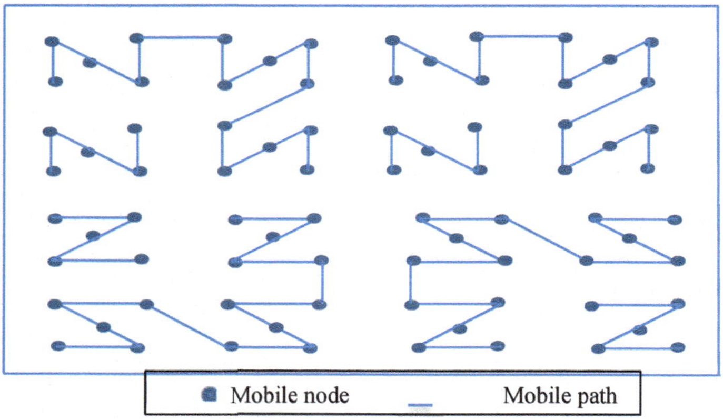

Fig. 8 *Z*-Path Model

If the area is very large, the mobile anchor path model does not cover the area. We divide the area into more than one level, and the *Z*-curve path model connects each level to the other.

In *H*-curve path model as in Fig. 9, the path model is formed using an *H*-curve path model. Each *H*-curve consists of deterministic number of stop points. The mobile anchor departs from the starting point at the angle of the network area and travels in the deterministic line from each anchor stop point to another in the fixed distance of 1 m which does not change through the simulation time. Then, through its trip, it sends packets which contain its information. The unknown nodes receive these packets and use the information to get their position. This path model can solve the problem

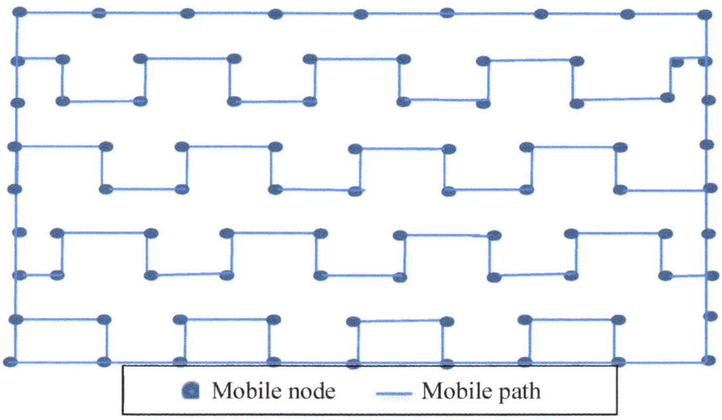

Fig. 9 *H*-Path Model

of collinearity, and it also guarantees that the area is fully covered by *H*-curve path model.

In SLMAT path model [24] as in Fig. 10, the stop points of mobile nodes form uniform triangles. The process of determining the unknown node position goes through four phases:

1. Set the communication domain.
2. The mobile node decides the locations of stop points, and then, it travels through the area based on its determined path and sends messages to the unknown nodes from its stop points. The message consists of the positions of the mobile node at each stop point and time of sending.

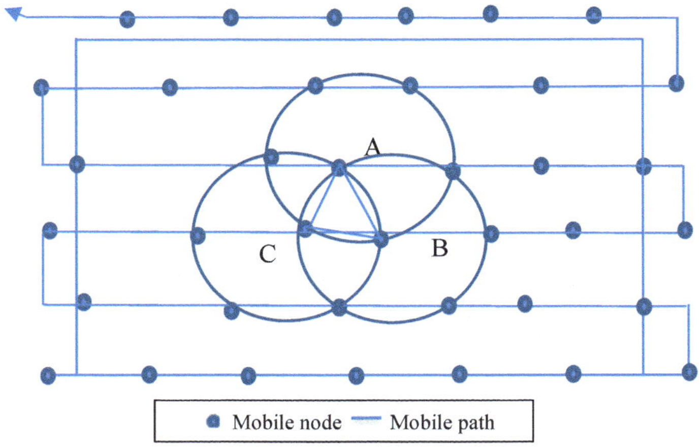

Fig. 10 SLMAT path model

3. Each unknown node saves the messages and determines the distances between it and stop points of the mobile node using RSSI technique.
4. Each unknown node checks if three of mobile node messages it saved can arrange a uniform triangle and if it is inside the triangle. If so, it calculates its position using trilateration.

This path model can solve collinearity and can handle the obstacles which it faces during its trip. This model can avoid the obstacles which it meets.

3.1.3 Dynamic Path Model

The third type of path models is a dynamic path model. The main difference between the dynamic path model and static path model is the nature of varying the movement path and the varying path shape which depends on the current status of the sensor nodes and their information. There are many dynamic paths examples such as Breadth-First (BRF), Backtracking Greedy (BTG) algorithms [25], and Dynamic Path of Mobile Beacon (DPMB) [26]. In [25], the authors present a technique based on managing the mobile anchor movement via reinforcement learning (RL) which is a machine learning approach. RL needs an uncountable number of tries to decide the position of unknown nodes with high accuracy. In [27], the authors proposed NLA_MB algorithm which relies on dividing the operation area into hexagonal subareas and build a model of mobile node localization error ratio related to its dynamic path and distance which it moved. Mobile beacon node resolves this model using an equation based on effective force theory and information of guide sensor nodes position and then creating a path which is compatible with the node arrangement. Based on beacon node position information, the unknown nodes use maximum likelihood estimation algorithm to calculate their position. In [28], the authors propose a mobile beacon supported localization (MBL), and based on that, the mobile node picks the unknown node with the greatest number of nodes around it as a bunch node. Then, it merges global path design for all bunch heads and local path design for every bunch member to design the mobile beacon path. The mobile beacon node walks through all the bunch head using the genetic theory. Then, the mobile node moves through hexagon paths in each bunch with a bunch head set in the middle of hexagon. This algorithm introduces a path design which obtains high accuracy results and minimizes energy consumption.

In [29], the authors propose a virtual force dynamic path relied on the force which is created by the interchanging force between the mobile node and fixed unknown node. This dynamic path is suitable for irregular networks such as, "*U*" form and "*L*" form. The mobile node has omnidirectional antenna, and each antenna has ID (1, 2, … 0.6). It uses these antennas to collect messages from sensor nodes and computes the sum of the force on mobile node. After that, the mobile node changes its position according to the total sum of the force and repeats this operation again (Fig. 11).

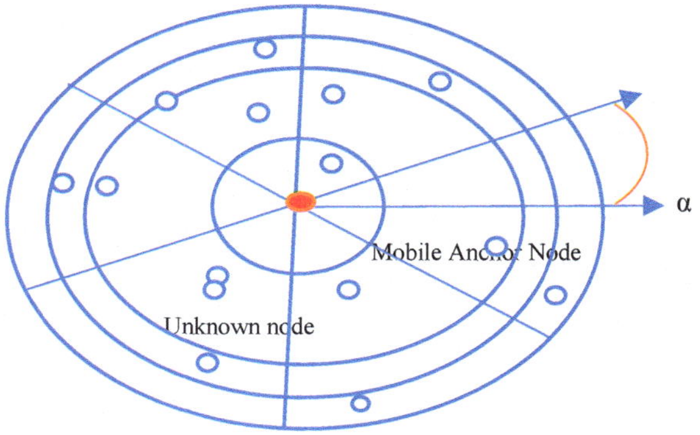

Fig. 11 Mobile node has omnidirectional antenna which collects messages in six directions

In [30], the authors propose a mobile beacon assisted localization algorithm (MBAL) which has three operations, namely moving operation, finding the locations of unknown nodes operation, and determining the path trajectory operation. First, the mobile node moves on the equal edges' triangle, and the unknown nodes which do not have their location information send messages to the mobile node to ask for their position. The mobile node collects these messages and forms a path with minimum length. At each process, the mobile node elects the closest node and accepts its messages to form the path.

4 Comparative Analysis

As shown in Table 1, the mobility model based on DV-hop in [17] is a random path model which achieves low position accuracy for the unknown nodes. It cannot solve the problem of collinearity and the full coverage area. This mobility model has no fixed design, because its parameters cannot be determined or measured. Also, it consumes high power more than static path model and dynamic path model. The advantage of this model is that it is easy to implement in a real environment. The Yu algorithm in [19] uses a random path model which is used to get the positions of the unknown nodes without having angle or distance parameters. The back draw of using this path is that it considers the path is direct lines; however, this is not the real. The ADO algorithm in [20] uses a random path model which does not need the distance parameter and does not consume much power. However, it needs a collection of anchor nodes to determine the position of unknown nodes. The DOL algorithm in [21] uses a random path model which does not need any extra hardware; however, it achieves low accuracy. The LMAT path in [23] is a static path which achieves high accuracy, and full coverage area, and it is suitable for any area size by setting the lengths of the triangle edges and the communication range to fit any area size; however, the energy consumption parameter has not been analyzed extensively in this model. The Z-Curve path in [3] is a static path which achieves high position

Table 1 Comparison of different types of paths

Algorithm	Type of anchor path	Full coverage area	Accuracy	Power consumption
Mobility model based on DV-hop	Random path	NO	Low	Medium
Yu	Random path	NO	Low	High
ADO	Random path	NO	Low	–
DOL	Random path	NO	Low	–
LMAT	Static path	YES	High	High
Z-Curve	Static path	YES	High	Medium
H-Curve	Static path	YES	High	Low
SLMAT	Static path	YES	High	Low
BRF & BTG	Dynamic path	NO	Low	–
NLA-BM	Dynamic path	NO	High	–
MBL	Dynamic path	NO	Medium	–
Virtual force	Dynamic path	NO	Medium	–
MBAL	Dynamic path	NO	Medium	–

accuracy results and can cover all the monitoring area. It can handle obstacles which the anchor node meets in its trajectory. The SLMAT path in [24] is a static path model which achieves higher results than the *Z*-Curve in [3] in accuracy, power, and path length. The *H*-Curve in [14] is a static path which achieves high accuracy, collinearity, and full coverage area. The strong point in this model is that it has a short path, shorter than the paths in [17, 23] and [3]. The BRF and BTG path in [25] is a dynamic path model which is suitable for non-uniform networks, and it achieves medium accuracy and cannot guarantee full coverage area. The NLA-BM path in [27] is a dynamic path model which uses force theory and positions of stop points of the mobile anchor node to draw the mobile node path, it can achieve high accuracy. However, it doesn't guarantee the full coverage area. The MBL model in [28] is a dynamic path which changes related to nodes' allocation; however, its complication increased with the quantity of nodes. The virtual force model in [29] is a dynamic path model which achieves high results in an irregular network; however, it needs extra hardware (omnidirectional antenna). The MBAL model in [30] is a dynamic path that achieves good results in an irregular network, but it does not take obstacles into consideration.

According to the applications, the random path model can be used in the applications which do not require high accuracy such as air pollution monitoring systems and circumstance monitoring. The static path models and the dynamic path models are used in applications which needed high accuracy such as military applications.

5 Conclusions

In this paper, localization techniques have been categorized based on mobility into four categories: the static anchors static unknown nodes, mobile anchors mobile unknown nodes, static anchors mobile unknown nodes, and mobile anchors static unknown nodes. Path models used in mobile anchors have been presented as follows: Static unknown nodes are categorized into three models, namely the random path model, the static path model, and the dynamic path model. We investigate through these models and chose the optimal, newest, and the most powerful ones to be presented to show the strong points and weak points in each one based on most suitable IoT application for each model type.

References

1. Faiz B, Luo J (2018) ELPMA: Efficient localization algorithm based path planning for mobile anchor in wireless sensor network. Wirel Person Commun Int J 721–744.
2. Rashid B, Mubashir HR (2016) Applications of wireless sensor networks for urban areas: a survey. J Netw Comput Appl 60:192–219
3. Rezazadeh J, Moradi M, Ismail AS, Dutkiewicz E (2014) Superior path planning mechanism for mobile beacon assisted localization in wireless sensor networks. IEEE Sens J 14(9):3052–3064
4. Mohamed E, Zakaria H, Abdelhalim MB (2017) An Improved DV-hop localization algorithm. Chap Adv Intell Syst Comput 533:332–341
5. Dong Q, Xu X (2014) A novel weighted centroid localization algorithm based on RSSI for an outdoor environment. J Commun 9(3):279–285
6. Luo RC, Chen O, Pan SH (2005) Mobile user localization in wireless sensor network using grey prediction method. In: 31st annual conference of IEEE industrial electronics society, IECON 2005, pp 6
7. Salem MA, Tarrad IF, Youssef MI, El-Kader SMA (2019) QoS categories activeness-aware adaptive EDCA algorithm for dense IoT networks. Int J Comput Netw Commun 11(3):67–83
8. Hussein HH, El-kader SMA (2018) Enhancing signal to noise interference ratio for device to device technology in 5G applying mode selection technique. In: ACCS/PEIT 2017–2017 International conference on advanced control circuits systems and 2017. International conference on new paradigms in electronics and information technology 2018, pp 187–192
9. Neuwinger B, Witkowski U (2009) Ad-hoc communication and localization system for mobile robots. In Kim JH et al (eds) Advances in robotics. Springer, Berlin
10. Guangjie H et al (2016) A survey on mobile anchor node assisted localization in wireless sensor networks. IEEE Commun Surveys Tutorials 18(3)
11. Santos F (2008) Localization in wireless sensor networks. ACM J 1–19
12. Alrajehm NA, Shams B (2013) Localization techniques in wireless sensor networks. Int J Distrib Sensor Netw 12(8):761–764
13. Han G, Yang X, Liu L, Zhang W, Guizani M (2020) A Disaster management-oriented path planning for mobile anchor node-based localization in wireless sensor networks. IEEE Trans Emerging Topics Comput 8:115–125
14. Hady AA, Abd El-Kader SM, Eissa HS (2013) Intelligent sleeping mechanism for wireless sensor networks. Egyptian Inform J 14(2):109–115
15. Abdel-Hady A, Fahmy HMA, El-Kader SMA, Eissa HS, Salem AM (2014) Multilevel minimized delay clustering protocol for wireless sensor networks. Int J Commun Netw Distrib Syst 13(2):187–220

16. Alomari A et al New Path Planning model for mobile anchor assisted localization in wireless sensor Networks. Wireless Networks 24(7):2589–2607
17. Han G, Chao J, Zhang C, Shu L, Li Q (2014) The impacts of mobility models on dv-hop based localization in mobile wireless sensor networks. J Netw Comput Appl 42:70–79
18. Johnson DB, Matz DA (1996) Dynamic source routing in ad hoc wireless networks. Mobile Comput 353:153–181
19. Yu G, Yu F, Feng L (2008) A three-dimensional localization algorithm using a mobile anchor node under wireless channel. In: IEEE international joint conference on neural networks. Hong Kong. pp 477–483
20. Xiao B, Chen H, Zhou S (2008) Distributed localization using a moving beacon in wireless sensor networks. IEEE Trans Parallel Distrib Syst 19(5)
21. Galstyan A, Kirshnamachari B, Lerman S (2004) Pattern distributed online localization in sensor networks using a mobile target. In: Information processing sensor networks (IPSN). Berkeley, pp 61–70
22. Koutsonikolas D, Das SM, Hu YC (2007) Path planning of mobile landmarks for localization in wireless sensor networks. Comput Commun 30(13):2577–2592
23. Han G, Xu H, Jiang J, Shu L, Hara T, Nishio S (2013) Path planning using a mobile anchor node based on trilateration in wireless sensor networks. Wireless Commun Mobile Comput 13(14):1324–1336
24. Han G, Yang X, Liu L, Zhang W, Guizani M (2017) A disaster management—oriented path planning for mobile anchor node- based localization in wireless sensor networks. IEEE Trans Emerging Topics Comput 8:115–125
25. Li H, Wang J, Li X, Ma H (2008) Real-time path planning of mobile anchor node in localization for wireless sensor networks. Int Conf Informat Automat ICIA 2008:384–389
26. Li X, Mitton N, Simplot-Ryl I, Simplot-Ryl D (2011) Mobile-beacon assisted sensor localization with dynamic beacon mobility scheduling. In: IEEE 8th international conference on mobile ad hoc and sensor systems. MASS, pp 490–499
27. Chen Y, Lu S, Chen J, Ran T (2016) Node localization algorithm of wireless sensor networks with mobile beacon node. Peer Peer Netw Appl 10(3):795–807
28. Zhao F, Luo H, Lin Q (2009) A mobile beacon- assisted localization algorithm based on network-density clustering for wireless sensor networks. In: 5th intentional conference on mobile ad-hoc and sensor networks. Fujian, pp 304–310
29. Fu Q, Chen W, Liu K, Wang X (2010) Study on mobile beacon trajectory for node localization in wireless sensor networks. In: IEEE international conference on information and automation. Harbin, pp 1577–1581
30. Kim K, Lee W (2007) MBAL: a mobile beacon- assisted localization scheme for wireless sensor networks. In: 16th International conference on computer communications and networks. Honolulu, pp 57–62

Chapter 6
Artificial Intelligence in 3D Printing

Fatma M. Talaat and Esraa Hassan

1 Introduction

Recently, artificial intelligence (AI) has become a new solution of our future life, and it can already be integrated into some advanced equipment. 3D printing technology can also make full use of artificial intelligence technology. It is an innovative new technology. Now, people are adding new technologies like artificial intelligence to 3D printers. This combination of artificial intelligence and 3D printing will enable 3D printing technology to develop more rapidly.

Artificial intelligence is a form of intelligence displayed by machines, sometimes called machine intelligence. This kind of machine can obtain information by itself and reasoning and analysis to draw conclusions (imitation of human learning thinking) in order to perform advanced tasks. AI machines simulate human intelligent behavior. The artificial intelligence automation process can be carried out autonomously in many different ways.

The operation of the 3D printer itself is a relatively complicated process, and the addition of artificial intelligence can greatly help it to improve, thereby making this new technology operate more efficiently. Nowadays, the 3D printing (3DP) is considered as one of the advanced manufacturing technologies which refer to the layer-basis manufacturing of 3D object using digital data [1, 2]. 3DP becomes an integral part of upper-limb prosthesis, resulting in a response to a number of tangible problems, including timely and, in some cases, restricted access to conventional prostheses.

The history of the plastic materials industry began in 1868 when John Wesley Hyatt tried to find a new material for billiard balls, and cellulose was discovered. In 1950, the company called Akerwerk made Germany's first type of injection molding machine (Mastro 2016).

F. M. Talaat (✉) · E. Hassan
Faculty of Artificial Intelligence, Kafrelsheikh University, Kafrelsheikh, Egypt

A. E. Hassanien et al. (eds.), *Enabling Machine Learning Applications in Data Science*, Algorithms for Intelligent Systems, https://doi.org/10.1007/978-981-33-6129-4_6

Manufacturing industry evolution came as plastics were replacing other materials. Today, plastics are multifunctional polymers that can be easily modified for various applications with the help of modern technologies. Artificial intelligence can be integrated into the design of 3D printing factories and thus change the future of manufacturing. AI Build is a London-based company that has developed an automated 3D printing platform based on artificial intelligence.

The device has a smart extruder that can detect any problems. It can also make autonomous response decisions. It is a large 3D printer using industrial robots and machine learning software. The use of robotic arms and robotic arms that can determine and print viable parts without any problems on their own is a real revolution. AI Build is working on machines that can view, create, and learn their own errors and create truly complex structures.

The rest of paper is organized as follows: Sect. 2 gives a background for some basic concepts. Section 3 introduces a comprehensive survey of various AI methods applied throughout 3DP. It includes AI in printability checking, slicing acceleration, nozzle path planning, cloud service platform, service evaluation, and security of attack detection. Our conclusion is discussed in Sect. 4.

2 Background and Basic Concepts

This section introduces some concepts in the field of additive manufacturing (AM), artificial intelligence (AI), and machine learning.

2.1 Additive Manufacturing (AM)

In the modern world, additive manufacturing is the most prospective and highly appraised technological field. 3D printing technologies have their origins in the stereolithographic printing technique invented in 1984 [3]. Additive manufacturing (AM) is used to create objects with complex forms that cannot be fabricated using traditional techniques.

With rapid prototyping and digital manufacturing, AM develops science evolution. In engineering applications, various polymer, metal, and biomaterials are used primarily to create prototypes and finished products with unique shapes, multifunctional compositions, reliability, and high quality.

Because of the development of AM technologies, the future will be different. Innovations will result in the supply chain being shortened, which means that consumers can print anything they want at home and remotely.

The digital manufacturing era has just begun, and ideas can be conveyed to the 3D models and then sent directly to the 3D printing machine, which can begin work and finish without human supervision. The only thing people will care about is the different food production materials, equipment, spare part replacement products, etc.

2.2 *Artificial Intelligence (AI)*

The key areas of potential work and consumer demand are artificial intelligence and software creation [4]. The familiar market should generally change to the digital market worldwide.

Countries will unite their local additive manufacturing markets to achieve this glorious future and collaborate on developing and exporting 3D printing technologies around the world. UK, USA, and most of Europe's leading countries are entering the digital manufacturing era. For example, the company "Airbus" uses most of the airplane parts built and manufactured with AM [3].

2.3 *Machine Learning (ML)*

The machine learning area is rising at an exponential pace due to data access from other channels. Machine learning is a sub-discipline of artificial intelligence, which studies algorithms that can automate learning from data to make data-based predictions [5]. It uses algorithms to locate patterns in large data volumes and extracts valuable knowledge/observations. Machine learning has big societal impacts throughout a broad variety of science, industry and market applications, big data for example, Internet of things (IoT), Web, psychology, health care, etc.

3 Artificial Intelligence Applications in 3D Printing

Machine learning and additive manufacturing (AM) have been influential in recent research years and have received even more attention from the industry, public, and scientists. AM is a woman promising and cutting-edge technology Various industries, with the goal of promoting the age of Future manufacturing industry named Industry 4.0, smart fabrication, automated fabrication. ML techniques are currently being used to solve problems at AM pre-manufacturing stage Production by generative nature and trial.

This paper session features a survey of recent research into the additive sector production using ML. In this case, the cameras were equipped with high resolution at the front, top, and left of the 3D desktop in the level printers to detect malicious infill faults during building product. The images were grabbed layer after layer from the top of a preview of the simulation. Throughout pictures of experimental, non-defect and defect infills were used for investigating malicious 3D printing flaws product. The images extracted the data, and two ML algorithms (Naïve Bayes) were applied: classification and decision tree J48.

The outcome of the analysis reveals 85.26% accuracy in Naïve Bayes In J48 decision trees, classifier and 95.51% rankings. However, compared with traditional

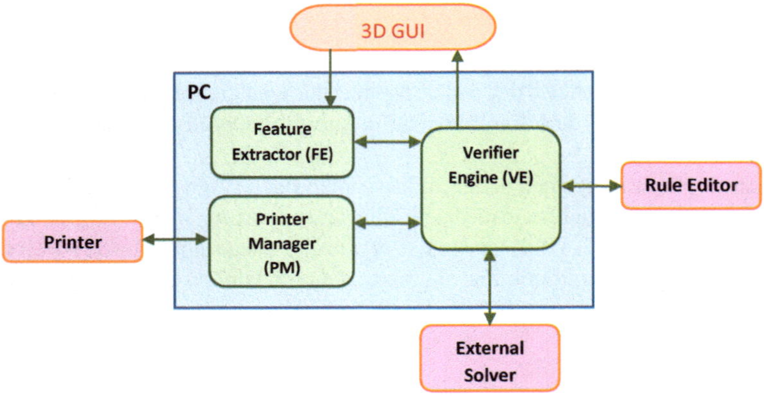

Fig. 1 Mechanism of original printability checker

manufacturing methods, the promotion and employment of 3DP are still limited due to the geometrical attribute, time consumption and specific material requirement.

3.1 Original Printability Checker

To minimize the difficulty of product manufacturing and ensure that the 3D model can be op mally produced, Lu proposed the printability checker (PC) scheme to determine whether an item is suitable for 3D printing or other manner of manufacturing [12] as shown in Fig. 1.

3.2 Improvement by ML

Although the experiments prove that the ability of the PC scheme to test printable model has been achieved, the practical implementation still lacks functionality and reliability, particularly when meeting a lot of different features or data. Hence, Lu suggested applying the ML approach to automatic modification of laws and parameters used, especially to check the printability.

Unlike the method above, the estimation model can be trained for the prediction of printability by using the support vector machine (SVM) instead of predefining any rules [6]. In this way, the optimum decision function may be achieved by using the same parameters for further classifications. This method was also proved by experiments, which effectively reduced the feature extraction time of 3D models without any negative impact on product precision.

3.3 Further Developments

In the realistic manufacturing sector, the measurement of printability or complexity is based not only on a single indicator but also on the integration of multiple indicators such as time, expense, raw material, model size, and geometry. The problems about how to test and determine multi-indicators and what effect proportion each indicator has should be converted into an optimization problem solved by genetic algorithms (GAs) and genetics-based ML method to achieve the optimal effect proportion and the minimum complexity value.

The reason is GA which is designed especially for large spaces or data that could be expressed in binary string format [7]. Compared to other methods, this probabilistic search method only requires few assumptions to build objective functions [8]. Furthermore, the PC scheme is developed on the basis of the current level of 3DP. To popularize 3DP, further research should focus on the optimization of printing technique to lower the complexity threshold under multi-indicator environment. By combining the improved printing method with the assist of printability checker, more and more products are possible to be classified from unprintable to printable.

3.4 Slicing Acceleration

To convert the sliced 3D model into a set of slicing planes in *z*-coordinates [9], the layer information requires being extracted from triangular mesh by using slicing algorithm. In [10], Wang et al. presents a slicing algorithm composed of three kernel modules, i.e., ray-triangle intersection (TRI), trunk sorting (TS), and layer extraction (LE), wherein TRI enables the slicing algorithm to calculate the intersection points between vertical rays on 2D image pixel centers and triangle meshes in STL format. A similar approach in [11, 12] utilizes the plane triangle intersection to calculate the intersection point.

TS is to sort the intersection points in the order in the trunk. According to the layer height and point position, LE is to calculate the binary value of each pixel and then generate the layer images for printing [10]. Though the slicing algorithm has established the foundation of subsequent layer-based addictive manufacture, it has a few weaknesses in computational complexity and difficulty of parallel implementation. In today's big data era, parallel computing has become a great potential for alleviating the computational demands of AI in the aspects of image processing, production rules, mechanization of logic, data filtering, data mining, etc. [13].

Hence, Wang et al. proposed two graphic processing unit (GPU) schemes to accelerate this prefabrication process with the assist of pixel-wise parallel slicing (PPS) and fully parallel slicing (FPS) methods [10]. The PPS aims to achieve the parallelism of slicing algorithm. As shown in Fig. 2, it enables GPU to support the threads and allocate them to each pixel ray, which means that all rays are operated by their specific threads. Also, the threads can help to store the intersection points

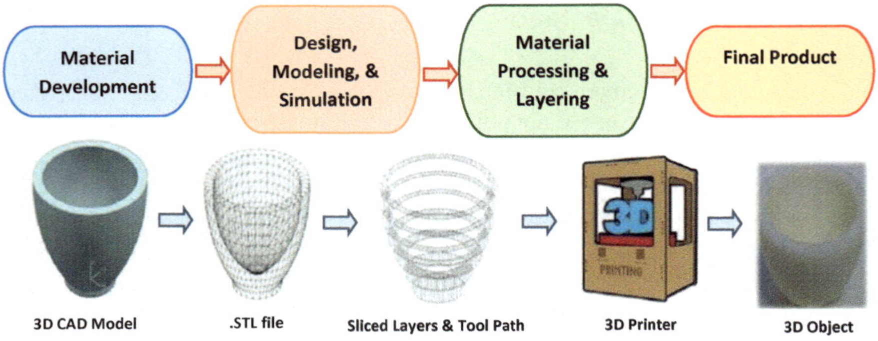

Fig. 2 Mechanism of original printability checker

and sorting results in a shared memory in GPU, which significantly reduces the time consumption. FPS further develops the parallelism of RTI, TS, and LE, especially for solving the large size slicing problems. Compared to PPS, FPS enables the multi-thread concurrency to operate three independent modules in a fully parallel method. [10].

3.5 Path Optimization

A feasible printing trajectory can not only direct the nozzle to form the desired shape but also significantly shorten the computational and printing time [1]. During the printing process, a printing nozzle basically spends time on traversing two types of segments, including print segment and transition segment [14]. Therefore, it is necessary to seek for an optimal path in the shortest traversing time or distance.

Similar to Traveling Salesman Problem (TSP), Fok et al. proposed a relaxation scheme of 3DP path optimizer to compute the nozzle traversing time. The path optimization problem is formulated as TSP by comparing each print segment to a city and finding the fastest or shortest tour to visit the whole country. In each layer, TSP implements optimization at inter-partition level (print area) and intra-partition level (blank area) based on the boundaries of these areas.

Between each inter-partition, TSP can compute the visit order and start the tour to visit each area. Then, the Christofides algorithm is used to find the shortest time of this visit. The transition segment in intra-partition level is the connection of two nearest end points located in two adjacent print areas [14].

3.6 Further Developments

Though the computational simulation has verified that the scheme in [14] is capable of simplifying and accelerating the printing process, there still exists a few defects which may affect the printing precision and optimal solution. Firstly, the Christofides algorithm utilizing the distance as a criterion can only prove that the nozzle traversing distance or time is optimal. But it fails to prove the total time consumption is the shortest under the circumstance of neglecting the retraction time.

In [14], it points out that the retraction method has to be considered due to the generation of excess filament leaking, which usually happens for a typical printer. It requires time to retract the filament back when the nozzle traverses between segments. Thus, this retraction time should not be simply regarded as a constant due to its positive correlation with the number of segments, which requires simulation and physical experiment to verify.

Secondly, to reduce the computing and printing time, researchers proposed a simplified method to consolidate the small connected print segments into integrated segments based on the consolidation threshold value [14]. However, this approach is a lack of an optimal threshold control. Because a large threshold is possible to affect the boundary shape of the print area, causing wrong transition segment between two adjacent inter-partitions and even the precision problem of the final 3D object.

Thus, how and what degree the threshold control has that can achieve the optimal printing speed without negative impact on precision are key directions in future. Finally, many researches have proved that parallel slicing acceleration and path optimization may offer an efficient performance. But there is a few researches on the integration of printability checking, slicing, and path planning. According to the layer-based printing process, the parallel computing to both convert the 3D model into layer image and automatically form an optimal path is reliable to further accelerate the prefabrication.

3.7 Service Platform and Evaluation

Service-oriented architecture (SOA), known as a core part of cloud manufacturing, refers to a computing paradigm that provides enabling technologies and services to fulfill the client requirements in an efficient and fast manner [15, 16]. A feasible SOA is able to intelligently realize the high flexibility, integration, and customization of 3DP [17]. To date, several researchers have studied the problem of correlation among virtual services under multiple demands and constraints [17].

For example, Li et al. proposed the impact of service-oriented cloud manufacturing and its applications. Ren Lei et al. presented the resource virtualization and allocation for cloud manufacturing [18]. Wu proposed the 3DP technique in cloud-based design and manufacturing system [7]. Y. Wu et al. also developed a conceptual scheme of

3DP service-oriented platform and an evaluation model on the basis of cloud service [19].

3.8 Cloud Service Platform

The cloud platform is an on-demand computing model composed of autonomous, hardware, and software resources. As an example, in [19], it provides clients with a convenient on-demand access to a shared collection of resources and then integrates the resources and capabilities into a virtualized resource pool [19]. The assist of service evaluation and demand matching algorithm enables the platform to intelligently make a comprehensive evaluation of terminal printers, providing an optimal resource allocation based on printing precision, quality, and cost and time [19]. Similarly, W. Wang et al. proposed the resource allocation algorithms to build a flexible and agile collaborative scheduling and planning of resources.

3.9 Service Evaluation

Based on above content, the method to evaluate and select the services of terminal printers was further developed by using ML method. In [19], Wu et al. presented a service evaluation model using principal indexes including time, cost, quality, trust, ability, and environment. With the help of fuzzy number, different quantizations based on hamming distance algorithm and the optimization algorithm are able to quantify the service quality and improve the accuracy of service selection [30]. Similarly, Dong, Y. F et al. also presented a quality-of-service acquisition method and a trust evaluation model for cloud manufacturing service using the genetic algorithm [20].

3.10 Further Developments

A comprehensive cloud platform is not only a collection of abundant resources. Most researches still focus on the technical innovation, resulting in the lack of attention to system security and safety for future deployment and adoption. Though the sharing platform may highly improve the resource usage and provides the opportunity to middle- or small-scale manufacturers to accomplish their production, it also leads to the malicious attacks from side channel, causing the risks of information steals or losses in the nowadays competitive environment.

For enhancing users' privacy, Tao et al. proposed the establishment of a private cloud platform, offering benefits and services from public platform environment but self-managed [18]. Although it may lower the risk to some extent, the specific criteria and standard to implement still require further research and simulation.

In addition, in spite of service and manufacturing intelligence, the cloud platform lacks real-time control during a printing process. Once a printing task begins, it will be difficult to interrupt or terminate the task already in progress, especially when connected with the online cloud platform. To reduce the massive time and material wastes, what intelligent method should be used is a principal objective to research.

D. Security problem in manufacturing industry has gained more attention in recent years. The cyber-physical attack, a new vulnerability to a cyber-manufacturing system including 3DP, may cause several defects of products including change of design dimensions, void infill, nozzle travel speed, heating temperature, and so on [21].

3.11 Attack Detection

For real-time detection of the malicious attacks in 3DP, Wu et al. proposed an ML method on physical data, covering k-nearest neighbor (KNN) algorithm, random forest algorithm, and anomaly detection algorithm. Specifically, each layer-based image with defects is converted into a grayscale plot. The features of grayscale mean, standard derivation, and number of pixels larger than the threshold are all extracted according to the grayscale value distribution.

Based on extracted features, the ML algorithms will in real time detect the outliers in defect areas and trigger alerts to the administrator, wherein KNN is to determine the classification of defect areas when the probability density of some parameters is unsure [21]. The anomaly detection is to detect unusual outliers that do not conform the predefined or accepted behaviors, such as increase of mean value, standard derivation, and number of pixels larger than a threshold [22].

Similarly, in a cyber-physical system, this unsupervised learning method is also applied to detect anomalies with low false rate by using a recurrent neural network and cumulative sum method [34]. The random forest can not only classify the defects by estimating the posterior distribution of each image layer but also build process-based patterns and use proximities to detect outliers according to the images. Through simulation and experiment, the anomaly detection algorithm was found to achieve the highest accurate detection [22].

3.12 Further Developments

The experiment has indicated the feasibility of ML in 3DP security. However, the research only works on the component surface with regular shapes or patterns. For a complex construction, it is unable to guarantee the detection system which will never identify a correct component as a defect by mistake. Hence, the current research on defect detection is only feasible for standardized components in mass production, requiring further development to fit into mass customization only with the capability to classify the defects accurately.

To solve the problem of collision detection while AM processes, and the need to control the absence of imprint head contamination and printing defects in AM machine during build process [23]. Had a universal system developed Visual control which performs a technical process Also a visual feedback and control was added. To inspect the 3D printing system visibly, two techniques are used [23], namely: (i) machine-based learning and (ii) tracking single point. These two techniques enhanced the quality of collision detection in 3D impression system PROTECTS. In this analysis, the methods used are commonly used for computer reconnaissance vision. The summary of AI methods for 3DP is shown in Table 1.

Table 1 Summary of AI methods for 3DP

3D printing process	Status	Application	Method	Solved problems
Printability checking) design and preparation)	Offline	(1) Original printability checker (2) Automatic checking	PC scheme (FE, PM, VE) ML (SVM)	1. Multi-indicator test and optimal effect proportion using GA 2. Lower complexity threshold
Prefabrication (planning)	Offline	Slicing acceleration path optimizer	Slicing algorithm (TRI, TS, LE) GPU (PPS and FPS parallelism) TSP-based optimization Christofides algorithm	1. Evaluation of timeconsumption closed to a practical situation 2. Consolidation of print segments and threshold control 3. Parallel computing of printability checking, slicing, and path planning
Service platform & evaluation (design, printing, service, and control)	Online	Cloud service platform evaluation model	Demand matching algorithm resource allocation algorithm ML (multi-criteria fuzzy decision based on hamming distance algorithm; GA)	1. Security enhancement 2. Real-time control 3. Design for printing
Security (control)	Online	Attack detection	kNN, anomaly detection, random forest	Defect detection for mass customization

4 Conclusion

This paper includes a description of the ML techniques, big data, and its recent uses in the field of additive manufacturing. The paper deemed a way to expand on the importance of ML with big data in industry AM. In this study, all the literature reviewed is related to the AM domain. The paper also lays out a total analysis of recent ML and major applications data used in the additive manufacturing industry, and possible future research in that field of research. This paper is theoretical in nature, but future research, these authors will use ML techniques to experiment with log AM machine files/data to detect defects, or abnormality. This paper encourages researchers to look into the advantages of the ML and big data analytics techniques out of novel AM technology research.

References

1. Yang F, Lin F, Song C, Zhou C, Jin Z, Xu W (2016) Pbench: a benchmark suite for characterizing 3D printing prefabrication. 2016 IEEE International Symposiumon Workload Characterization (IISWC). Providence, RI, pp 1–10
2. I. Gibson, D. W. Rosen, B. Stucker et al (2010) Additive manufacturing technologies. Springer
3. Attaran M (2017) The rise of 3-D printing: The advantages of additive manufacturing over traditional manufacturing. Bus Horiz 60:677–688
4. Raftery T Artificial intelligence and the future of jobs, [Online] Availableat:https://www.digitalistmag.com/iot/2017/11/29/artificial-intelligence-future-of-jobs-05585290. Accessed 22 Oct 2018
5. Özcan, E (2019) Scope for machine learning in digital manufacturing. J
6. Lu T (2016) Towards a fully automated 3D printability checker. In: Proceedings of IEEE International Conference Industrial Technology vol. 2016, pp 922–927
7. Goldberg DE, Holland JH (1988) Genetic algorithms and machine learning. Mach Learn 3(2):95–99
8. Weise T (2009) Global optimization algorithms–theory and application, vol. 1, p 820. http//www. it-weise. de, Abrufdatum
9. Zhou C, Chen Y, Yang Z, Khoshnevis B (2013) Digital material fabrication using mask-image-projection-based stereolithography. Rapid Prototyp J 19(3):153–165
10. Wang A, Zhou C, Jin Z, Xu W Towards scalable and efficient GPU-enabled slicing acceleration in continuous 3D printing, pp 623–628.
11. Minetto R, Volpato N, Stolfi J, Gregori RMMH, Silva MGV An optimal algorithm for 3D triangle mesh slicing and loop-closure, pp 1–31
12. Gregori RMMH, Volpato N, Minetto R, Silva MVDG (2014) Slicing triangle meshes: an asymptotically optimal algorithm. In: 2014 14th international conference on computational science and its applications. Guimaraes, pp 252–255
13. Kanal LN, Kitano H, Kumar V, Suttner CB (1994) Parallel processing for artificial intelligence 1. vo. 14, 1st edn. North Holland, pp 1–10
14. Fok K, Cheng C, Tse CK, Ganganath N (2016) A relaxation scheme for TSP-based 3D printing path optimizer
15. Hassan QF (2011) Demystifying cloud security. CrossTalk, pp 16–21
16. Tao F, Zhang L, Venkatesh VC, Luo Y, Cheng Y (2011) Cloud manufacturing: a computing and service-oriented manufacturing model. Proc Inst Mech Eng Part B J Eng. Manuf 225:1969–1976

17. Da Silveira G, Borenstein D, Fogliatto HS (2001) Mass customization: Literature review and research directions. Int J Prod Econ 72(49):1–13
18. Lei R, Lin Z, Yabin Z et al (2011) Resource virtualization in cloud manufacturing. Comput Integrat Manuf Syst 17(3):511–518
19. Wu Y, Peng G, Chen L, Zhang H (2016) Service Architecture and evaluation model of distributed 3D printing based on cloud manufacturing, pp 2762–2767
20. Dong YF, Guo G (2014) Evaluation and selection approach for cloud manufacturing service based on template and global trust degree. Comput Integrat Manufact Syst 20(1):207–214
21. Wu M, Song Z, Moon YB (2017) Detecting cyber-physical attacks in Cyber Manufacturing systems with machine learning methods. J Intell Manuf 1–13
22. Xie H, Liang D, Zhang Z, Jin H, Lu C, Lin Y (2016) A novel pre-classification based KNN algorithm. In: 2016 IEEE 16th international conference on data mining workshops (ICDMW), Barcelona, pp 1269–1275
23. Makagonov N, Blinova E, BezukladnikoIgor (2017) Development of visual inspection systems for 3D printing

Chapter 7
Finding Suitable Threshold for Support in Apriori Algorithm Using Statistical Measures

Azzeddine Dahbi, Siham Jabri, Youssef Balouki, and Taoufiq Gadi

1 Introduction

Data growth on the Internet is steadily increasing. The size of the data is so large that new prefixes have been introduced: peta, exa, and zettabyte. We are drowning in data, but starving for knowledge and information. Data mining has arisen in response to these shortcomings [1].

The investigation of association rules is one of the most crucial tasks of data mining, and it strives to reveal precious patterns and exciting relationships for a large volume of data. It was first advocated for the analysis of client data, but is now being envisioned as general data that can be used in various areas such as finance and business, scientific research and government security. Currently, many algorithms are proposed for mining association rules. The most prominent and straightforward is the Apriori algorithm [2] provided by Agrawal in 1993.

The application of the Apriori algorithm in DM permits to test the different combinations between items (Data_Atributes) to recover attribute values that are frequently retrieved in a database, which will be displayed in the form of association rules.

Exploring association rules is based on the request of a minimum support and minimum confidence thresholds, and these parameters are intuitively estimated by users. These constraints are used to tailor the research space and make the computation process possible. However, this can be restrictive because support and confidence influence directly the number and quality of association rules discovered.

According to the selection of these thresholds, association rule mining algorithms are able to produce a high number of rules that require algorithms to suffer from long execution times and a large memory consumption, or they are able to produce a low number of rules and thus can discard valuable information.

A. Dahbi (✉) · S. Jabri · Y. Balouki · T. Gadi
Faculty of Science and Technology, Laboratory Mathematics Computer Science and Engineering Sciences (MISI), Hassan 1st University Settat, Settat, Morocco

A. E. Hassanien et al. (eds.), *Enabling Machine Learning Applications in Data Science*, Algorithms for Intelligent Systems, https://doi.org/10.1007/978-981-33-6129-4_7

In an attempt to reconcile this gap, there are considerable researches in the literature to assist the process user in setting the most appropriate and suitable support and confidence thresholds for the scope of the decision-making process.

Liu et al. [3] address the problem of using one minimum support for the whole database, which is not logical because the data items are not of the same nature and/or do not have similar frequencies in the data. And it offers a technique that allows the user to specify several minimum supports to reflect the nature of the components and their various frequencies in the database.

Lee et al. [4] suggest a new paradigm to reach a minimum support in the case of a different minimum support and provide a straightforward algorithm based on Apriori under maximum constraint.

Salam et al. [5] introduce a new approach which is based on the association ratio graph to investigate efficiently the first few patterns of maximum frequency without using the minimum support parameter.

Fournier-Viger [6, 7] redefine the problem of deriving association rules as the deriving of top*k* association rules by providing an algorithm to find the top*k* rules with the highest support, where *k* is the number of rules to be generated and defined by users.

Kuo et al. [8] propose a new method to automatically determine the best values of support and confidence parameters, in particular to find association rules using binary particle swarm optimization, and proposed a new approach for association rules to enhance computational performance and to automatically define appropriate threshold values.

Other works [9, 10] have presented an approach focusing on multi-criteria optimization in order to identify the preferred association rules. Without any need to specify any parameter, the principle is to discover patterns that are not dominated by any other pattern by using a set of interesting criteria. In our previous work, we proposed an innovative methodology to discover association rules using an automatic multi-level minimum support [11].

In the same context, and to automatically adjust the choice of the minimum support according to each dataset, we propose an approach based on different statistical measures of the datasets.

The document is structured as follows. In Sect. 2, we present the indispensable scientific context: the exploitation of association rules and statistical measures. Section 3 presents our approach based on the Apriori algorithm for mining association rules with auto-adjusting threshold support with multiple minimum support. Section 4 is devoted to the study of experiences, which gives credibility to our approach. And the last section concludes our work and presents the prospects for the future.

2 Background of Association Rules Mining

2.1 Process Association Rules Mining

Association rules are one of the main themes of data mining. This technique is a way of pooling elements that usually belong together [3]. It is formalized as follows:

Let $I = \{i_1, i_2, \ldots, i_n\}$ being a package of all the elements, the association rules are produced across a large set of transactions, symbolized by T with $T = \{t_1, t_2, \ldots, t_m\}$, every transaction t_i is an item set and meet $t_i \subseteq I$.

An association rule is a statement of the form $X \rightarrow Y$, where $X, Y \subset I$ with the fact that $X \cap Y = \emptyset$. We designate by the set X the antecedent of the rule and by the set Y the consequence of the rule where I is a non-empty set.

An association rules may be considered meaningful if the elements in question often occur together and if it can be assumed that one of the sets could somehow lead to the presence of the other set. The strength of an association rule can be judged by mathematical notions called 'support' and 'confidence.' The designation $P(X)$ is intended to indicate the proportion of the time that the set X appears in the transaction set T.

The support of the rule $X \rightarrow Y$ is the percentage of transactions in a database D that contain $X \cup Y$ and is represented as:

$$\text{Support}\,(X \rightarrow Y) = P(X, Y) = \frac{n(X \cup Y)}{n}$$

The confidence of a rule $X \rightarrow Y$ describes the percentage of transactions containing X which also contain Y and is represented as:

$$\text{Confidence}\,(X \rightarrow Y) = \frac{P(X, Y)}{P(X)} = \frac{n(X \cup Y)}{n(X)}$$

where $n(X \cup Y)$ is the number of transactions that contain items (i.e., $X \cup Y$) of the rule $n(X)$ is the number of transactions containing itemset X and n is the total number of transactions.

In general, the generation of association rules can be broken into two steps:

- Discover the list of the most frequent (or most important) items: The frequent item set is a set of items that check a predefined threshold of minimum support.
- Generate all strong association rules from the frequent itemsets: The strong association rule is a rule that checks a predefined threshold of minimum confidence.

There are multiple models for AR generation in the literature: Apriori [1], Close, Close+ [12] or Charm [13], etc.; The most modest and well-known algorithm is the Apriori algorithm for finding frequent items and generating all association rules. It is the most powerful method to generate $k + 1$-itemset candidates from k-frequent items according to the principle of the Apriori algorithm that any subset of frequent itemsets are all frequent itemsets. Foremost, find the frequent 1-itemsets L_1. Then L_2 is generated from L_1 and so on, until no more frequent k-itemsets can be found and then algorithm desists. Every L_k generated should scan database once. Then C_k is generated from $L_k - 1$.

2.2 Statistical Measures

Collecting a large amount of data can be easy, but understanding what you've found is often complicated. For this reason, we apply statistics to describe and analyze the data [14]. Several characteristics are used to describe and define the characteristics and structure of a dataset. They cover the number and types of variables or attributes, as well as many statistical measures applicable to them. They can be divided into two types: measures of central tendency and measures of dispersion.

They are often called descriptive statistics because they can help you describe your data.

The measure of central tendency is a basic measure that attempts to describe an entire dataset within a single value that describes the middle or center of its distribution. These measures are useful because they:

- Summarize a large amount of data into a single value
- Indicate that there is some variability around this single value in the original data.

There are three major measures of central tendency: mode, median, and mean. Each measure illustrates a distinct indication of the common or central value in the distribution. The mean, median, and mode are all determined using different techniques and, when used for the same original dataset, they often yield different average values. It is necessary to know what each mathematical measure of the average informs you about the original data and to consider which measure, the mean, median, or mode, is most appropriate to calculate if you need to apply a mean value to describe a dataset.

Mode is the most common value that occurs with the highest frequency in a dataset or distribution. It is representative. There may be more than one mode in a dataset if many values are also common; otherwise, there may be no mode at all and it can be found for both numeric and categorical (non-numeric) data, but there are some limitations to using the mode. In some datasets, the mode may not reflect the center of distribution very well when the distribution is organized from the lowest to the highest value.

The median indicates the middle value in a dataset or distribution when the values are organized in order of size from smallest to largest order or vice versa. The median divides the distribution in half (there are 50% of observations on either side of the median value) When there is an odd number of values in the dataset, the middle value is easy to find. When there is an equal number of values, the midpoint between the two central values is the median. The median is a useful measure of the average value when the data involve particularly high or low values because these have a slight impact on the result. The median is the most proper measure of average for data classified on an ordinal scale. The median is less affected by outliers and skewed data than the mean and is habitually the favored measure of central tendency when the distribution is not symmetrical. The median is also easy to calculate but this does not imply that it is an inferior measure to the mean—what is important is to use an appropriate measure to determine the average. Another field where the median is helpful is with frequency data. Frequency data give the numbers of people or things in particular categories but it cannot be identified for categorical nominal data, as it cannot be logically ordered.

The mean is the most usually used mathematical measure of average and is generally, what is being referred to when people use the term average in everyday language. The mean is the total of the value of each observation in a distribution divided by the number of values that make up the dataset. This is also called as the arithmetic average. The mean can be applied to both discrete and continuous numeric data, but it cannot be calculated for categorical data, as the values cannot be summed, and if a dataset includes one or two very high or very low values the mean will be less common as it will be unfavorably impacted by these exceptional value(s).

The measures of dispersion refer to the diffusion of the data series around it average. They are the degree to which value in a dataset deviates from the average of distribution. There are three principal measures of dispersion: the range, the interquartile range, the variation/the standard deviation.

The range is estimated by applying the difference between the greatest value and the lowest score in the set of data.

Quartiles in statistics are values that split your data into quarters. They divide your data into four portions according to where the numbers fall on the number line.

The lower quartile (Q_1) is the median of the lower half of the dataset. It is the number separating the first and second quartile. The lower quartile is also called the 25th percentile; The upper quartile or Q_3 is the number separating the third and fourth quartile. The upper quartile is also called the 75th percentile; it divides the lowest 75% of data from the highest 25%.

The interquartile range is explained as the difference between the upper quartile (Q_3) and the lower quartile (Q_1). It represents the dispersion of the middle 50% of the data values. The interquartile range is a more helpful measure of the dispersion than the range.

The variance estimates how distant a dataset is spread out. The scientific definition is 'The average of the squared differences from the mean,' but all it really does is to give you a very general idea of the spread of your data. A value of zero means that there is no variability; all the numbers in the dataset are the same.

The standard deviation is the square root of the variance. While variance provides you a strict view of spread, the standard deviation is more accurate, providing you precise distances from the mean.

3 Proposed Approach

One of the main challenges of the Apriori algorithm is the choice of threshold for support and confidence. Apriori finds the set of frequent candidate items by generating all possible candidate items that check a user-defined minimum threshold. This choice influences the number and qualities of AR.

In the proposed algorithm, the description of the association rules remains the same as in the Apriori algorithm, but the description of the minimum support is revised. We rely on many statistical measures to determine the minimum support threshold and this constraint is adopted to find a large set of items. Below the minimum support threshold, the level-by-level processing characteristic is retained, so that the original Apriori algorithm can be easily extended to find frequent item sets.

Our algorithm requires a minsup threshold that is established according to the transactional dataset, which is consistent and logical.

The main innovation of our paper is to calculate automatically the minsup threshold based on each dataset instead of using a constant value predefined by the users. To achieve this goal, we use a set of statistical measures:

Central tendency measures such as mean, mode, and median and dispersion measures for which we choose the range, standard deviation, quartile 1, quartile 3, and interquartile range.

The majority of existing methods for generating association rules apply a single, uniform minimum support threshold value for all elements or sets of elements; however, all elements in a set of elements do not operate according to the same process, some appear very frequently and often, and others appear infrequently and very rarely. Therefore, the minsup threshold should be changed according to the different levels of the element set. The second intervention in the proposed approach is therefore the updating of the minsup threshold at each level, and we make this minsup change (update) dynamically according to each level, still using the statistical measures of each level.

Like the original algorithm Apriori, our proposed algorithm is also based on levelwise search. It generates all frequent itemsets by making multiple passes over the dataset. In the first pass, it counts the supports of individual items and determines different statistical measures and choose one of them to take as the selected threshold then determine whether they are frequent.

In each subsequent pass, it starts with the former set of itemsets found to be frequent in the previous pass. It uses this set by join it with itself to generate new possibly frequent itemsets, called candidate itemsets. We compute the actual supports for these candidate itemsets during the pass over the dataset. At the end of the pass, it determines which of the candidate itemsets are actually frequent by using new threshold for each level.

- Input:

Transactional dataset containing a set of n transactions.

- Step 1:

Calculate the count ci of each item 1-itemset$_i$, $i = 1$ to m, as its occurrence number in the dataset then derive its support value sup-1itemset$_i$ as: $\sup - 1\text{itemset}_i = \frac{c_i}{n}$

- Step 2:

Determine the minimum support for the first level for 1-itemset: minsup-1 by using the selected statistical measures for example: The means or the median.... By using all itemset with one item.

Minsup: is a minimum support and 1-itemset is a set of items composed of 1item.

- Step 3:

Verify whether the support sup-1itemset$_i$ of each item$_i$ is large than or equal to minsup-1 defined using the selected statistical measures. If i satisfies the above condition put in the set of large or frequent 1-itemset (L_1).

$$L_1 = \{1 - \text{itemset}_i / \sup - 1\text{itemset}_i \geq \text{minsup1} \\ \text{with } i = 1.....N \text{ number of all } 1 - \text{itemset}\}$$

- Step 4:

Generate the candidate C_2 from L_1 with the same way to the Apriori algorithm, by join L_1 with itself. k-itemset: set of items composed of k items.

- Step 5:

Compute the new minsup of the 2-itemset level by using the selected statistical measures of support of the generated C_2 then generate the L_2.

$$L_2 = \left\{ \begin{array}{c} 2 - \text{itemset}_i / \sup - 2 - \text{itemset}_i \geq \text{minsup2} \\ \text{with } i = 1 \ldots N \text{ number of all 2 - itemset} \end{array} \right\}$$

- Step 6:

Check whether the support sup k-itemset$_i$ of each candidate k-itemset$_i$ is larger than or equal to minsup k obtained in step5. If it satisfies the above condition put in the set of large k-itemset(L_k).

$$L_k = \left\{ \begin{array}{c} k - \text{itemset}_i / \text{sup} - k - \text{itemset}_i \geq \text{minsupk} \\ \text{with } i = 1 \ldots N \text{ number of all kitemset} \end{array} \right\}.$$

- Step 7:

Repeat steps 4–6 until Li is null.

- Step 8:

Construct the association rules for each large k-itemset$_i$ with items: $\{Ik_1, Ik_2, \ldots Ik_q\}$ which verify the threshold of confidence, i.e., the association rule whose confidence values are larger than or equal to the threshold of confidence defined by the mean of support of all large q itemset Ik.

- Output:

A set of association rules using an automatic threshold of support in multi-level.

Let us now see how we compute the threshold of minimum support using different statistical measures. The first one is the means, the minimum support using the means is calculated as:

$$\min \text{sup} -l = \sum_{i=1}^{S} \frac{\text{sup} -l - \text{itemset}_i}{S}$$

with l is the level and S is the number of the frequent l-itemset in the level l.

In the same way, we calculate the various threshold using other statistical measures defined above except that the approaches using standard deviation we take as threshold of minimum support a score take into account the level and the expression of this is formulated as:

$$\min \text{sup} -l = \exp(-2.l.\sigma)$$

With l is the levelwise, and σ is the standard deviation correspondent to each dataset.

4 Experiment Study

In this part, we will evaluate the efficiency and effectiveness of the proposed approach; several experiments on various kinds of datasets are tested. We use different public datasets: (mushroom, flare1, flare2, Zoo, Connect) which are real datasets and widely tested by other approaches, they are taken from UCI machine learning repository [15]. T10I4D100K(T10I4D) commonly used as benchmarks for testing algorithms, and it is also generated using the generator from the IBM Almaden Quest research group [16] and Foodmart is a dataset of customer transactions from a retail

Table 1 Characteristics of the used datasets

Dataset	Items	Transactions
Mushroom	22	8124
Flare1	32	323
Flare2	32	1066
Zoo	28	101
Connect	42	67,557
Chess	36	3196
Foodmart	1559	4141
T10I4D	1000	100,000

store, obtained and transformed from SQL-Server 2000 [17]. Table 1 summarizes the properties of the used datasets.

Our objectives in this section are multiple, the first, we show through many experiments the number of the generated association rules given by different approaches which took as minimum support different statistical measures explained above to illustrate the advantage of those approaches in front of the original Apriori.

Secondly, the aim of this experiment is to compare the quality of the generated association rules given by these approaches and comparing them with the original Apriori and another algorithm Topkrule. Same thing for the third experiment, we will examine and study the runtime.

All experiments were performed on an Intel(R) core i3-3210 M CPU @ 1.70 GHz, with 4 GB RAM running on Windows 7(64 bits) Enterprise edition. The algorithm was written in Java programming language using Eclipse Neon IDE.

Table 2 shows different values of threshold of confidence chosen and different values of minimum support found by different statistical measures in the first level for different dataset.

Table 2 Threshold values of support and confidence

Dataset	Moy	Med	Q1	Q2	InterQ	STD	Minconf
Mushroom	20	45	26	63	37	30	70
Flare1	32	63	27	90	63	23	50
Flare2	32	59	31	94	65	21	50
Zoo	30	43	35	74	38	29	50
Connect	33	89	55	96	41	15	80
Chess	49	79	61	94	32	18	70
Foodmart	0.02	0.003	0.02	0.0036	0.001	–	70
T10I4D	0.01	0.018	0.011	0.02	0.018	–	70

Table 3 Number of Ar generated for each dataset

	Mushroom	Flare1	Flare2	Zoo	Connect	Chess	Foodmart	T10I4D
Moy	8353	106	127	362	422,874	55,700	2037	111,482
Med	337	26	26	28	5860	2750	1800	2671
Q1	65,996	811	998	1137	–	2,860,531	11,486	1,490,265
Q3	16	2	2	2	49	66	370	9
InterQ	1,632,735	602	180	12,457	–	–	25,882	6351
Std	31,031	6384	14,277	241	–	–	–	–
APR	14,965,927	1430	1298	3716	220,224,860	1,776,698	21,982	206,772

4.1 Reduction of a Number of Rules

In this experiment, we show the ability of our proposed approach to reduce the number of AR generated from the chosen datasets. Our experiment compares our approaches to APRIORI based on thresholds.

Table 3 compares the size of AR generated by different methods using statistical measures to the APRIORI. For Apriori algorithm, we took as threshold of support the value obtained by the approach of the means in the first level.

As we see in the result, the number of rules generated by Apriori is greater than the number given by all the other approaches and in all the datasets, which ensures their superiority over him. Another remark is that the approach of Q3 keep only a small number of rules which means, this approach removes rules may contain a valuable information which imply that there is a loss of information.

On the other hand, the proposed approach Sup-moy can significantly reduce the huge number of rules generated from datasets, which can facilitate interpretation and help users to see the most interesting ones and thus make the right decision.

4.2 The Running Time Analysis

We realized an implementation for traditional Apriori from [1] and different proposed algorithms Sup-moy, Sup-med, Sup-Q1, Sup-Q3, Sup-interQ, Sup-std, and we compare the time wasted of original Apriori (APR) and Topkrule algorithm (Topk), and the time wasted by different approaches proposed by applying many datasets, various values for the minimum support given in the implementation. The result is shown in Table 4.

As we see in Table 4, that the time-consuming in our entire proposed algorithm in each dataset is less than it is in the original Apriori, and the difference grows more and more as the number of transactions of datasets increases.

Table 4 Time-consuming in different datasets (in Ms)

	Muchroom	Flare1	Flare2	Zoo	Connect	Chess	Foodmart	T10I4D
Moy	3746	10	73	24	224,335	1852	20,660	1,283,038
Med	1212	48	28	28	24,003	2999	17,110	106,657
Q1	10,440	178	126	110	–	64,919	56,884	8,582,360
Q3	290	30	22	20	3963	320	5530	11,348
intQ	30,061	55	40	154	–	–	57,240	27,555
std	6868	110	261	61	–	–	–	–
APR	86,984	100	81	60	888,383	9432	62,152	340,642
Topk	11,419	10	20	12	–	41,501	18,600	–

Table 5 Average of confidence for different datasets

	Muchroom	Flare1	Flare2	Zoo	Connect	Chess	Foodmart	T10I4D
Moy	0.89	0.85	0.9	0.86	0.96	0.94	1	0.87
Med	0.93	0.92	0.95	0.86	0.98	0.96	1	0.87
Q1	0.87	0.72	0.75	0.86	–	0.88	1	0.82
Q3	0.99	0.96	0.99	0.94	0.99	0.99	1	0.87
inteQ	0.85	0.70	0.89	0.77	–	–	1	0.86
std	0.88	0.67	0.66	0.81	–	–	–	–
APR	0.80	0.75	0.85	0.84	0.84	0.86	0.98	0.78
Topk	0.89	0.85	0.9	0.88	–	0.94	0.99	–

On the other hand, we see that this time-consuming in our approach is the same as the time-consuming in Topkrule in some datasets and it is less than it in other datasets.

We can add another advantage to some approaches, which is the use of memory space. As we see, we did not obtain the result of Topkrule in Connect and T10I4D100K datasets, the Topkrule and interQ algorithm cannot run on the machine and this is due to the memory problem, while the other algorithms run without any problem.

4.3 *The Quality of the Extracted Rules*

In order to analyze the performance of the proposed algorithms, we have compared the average value of confidence in each dataset of the methods using statistical measures as threshold of minimum support to the original Apriori with the threshold and to the Topkrule algorithm.

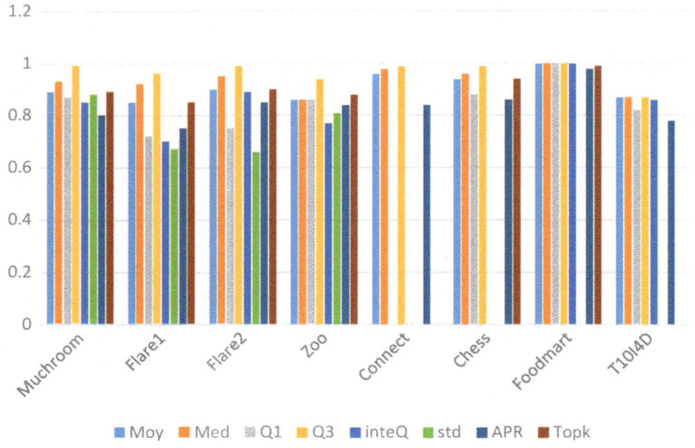

Fig. 1 Histogram of the average of confidence

From Table 5, we can observe that the method Q_3, means the median has found rules with high value of confidence for the generated association rules in the majority of used datasets, followed by Topkrule algorithm, then Q_1, and interQ and std approach provide us with medium confidence and in the last position we found the Apriori algorithm (Fig. 1).

5 Conclusion

In this paper, we proposed a new approach based on Apriori algorithm for discovering the association rules to auto-adjust the choice of the threshold of support. The main advantage of the proposed method is the automatism of the choice of support in multi-level; we get results containing desired rules with maximum interestingness in a little time. The numbers of rules generated by proposed algorithm are significantly less as compared to Apriori algorithm. Hence, we can say our algorithm answer the problem of the choice of the threshold of the support efficiently and effectively. As future works, we plan to ameliorate our approach to be able to select the interesting association rules without using any predefined threshold.

References

1. Zhou S, Zhang S, Karypis G (eds) (2012) Advanced data mining and applications. In: 8th International conference, ADMA 2012, December 15–18, 2012, Proceedings vol. 7713. Springer Science & Business Media, Nanjing, China

2. Agrawal R, Srikant R (1994) Fast algorithms for mining association rules. In: Proceeding VLDB '94 proceedings of 20th international conference very large data bases, vol. 1215, pp 487–499
3. Liu B, Hsu W, Ma Y (1999) Mining association rules with multiple minimum supports. In: Knowledge discovery and databases, pp 337–241
4. Lee YC, Hong TP, Lin WY (2005) Mining association rules with multiple minimum supports using maximum constraints. Int J Approximate Reasoning 40(1):44–54
5. Salam A, Khayal MSH (2012) Mining top-k frequent patterns without minimum support threshold. Knowl Inf Syst 30(1):57–86
6. Fournier-Viger P, Wu C-W, Tseng VS (2012) Mining top-k association rules. In: Proceedings of the 25th Canadian conference on artificial intelligence (AI 2012). Springer, Canada, pp 61–73
7. Fournier-Viger P, Tseng VS (2012) Mining top-k non-redundant association rules. In: Proceedings of 20th international symposium, ISMIS 2012, LNCS, vol. 7661. Springer, Macau, China, pp 31–40
8. Kuo RJ, Chao CM, Chiu YT (2011) Application of particle swarm optimization to association rule mining. In: Proceeding of applied soft computing. Elsevier, pp 326–336
9. Bouker S, Saidi R, Ben Yahia S, Mephu Nguifo E (2014) Mining undominated association rules through interestingness measures. Int J Artif Intell Tools 23(04):1460011
10. Dahbi A, Jabri S, Ballouki Y, Gadi T (2017) A new method to select the interesting association rules with multiple criteria. Int J Intell Eng Syst 10(5):191–200
11. Dahbi A, Balouki Y, Gadi T (2017) Using multiple minimum support to auto-adjust the threshold of support in apriori algorithm. In: Abraham A, Haqiq A, Muda A, Gandhi N (eds) Proceedings of the ninth international conference on soft computing and pattern recognition (SoCPaR 2017). SoCPaR 2017. Advances in intelligent systems and computing, vol 737. Springer, Cham
12. Pei J, Han J, Mao R (2000) CLOSET: an efficient algorithm for mining frequent closed itemsets. In: Proceeding of the 2000 ACM-SIGMOD international workshop data mining and knowledge discovery (DMKD'00), 2000, TX. ACM, Dallas, pp 11–20
13. Zaki MJ, Hsiao CJ (2002) CHARM: An efficient algorithm for closed itemset mining. In: SDM'02. Arlington, VA, pp 457–473
14. Tamhane AC, Dunlop DD (2000) Statistics and data analysis. Prentice Hall
15. UCI machine learning repository, https://archive.ics.uci.edu/ml/index.php. Last Accessed 31 March 2020
16. Frequent Itemset Mining Implementations Repository, http://fimi.ua.ac.be/data/. Last Accessed 31 March 2020
17. An Open-Source Data Mining Library, http://www.philippe-fournier-viger.com/index.?li-nk=datasets.php. Last Accessed 31 March 2020

Chapter 8
Optimum Voltage Sag Compensation Strategies Using DVR Series with the Critical Loads

Alaa Yousef Dayoub, Haitham Daghrour, and Amer Nasr A. Elghaffar

1 Introduction

In modern times, as we known, the global demand for electric power has increased dramatically. In addition, those concerned with environmental cleanliness have begun to think seriously about the negative effects of fossil fuels, and since that time, governments have begun to create energy resources that are clean and highly reliable [1].

In the ideal electrical system, the power distribution companies must provide their customers with a sinusoidal voltage with a constant amplitude and frequency. However, it is also known that power systems, especially at the distribution level, consist of a number of nonlinear loads that negatively affect the quality of energy and their negative impact increases with the increase in their number [2]. The energy released by most renewable energy sources is unfortunately very variable, as it results in a decrease in the quality of energy and fluctuations in voltage, both of which have major negative effects on the energy system as a whole and this effect has increased with the increase of the impact of the large influx of renewable energy confiscations associated with the power system [3]. Clear examples of voltage fluctuations that have significant damage to the power system are spikes, sags, inflation, harmonic distortions, spikes, swells, and momentary disturbances. Among the most common disturbances that affect energy quality is voltage sag, and among the causes of its occurrence are storms, animals, which are external factors, and also the cause of its occurrence may be internal factors such as overloading and grounding problems.

It is known that each electrical device has a specific and fixed voltage to operate on, and thus, the voltage sag can lead to damage to these equipment, and given that most electrical equipment, especially sensitive ones, are of high cost, which

A. Y. Dayoub (✉) · H. Daghrour
Electrical Engineering Department, Tishreen University, Lattakia, Syria

A. N. A. Elghaffar
Electrical Engineering Department, Minia University, Minya, Egypt

A. E. Hassanien et al. (eds.), *Enabling Machine Learning Applications in Data Science*, Algorithms for Intelligent Systems, https://doi.org/10.1007/978-981-33-6129-4_8

forced the implementation of new strategies and control devices that will protect these equipment from differences Harmful in voltage and maintaining the quality of electrical energy [4]. Low energy quality and variation in voltage levels are never desirable in the power system because of its negative impact on causing large losses on the power system as a whole, which results in high costs for both system operators and customers [5].

2 DVR Operation Characteristics

The dynamic voltage regulator (DVR) is one of the most important discoveries in the electrical equipment industry used as solutions to problems of energy quality and improvement. DVR is a recently proposed series compensation device whose main objective is to protect the sensitive load associated with it from voltage disturbances, the most important of which are voltage sag and swell. The DVR consists of voltage source converter (VSC), injection transformer, energy storage device, harmonic filter, and DC charging circuit [6]. A simple diagram of a DVR system including these components can be seen in Fig. 1 [7]. The DVR model discussed in this research paper will abide to the IEEE1547 standard of regulating the protected load's bus voltage to within 5% of nominal value. Figure 2 [8] shows a simplified diagram of the DVR.

The voltage of the converter (V_{conv}) is connected in series between the supply voltage (V_s) and the load voltage (V_l) having impedances Z_S, Z_L, respectively [9]. The term DVR losses is denoted by the symbol R_{DVR}, while the reactance of the injection transformer and DVR filters is denoted by the symbol X_{DVR}. The value of

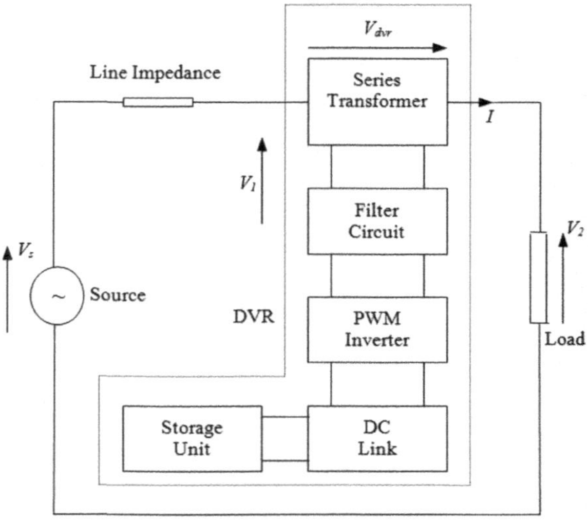

Fig. 1 Simplified DVR basic operation system

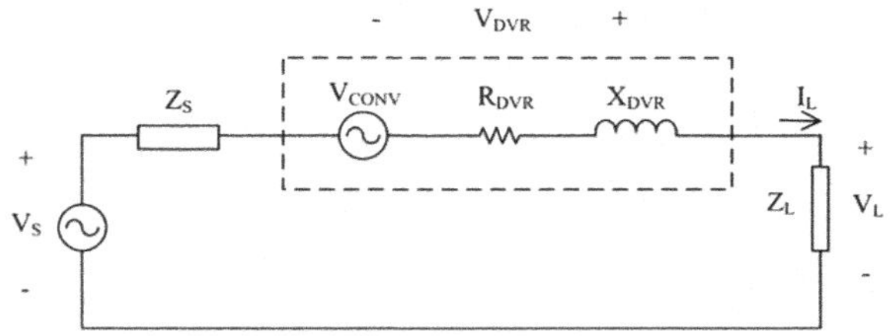

Fig. 2 Simplified diagram of the DVR

X_{DVR}, R_{DVR} is related to both the power of the DVR (S_{DVR}) and the voltage of the DVR (V_{DVR}).

$$X_{\mathrm{DVR}} = \frac{V_{\mathrm{DVR}}{}^{2}}{S_{\mathrm{DVR}}} u\mathrm{DVR}, X \tag{1}$$

$$R_{\mathrm{DVR}} = \frac{V_{\mathrm{DVR}}{}^{2}}{S_{\mathrm{DVR}}} u\mathrm{DVR}, R \tag{2}$$

$$Z_{\mathrm{DVR}} = \frac{V_{\mathrm{DVR}}{}^{2}}{S_{\mathrm{DVR}}} u\mathrm{DVR}, Z \tag{3}$$

$$U_{\mathrm{DVR}}, Z = U_{\mathrm{DVR}}, R + jU_{\mathrm{DVR}}, X \tag{4}$$

The following symbols represent U_{DVR}, R, U_{DVR}, X, U_{DVR}, Z; The p.u value of the resistance (R), reactance (X), and impedance (Z) of the DVR, respectively. The ability to handle both voltage (V_{DVR}) and current (I_{DVR}) of the DVR is given as a percentage according to the relationships [10]

$$\mathrm{iDVR\%} = \frac{I_{\mathrm{DVR}}}{\mathrm{IL, rated}} 100\% \tag{5}$$

whereas I_{DVR} is the current rating of the DVR and the rated supply voltage and rated load current are represented by Vs, rated and IL, rated, respectively.

3 Tests and Results

The DVR injects an in-phase voltage to the source voltage, using an in-phase compensation strategy. Several tests have been developed using PSCAD software in order to test the DVR's ability to mitigate voltage disturbances and also regulate voltage on the sensitive load side. These tests are designed to show how flexibility in the DVR will prevent the weak load from voltage sag and swell. During testing, the DVR was connected to the end of the test circuit with a 100 kW sensor load. The DVR and sensor load replace the photoelectric load attached to bus 310 in Fig. 3 from the

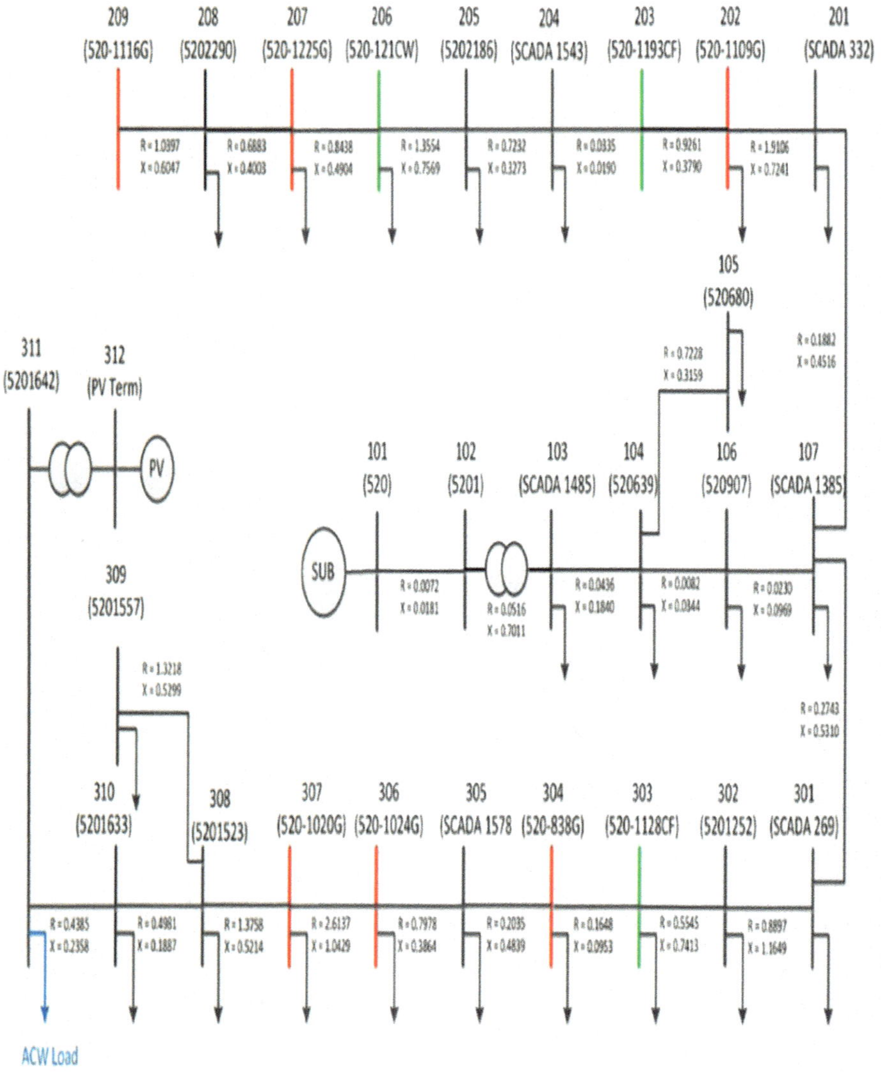

Fig. 3 Test circuit diagram

previous section. The load volume was up to 100 kW due to its similarity to the loads in the presented test circuit. Figure 4 shows the initial setup before performing any tests.

Fig. 4 DVR with the load sensor before starting the test

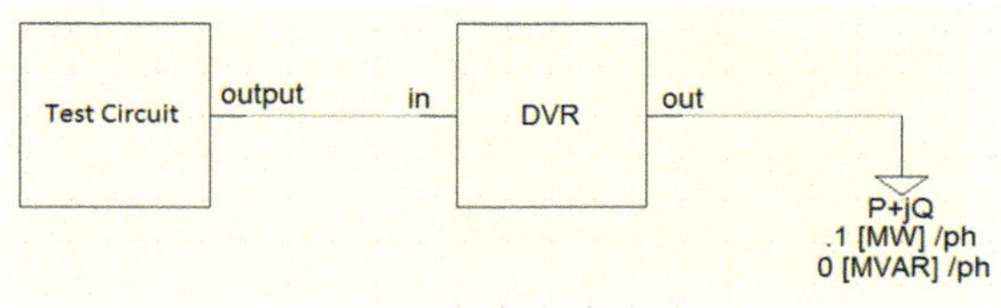

3.1 Load Side Voltage Sag Protection

In addition to the main function of the DVR in protecting against disturbances from the side of the feeding source, it also corrects light voltage disturbances from the side closest to the pregnancy. To test this DVR capability, a purely inductive load of 1 MVAr was added. Figure 5 shows inductive loads attached in parallel to the sensitive load through a small transmission line impedance.

So that we can see the response of the DVR after steady state occurs, the DVR will not do any voltage magnitude correction until the simulation runtime reaches 0.3 s. In this test, the inductive load is added at start up, so the steady-state voltage sag is evident just before 0.2 s run time. Figure 6 shows the DVR started to correct the voltage sag at 0.3 s and the three-phase rms voltage of the load and the source during the drop. The source voltage remains fairly constant at around 0.93 per unit, but the load voltage is corrected from 0.89 per unit to 0.975 per unit and is kept there by the DVR. The DVR successfully protects the load without fixing the size of the source voltage. Figure 7 shows the effect of voltage sag on the phase voltages of the source, load, and DVR. Note that the DVR injects a phase voltage at the source voltage to restore the load voltage size to 5% of the rated value. The loading angle is not preserved, as it is forced to be in phase with the source during DVR compensation. In addition, the source voltage, load, and DVR of phase A are superimposed in Fig. 8, showing the in-phase compensation more clearly. This same test was performed for a small number of loads in order to gather sufficient information about the DVR's response to the load side voltage sag. The acceptable range for static load bus voltage is between 0.95 and 1.05 per unit. The data in Table 1 show the ability of the DVR to regulate the load bus voltage in the range of 5% (0.05 per unit) of the rated value

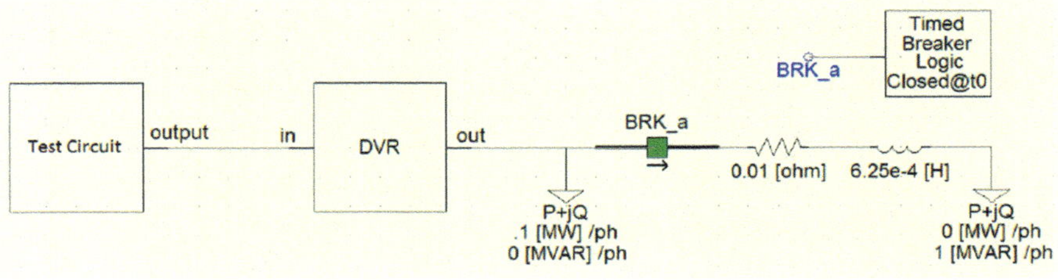

Fig. 5 Circuit simulated voltage sag on the load side

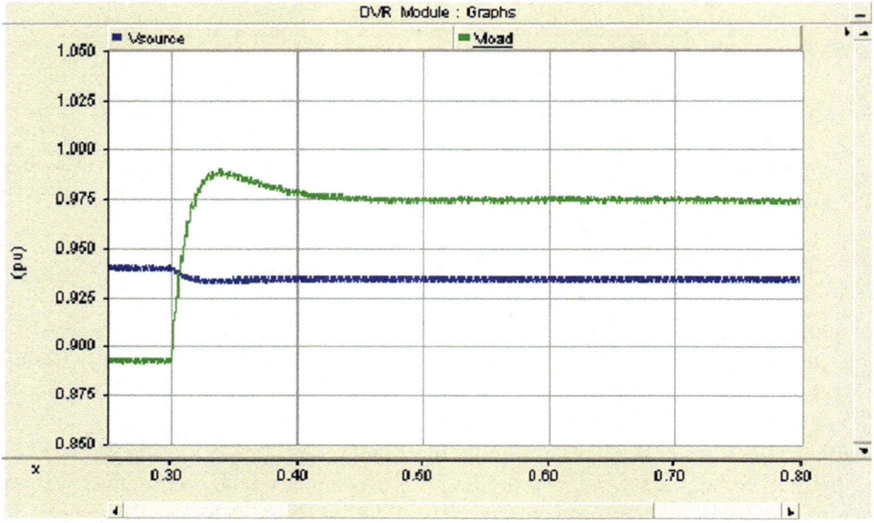

Fig. 6 Effect of load side voltage sag on load and source voltage

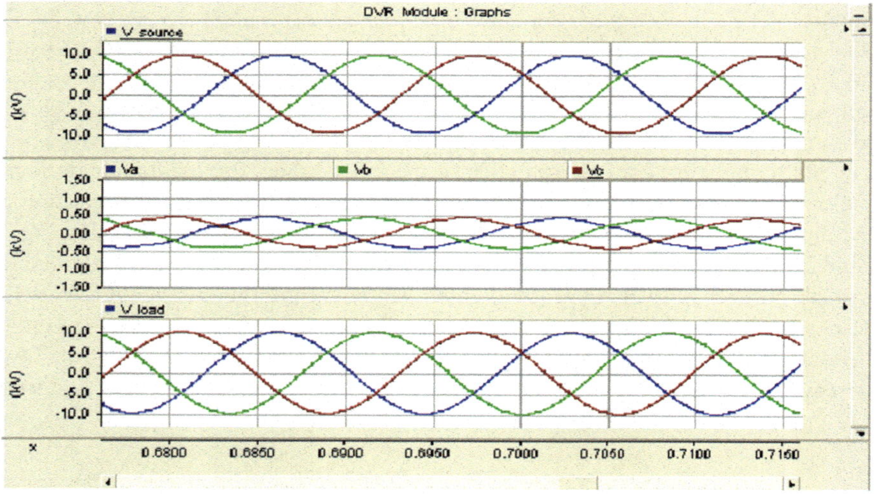

Fig. 7 Effect of load side sag on both source, load, and DVR phase voltages

up to the inductive load 2 MVAR. Figure 9 also shows how the DVR can protect the sensitive load from loading side sag.

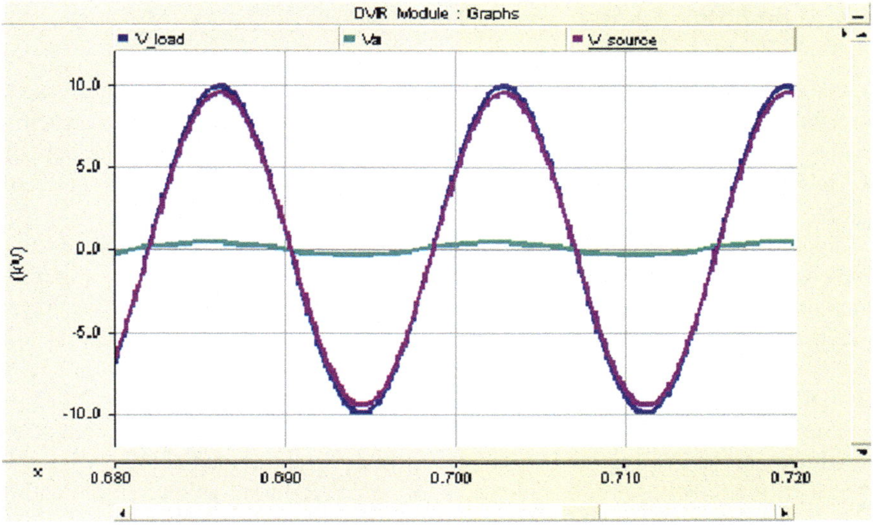

Fig. 8 Compensation in phase A load, source, and DVR voltages during load side sag

Table 1 Correction with added load provided by the DVR when voltage sag occurs

Added load (MVAr)	Load voltage w/out DVR (pu)	Load voltage w/DVR (pu)
0.5	0.94363	0.98670
1	0.89336	0.97547
1.5	0.84748	0.96192
2	0.80683	0.95015

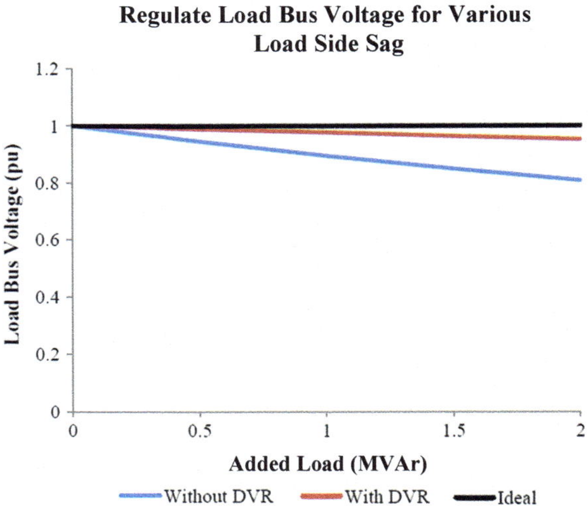

Fig. 9 Regulate load bus voltage for various load side sag

3.2 Load Side Voltage Swell Protection

A test was performed similar to the one performed to simulate a voltage sag to simulate voltage swells on the load side of the DVR. However, a purely capacitive load of (−1 MVAR) was added instead of adding an inductive load in parallel with the sensitive load. Figure 10 shows the additional load in parallel with the protected load.

Again, the comparison was designed in calculating the modification index to prevent the DVR from injecting a voltage of up to 0.3 s into the simulation. Figure 11 shows the load and source voltage in per unit as the DVR begins to inject a compensation voltage to counteract the voltage swell. Prior to correction, the capacitive load increases the load and the source voltage, with the load voltage normalizing at 1.132 per unit. Once the DVR is turned on, the load voltage is rapidly reduced to 1.025 per unit and this value is within the acceptable range of 5% of the rated value, and we conclude that the DVR has the ability to regulate the load carrier voltage. Figure 12 shows the phase voltages of the source, load, and DVR during voltage amplification.

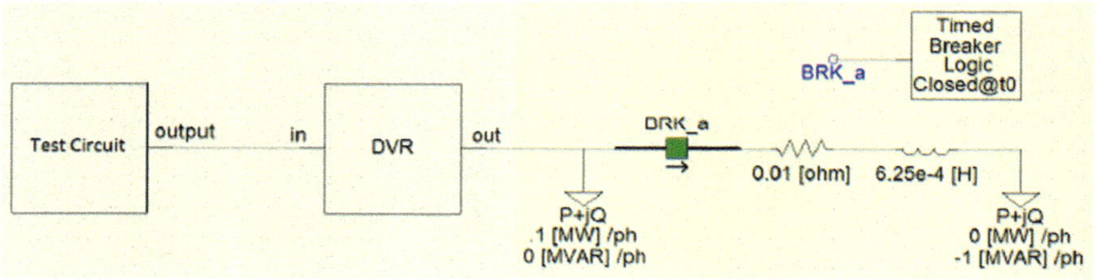

Fig. 10 Circuit simulated voltage swell on the load side

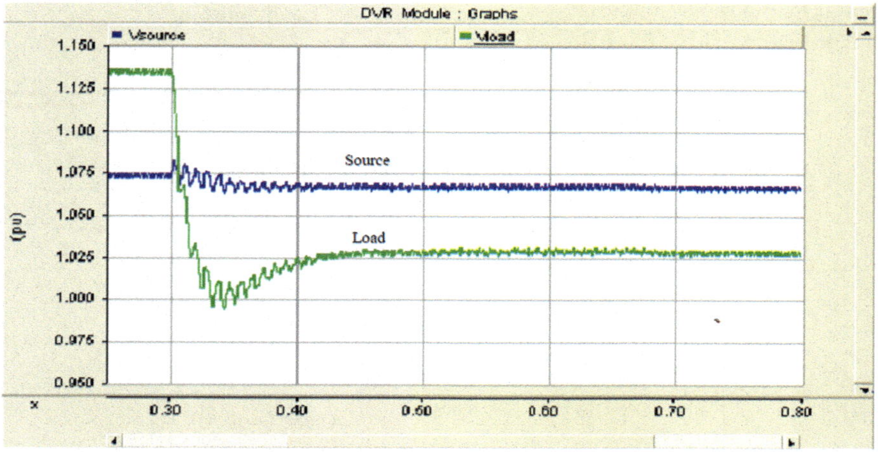

Fig. 11 Effect of load side voltage swell on load and source voltages

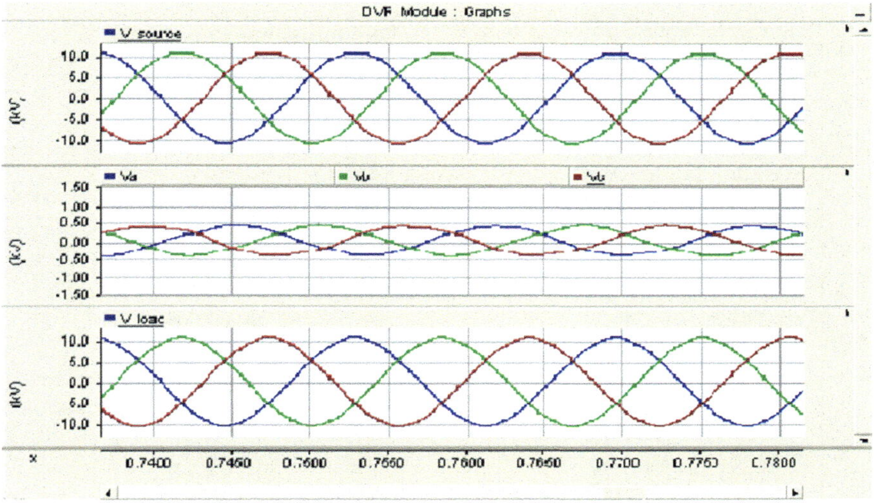

Fig. 12 Effect of load side swell on both source, load, and DVR phase voltages

In this case, the DVR injects a 180° off-phase voltage at the source voltage in order to restore the load voltage magnitude to 5% of the rated value. Figure 13 shows the phase A voltages of the source, load, and DVR. The DVR injected voltage is 180° out of phase with the source so that the size of the load voltage is reduced to the rated value.

Exactly like this test done in case of voltage sag we did to find out the ability of the DVR to correct voltage swells of varying volumes at different values of capacitive

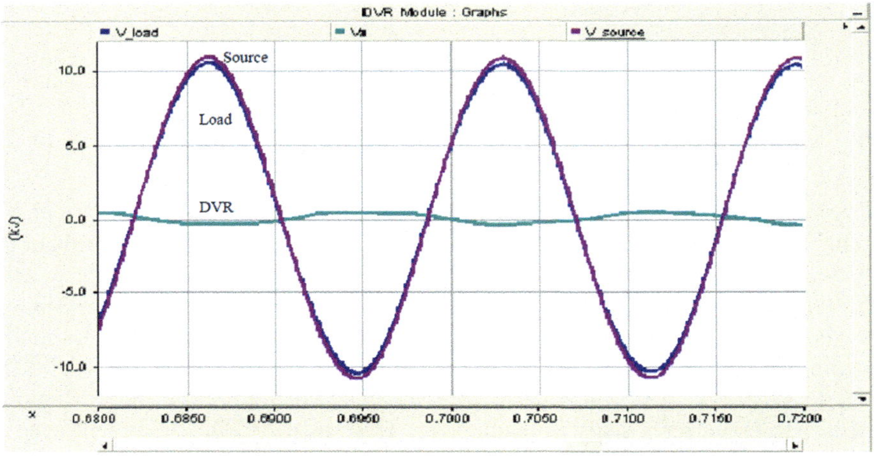

Fig. 13 Compensation in phase A load, source, and DVR voltages during load side swell

Table 2 Correction with added load provided by the DVR when voltage swell occurs

Added load (MVAr)	Load voltage w/out DVR (pu)	Load voltage w/DVR (pu)
−0.5	1.06023	1.01387
−1	1.13183	1.02519
−1.2	1.16307	1.03552
−1.4	1.20229	1.05000

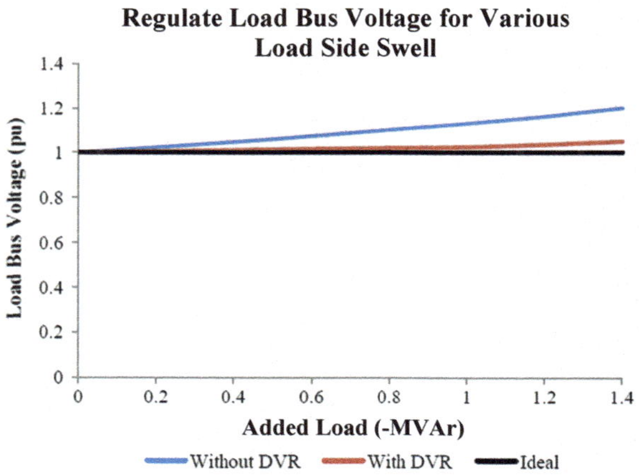

Fig. 14 Regulate load bus voltage for various load side swell

load. Table 2 shows the load bus voltage in per unit with and without the DVR for various capacitive loads. The DVR demonstrates the ability to keep the bus voltage at the load to within 0.05 per unit (5%) of rated for loads up to −1.4 MVAr. Figure 14 clearly shows the effectiveness of the DVR in protecting the sensitive load from load side swell.

4 Conclusion

The annual increased government investment in research and development of highly reliable clean energy resource technologies has greatly increased the demand for them, and renewable energy sources have become interconnected with the electric grid in large numbers and their economic benefits have become widely known, but the large influx of renewable energy resources has negatively affected on the quality of the electrical energy and also on the operation of the power system as a whole. The energy output of these sources is highly variable, which results in serious voltage disturbances on most sensitive equipment. This has called for the urgent need to mitigate these dangerous fluctuations in voltage and maintains the quality of electrical power, especially with the increasing investment of renewable energy sources.

Recent developments in the world of power electronics have created distinct and novel solutions for different voltage disturbances in power systems. One of the most important of these modern electronic devices is the dynamic voltage regulator, a device used to protect the sensitive load connected to it through sequential compensation from voltage sag and swell. This paper provides an analysis and modeling of the DVR working system using the PSCAD program, as it showed the simulation results in addition to protecting the sensitive load from voltage sag and swell, showing the ability of the DVR to regulate the load voltage to its reference value. Using a phase compensation strategy for sequential voltage injection, the DVR mechanically corrects for voltage sag and swell over only a few cycles. The developed DVR model is a useful engineering benefit for future analysis and development of newer DVR systems.

References

1. Fung C-C, Thanadechteemapat W, Harries D (2009) Acquiring knowledge and information on alternative energy from the world wide web. In: Power & energy society general meeting
2. Benachaiba C, Ferdi B (2008) Voltage quality improvement using DVR. Electr Power Qual Utilisation 14(1):39–45
3. Lazaroiu GC (2013) Power quality assessment in small scale renewable energy sources supplying distribution systems. Energies 6(1):634–645
4. Voltage Sags (2004) Platts, a division of The McGraw Hill Companies, Inc., Moncks Corner, SC
5. Pal S, Nath S (2010) An intelligent on line voltage regulation in power distribution system. In: 2010 international conference on power, control and embedded systems (ICPCES), Allahabad
6. Zhan C, Ramachandaramurthy VK, Arulampalam A, Fitzer C, Kromlidis S, Bames M, Jenkins N (2001) Dynamic voltage restorer based on voltage-space-vector PWM control. IEEE Trans Ind Appl 37(6):1855–1863
7. Ghosh A, Jindal AK, Joshi A (2004) Design of a capacitor supported dynamic voltage restorer (DVR) for unbalanced and distorted loads. IEEE Trans Power Delivery 19(1):405–413
8. Ghosh A, Ledwich G (2001) Structures and control of a dynamic voltage regulator. In: IEEE power engineering society winter meeting, 2001, Columbus, Ohio
9. Nazarpour D, Farzinnia M, Nouhi H (2017) Transformer-less dynamic voltage restorer based on buck-boost converter. IET Power Electron 10(13):1767–1777
10. Nielsen JG (2002) Design and control of a dynamic voltage restorer, Institute of Energy Technology, Aalborg University, Aalborg

Chapter 9
Robust Clustering Based Possibilistic Type-2 Fuzzy C-means for Noisy Datasets

Abdelkarim Ben Ayed, Mohamed Ben Halima, Sahar Cherif, and Adel M. Alimi

1 Introduction

In machine learning, a wide variety of clustering algorithms are employed in several applications. Among these algorithms, four categories are distinguished which are hierarchical methods, centroid methods, distribution-based methods, and density-based methods. Hierarchical methods consist in dividing or merging data points until a certain stop criterion will be satisfied. Despite these methods achieve good clustering accuracy results, they still have several problems. They are sensitive to outliers and inadequate for the large dataset. On the other side, centroid methods randomly initialize the cluster centers which are then optimized based on the distance between the centers and the closest data points. These methods are fast, effective for simple datasets and have a simple implementation. However, centroid methods result in local optimum and depend greatly on the choice of initial centers. Even though centroid methods create clusters with regular shapes and similar size, they cannot cluster density-based clusters or irregular clusters and they require a manually fixed number of clusters. Distribution-based methods consist of clustering data points based on their distribution. The results of these methods depend greatly on the choice of data distribution model and result in a local optimum. Density-based methods cluster data are based on their density. These methods are robust to outliers and allow

A. Ben Ayed (✉) · M. Ben Halima · S. Cherif · A. M. Alimi
REGIM-Lab.: Research Groups in Intelligent Machines, University of Sfax, National Engineering School of Sfax, LR11ES48, 3038 Sfax, Tunisia
e-mail: abdelkarim.benayed.tn@ieee.org

M. Ben Halima
e-mail: mohamed.benhlima@ieee.org

S. Cherif
e-mail: sahar.cherif.tn@ieee.org

A. M. Alimi
e-mail: adel.alimi@ieee.org

A. E. Hassanien et al. (eds.), *Enabling Machine Learning Applications in Data Science*,
Algorithms for Intelligent Systems, https://doi.org/10.1007/978-981-33-6129-4_9

them to find arbitrary shape clusters but require to fix manually many parameters. A detailed review of clustering algorithms was carried out in [1, 2].

Conventional data clustering algorithms cannot handle the new aspects of data coming from social media, personal devices, sensors, medical scanners, etc. Specifically, these data may contain outliers, missing values, a large number of irrelevant features, etc. To deal with such complex data, many recent works have applied advanced soft computing techniques such as unsupervised feature selection to reduce redundant and noisy features [3]. It proved to be robust in improving a large variety of systems, such as image segmentation [4], bioinformatics [5], pattern recognition [6, 7], and medical applications [8]. Other interesting works are based on fuzzy logic in order to handle the uncertainty concept such as fuzzy C-means clustering [9], possibilistic C-means [10, 11], and type-2 fuzzy clustering [12].

Ensemble and fuzzy clustering techniques have acquired a lot of attention recently. Ensemble learners are a branch in machine learning using a set of weak learners on all or subset of data. After that, the results of each learner are merged in order to get the final result and remove the outliers [13–17]. Cluster forests [14] are an effective ensemble clustering algorithm inspired from random forests [13] which is a popular ensemble classification algorithm. The main idea of random forests algorithm is to run decision trees on a randomly chosen subset of data features. Then, a vote technique is used to aggregate the results. Cluster Forests apply K-means clustering algorithm on a part of data features chosen using a certain feature competition algorithm, then aggregates the intermediate clustering results using the N-cut spectral clustering algorithm [18]. The cluster forest algorithm is used as an ensemble learner in our work to deal with outliers within real datasets. Janoušek et al. [17] proposed bio-inspired cluster forests which is a modified version of the cluster forests using differential equation as the base clustering algorithm to further enhance clustering results. Inspired from this work, we replaced the base clustering algorithm (K-means) in cluster forest algorithm by the improved type-2 fuzzy possibilistic C-means clustering algorithm as the base clustering algorithm.

In this paper, we proposed a new method called improved possibilistic type-2 fuzzy C-means clustering-based cluster forests (IP-T2-FCM-CF) considering the difficulty of the clustering task and taking into consideration the challenging nature of data. The contributions of our work are as follows:

- Cluster forests based on improved possibilistic type-1 fuzzy C-means.
- Cluster forests based on type-2 fuzzy C-means.
- Cluster forests based on improved possibilistic type-2 fuzzy C-means.

The impact of feature selection, the possibilistic aspect, the transition from type-1 fuzzy to type-2 are all studied on real-world data for measuring the clustering quality of the hybrid proposed method IP-T2-FCM-CF.

The rest of the paper is organized as follows. Section 2 describes possibilistic concept, feature selection, and cluster forests adopted in our work. Next, Sect. 3 presents our new proposed method and its three variants. Then, Sect. 4 illustrates the experimental results and discussion using two external indices. Finally, Sect. 5 concludes the paper and outlines our future work.

2 Materials and Methods

In this section, we present theoretical concepts of the techniques used in our new clustering method.

2.1 A Brief Review of the Baseline Methods

The possibilistic C-means clustering algorithm or PCM is proposed in [10] to improve clustering performance for dataset with noise. The PCM clustering algorithm steps are similar to FCM steps. The authors published a modified version in [11] with an improved objective function. PCM produces dense clusters and is less sensitive to outliers. PCM can ignore isolated outliers' data points, while a single outlier can alter all FCM membership matrix. However, PCM has the drawback to be very sensitive to initialization and can produce coincident clusters. PCM is used in many clustering systems [19]. Other modified versions have been published lately including modified FPCM [20] and improved PFCM [21]. In this work, we decided to use the improved PFCM algorithm which is proved to have better performance than the other versions [21]. Unlike FCM which is based on fuzzy membership degree, PCM is based on the concept of typicality membership degree. Compared to original PFCM [21], IPFCM determines dynamically the constants a and b representing the weights of membership and typicality, respectively [21]. They are defined by Eq. (1).

$$a_{ij} = \exp\left(\frac{-\|x_i - v_j\|^2}{\sum_{i=1}^{n} \|x_i - \overline{x}\|^2} \times \frac{n}{\sqrt{c}}\right);\ b_{ij} = \exp\left(-\frac{n \times \|x_i - v_j\|^2}{\sqrt{c}}\right) \quad (1)$$

The correspondent objective function is expressed in (2).

$$J_{\mathrm{IPFCM}} = \sum_{j=1}^{c}\sum_{i=1}^{n}\left(\mu_{ij}^{m} a_{ij}^{m} d^2(x_i, v_j) + t_{ij}^{\eta} b_{ij}^{\eta} d^2(x_i, v_j)\right) \quad (2)$$

where m and η represent the fuzzy exponent and the typicality exponent, respectively. Besides, μ_{ij} and t_{ij} represent the degree of membership and the degree of typicality as in Eq. (3).

$$\mu_{ij} = \left[\sum_{k=1}^{c}\left(\frac{d^2(x_i, v_j)}{d^2(x_i, v_k)}\right)^{\frac{2}{(m-1)}}\right]^{-1};\ t_{ij} = \left[\sum_{k=1}^{n}\left(\frac{d^2(x_i, v_j)}{d^2(x_i, v_k)}\right)^{\frac{2}{(\eta-1)}}\right]^{-1} \quad (3)$$

Note that, v_j represents the vector of centers and is defined in (4).

$$v_j = \frac{\sum_{i=1}^{n}\left(\mu_{ij}^{m} a_{ij}^{m} + t_{ij}^{\eta} b_{ij}^{\eta}\right) x_i}{\sum_{i=1}^{n}\left(\mu_{ij}^{m} a_{ij}^{m} + t_{ij}^{\eta} b_{ij}^{\eta}\right)} \tag{4}$$

Feature selection methods are widely used in data mining applications and mainly in classification algorithms for high-dimensional data sets to reduce the number of similar features or to remove redundant and irrelevant features. These methods have the advantage of reducing the computing time while improving the clustering accuracy of the classification task. Feature selection methods are different for supervised classification and unsupervised classification algorithms. The traditional methods of feature selection were statistical-based, similarity-based, and information-based while the modern methods are sparse-learning-based, deep-learning-based, and ensemble-based methods. A recent state of the art on both supervised and unsupervised feature selection methods is presented in [22]. In this work, we use an ensemble-based feature selection method inspired from [14] that build randomly different vector subsets of features then choose the best vector of features using a cluster quality measure.

Spectral clustering (SC) has acquired a lot of attention in the recent years. SC is applied in many fields including mainly image segmentation and pattern recognition problems. SC became popular since 2001 with the works of [18] and [23]. Recent applications of SC are presented in [24]. SC main idea is to reformulate the problem of clustering to graph partitioning problem. The solving procedure includes the use of similarity matrix and generalized eigenvalues.

Cluster forests clustering algorithm (CF) [14] is an unsupervised ensemble classification algorithm inspired from the popular random forests ensemble classification algorithm [13]. CF algorithm is composed of three major steps. In the first step, subsets of strong feature vectors, called trees, are constructed based on subsets of the data set, the set of trees from a forest. In the second step, the same clustering algorithm is applied to all feature vectors. Finally, n-cut spectral clustering is used to aggregate all the vector labels resulting from the previous step [14].

First, a feature selection step is applied. It consists in building a strong feature vector to improve clustering accuracy and reduce its complexity. A cluster validity measure κ expressed in Eq. (5) is employed to choose the strong subset of features [14].

$$\kappa = \frac{\text{Within Cluster Distance}}{\text{Between Cluster Distance}} \tag{5}$$

2.2 Discussion

The most important issue for K-means clustering is that the clustering results depend strongly on the initialization of the cluster centers and their number. K-means algorithm is convenient only for numerical data. Besides, K-means do not converge to global minimum but converge to a local minimum. To overcome these limitations, it is recommended to iterate K-means several times and then choose the one with the best objective function value.

Unlike traditional clustering such as K-means, SC has the advantage to be robust against initial cluster initialization and non-convex clusters which makes SC has better clustering performance than most traditional algorithms [25].

The IPFCM is proved to find better clusters and to detect more effectively unequal size clusters [21]. In our work, we extend IPFCM to type-2 IPFCM to support more dataset uncertainties.

To choose the best feature vector, Yan et al. [14] apply K-means clustering algorithm on the candidate features to determine the related value of κ. Strong feature vectors correspond to low value of κ. Feature vectors are initialized by making feature competition between different couples of random features. The couple of features having the lowest κ value is chosen. After that, additional random features are added iteratively to the chosen couple of features at the condition of improving κ value. The growth of feature vector stops after three failed attempts to add additional features.

In the second step, K-means the base clustering algorithm is applied to the selected feature vectors (trees). K-means is used to cluster each of the T feature vectors. The resulted T labels vector are used in the next step to create an affinity matrix.

The first two steps are repeated T times, where T represents the size of forest or the number of trees. All affinity matrices are summed then regularized. Finally, n-cut spectral clustering is employed to extract the final labels vectors from the resulted affinity matrices.

3 Improved Possibilistic Type-2 Fuzzy C-means Clustering-Based Cluster Forests (IP-T2-FCM-CF)

In this section, we present the three variants of fuzzy clustering method by presenting the corresponding algorithm. We recall that our method consists in changing the base clustering algorithm of the cluster forests algorithm, namely IPFCM, T2-FCM, and IP-T2-FCM, respectively. In this regard, the algorithm IPFCM was changed to handle the fuzzy type-2. Therefore, the objective function (2) become as in (6).

$$J_{\mathrm{IP-T2-FCM}} = \sum_{j=1}^{c}\sum_{i=1}^{n}\begin{pmatrix} \mu_{ij}^{(m_1+m_2)/2} a_{ij}^{(m_1+m_2)/2} d^2(x_i, v_j) + \\ t_{ij}^{(\eta_1+\eta_2)/2} b_{ij}^{(\eta_1+\eta_2)/2} d^2(x_i, v_j) \end{pmatrix} \tag{6}$$

Lower and upper fuzzy exponents are calculated using Eq. (7).

$$\underline{v}_j = \frac{\sum_{i=1}^{n}\left(\mu_{ij}^{m_1} a_{ij}^{m_1} + t_{ij}^{\eta_1} b_{ij}^{\eta_1}\right)x_i}{\sum_{i=1}^{n}\left(\mu_{ij}^{m_1} a_{ij}^{m_1} + t_{ij}^{\eta_1} b_{ij}^{\eta_1}\right)}; \overline{v}_j = \frac{\sum_{i=1}^{n}\left(\mu_{ij}^{m_2} a_{ij}^{m_2} + t_{ij}^{\eta_2} b_{ij}^{\eta_2}\right)x_i}{\sum_{i=1}^{n}\left(\mu_{ij}^{m_2} a_{ij}^{m_2} + t_{ij}^{\eta_2} b_{ij}^{\eta_2}\right)} \tag{7}$$

Type-2 lower and upper typicality memberships are calculated using (8) and (9).

$$\underline{T}_i(x_j) = \min\left\{\left[\sum_{k=1}^{c}\left(\frac{d_{ij}^2}{d_{ij}^2}\right)^{\frac{2}{\eta_1-1}}\right]^{-1}, \left[\sum_{k=1}^{c}\left(\frac{d_{ij}^2}{d_{ij}^2}\right)^{\frac{2}{\eta_2-1}}\right]^{-1}\right\} \tag{8}$$

$$\overline{T}(x_j) = \max\left\{\left[\sum_{k=1}^{c}\left(\frac{d_{ij}^2}{d_{ij}^2}\right)^{\frac{2}{\eta_1-1}}\right]^{-1}, \left[\sum_{k=1}^{c}\left(\frac{d_{ij}^2}{d_{ij}^2}\right)^{\frac{2}{\eta_2-1}}\right]^{-1}\right\} \tag{9}$$

IP-T2-FCM combines both possibilistic and type-2 fuzzy set theory set to handle both noisy data and outliers. In our work, we introduced the IP-T2-FCM as a base clustering algorithm in cluster forests to be IP-T2-FCM-CF. This hybridization makes our method robust and suitable for real-world data where different sources of information are provided and big data are more and more emergent. The process of IP-T2-FCM is also shown in Fig. 1 using the flowchart notations (see Fig. 1).

The proposed method starts with a feature selection step using the chosen base clustering algorithm. This step includes a feature competition technique ensuring the composition of a first feature vector and κ cluster quality measure for the selection of strong features f'. At each feature selection iteration, a randomly selected feature f from the feature space $\mathcal{F}$ is added to f' and IP-T2-FCM is applied to the data induced by the concatenation of (f', f). Again, κ measure is calculated to decide if this new added feature f. will be kept or discarded. The feature selection step takes end if the maximum number of discards is reached. The result is the features considered as the best feature vector fb*.

Then, IP-T2-FCM clustering algorithm is applied to fb* to get data partition. Giving n the number of samples, the next step is to calculate the $n \times n$ co-cluster indicator matrix which is also called affinity matrix as expressed in Eq. (10).

$$P_{ij} = \begin{cases} 1 & X_i \text{ and } X_j \text{ are in the same cluster} \\ 0 & \text{otherwise} \end{cases} \tag{10}$$

Previous steps are repeated T times, then add P_{ij} matrices using Eq. (11).

$$P = \sum_{i=1}^{T} P_{ij} \tag{11}$$

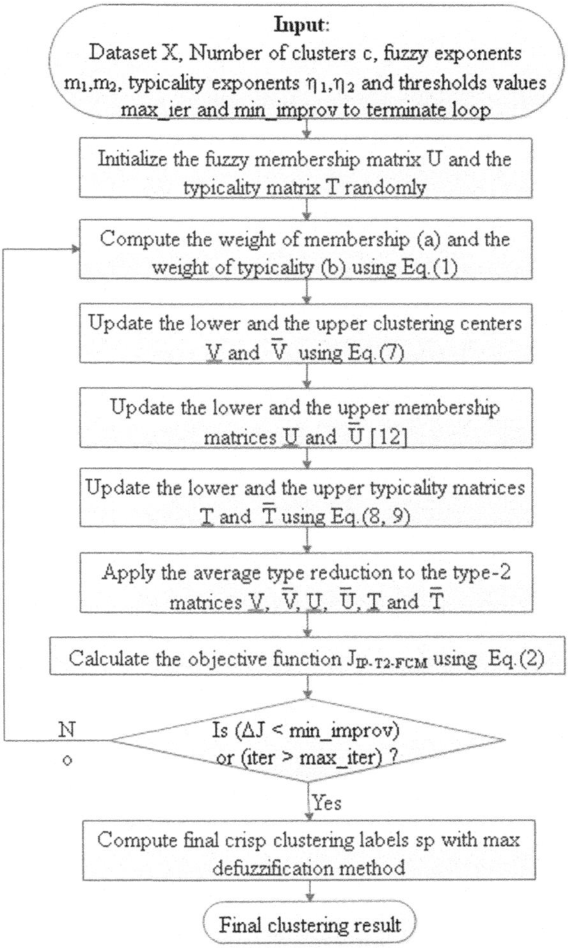

Fig. 1 Flowchart of IP-T2-FCM clustering algorithm

The next step is to apply thresholding and scaling to the resultant sum P matrix [14]. Spectral clustering is the final step of the IP-T2-FCM-CF method. Here, we recall that we chose the n–cut clustering algorithm to get final clustering result as recommended in [14]. Figure 2 explains the steps of the whole process showing the transition from a step to another.

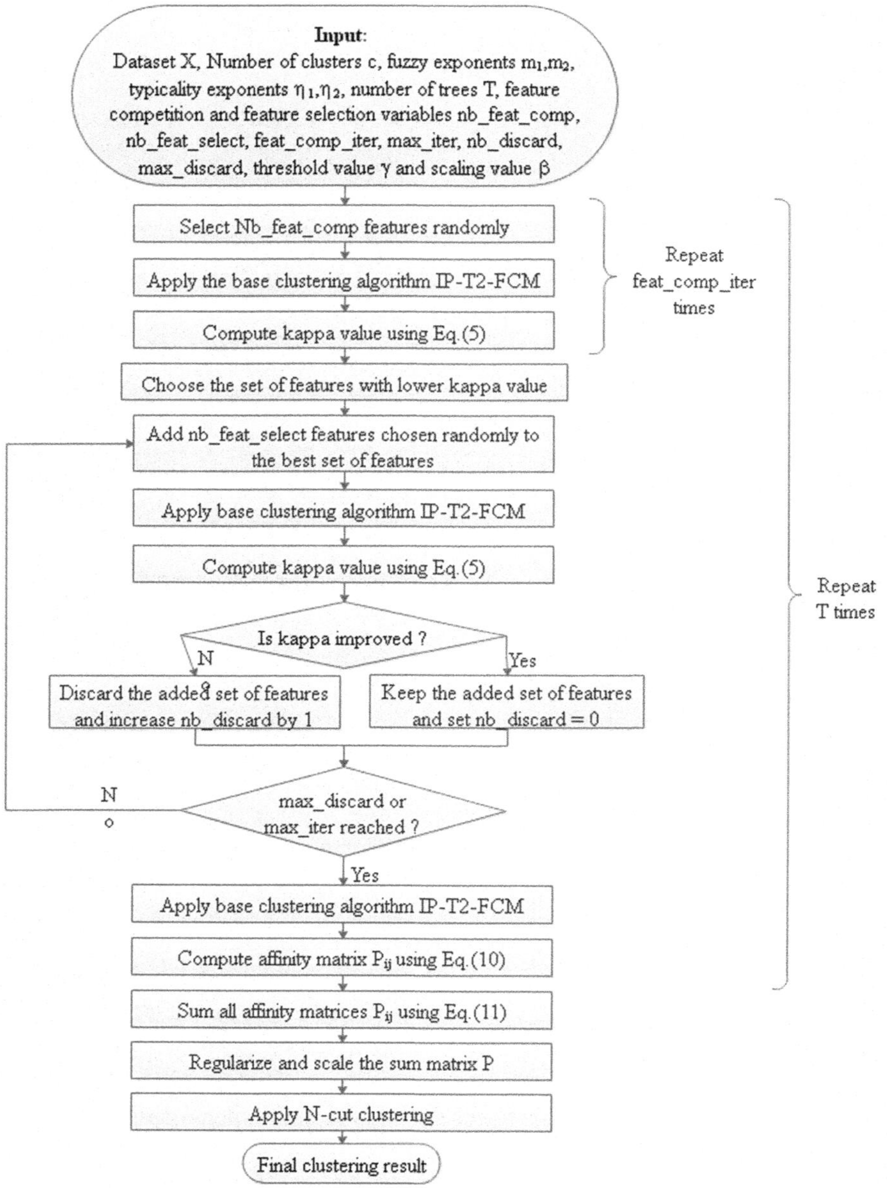

Fig. 2 Flowchart of cluster forests-based IP-T2-FCM-CF clustering algorithm

4 Evaluation of the Proposed IP-T2-FCM-CF Algorithm

4.1 Experiments Setup

Eight real-world UC Irvine datasets [26] varying from small to large (Soybean small, Robot, Wine, Heart, SPECT, WDBC, Image Segmentation, and Madelon) are used to validate the proposed method. The datasets details are presented in Table 1.

Two external evaluation metrics ρ_r (clustering quality) and ρ_c (clustering accuracy) [14] are considered to measure the performance of the clustering algorithms by comparing the labels resulted from clustering algorithms with the true labels of the eight datasets. For more details about clustering evaluation metrics, the reader can refer to the review presented in [27].

The clustering quality metric ρ_r compares the calculated pairs of data points with those in true labels and is defined by Eq. (12).

$$\rho_r = \frac{\text{Number of correctly clustred pairs}}{\text{Total number of pairs}} \tag{12}$$

The clustering accuracy metric ρ_c is inspired from the classification accuracy metric and is defined by Eq. (13).

$$\rho_c = \max_{\text{all permutations}} \left\{ \frac{\text{Number of correctly clustred pairs}}{\text{Number of data points}} \right\} \tag{13}$$

In particular, clustering algorithm with the highest values of these metrics implies that the data are well clustered and it is the best algorithm.

Table 1 Testing datasets

Dataset	Number of features	Number of instances	Number of classes
Soybean	35	47	4
SPECT	22	267	2
ImgSeg	19	2100	7
Heart	13	270	2
Wine	13	178	3
WDBC	30	569	2
Robot	90	164	5
Madelon	500	2000	2

4.2 Experimental Results

In the experiment settings, 40 clustering trees were created and the results reported on the average of 20 runs. For the number of clusters, it was initially fixed equally to the number of classes for each dataset. Table 2 details all the aforementioned parameters in Sect. 3. Note that, the values of these parameters were considered following previous work from the literature.

Handling noisy data, sensitivity to the outlier's samples, and uncertainty of real-world data are the major issues that our proposed method tries to overcome in this work. Three variants are considered which are T2-FCM-CF, IPFCM-CF, and IP-T2-FCM-CF, respectively. The aim, here, is to study the impact of each improvement, to know, the introduction of the type-2 fuzzy concept, the possibilistic aspect in the membership function, and the ensemble clustering with the unsupervised feature selection.

Furthermore, we cannot present our results only, but also it is interesting to compare them to traditional clustering algorithms which are K-means [2], FCM [9], Type-2 fuzzy C-means (T2-FCM) [12], improved possibilistic fuzzy C-means (IPFCM), improved possibilistic type-2 fuzzy C-means (IP-T2-FCM), and K-means-based cluster forest (KM-CF) [28, 29] algorithms using clustering quality (see Table 3) and clustering accuracy (see Table 4) metrics. The highest achieved results are highlighted in both tables.

Table 2 Details of the experimental settings

Methods	Parameters	Values
K-means	max_iter	100
FCM	m	2
T2-FCM	m_1 m_2	1.5 2.5
IPFCM	m η	2 4
IP-T2-FCM	m_1 m_2 η_1 η_2	1.5 2.5 3.5 4.5
CF	T feat_comp_iter nb_feat_comp nb_feat_select max_discard gamma beta	40 5 $\sqrt{}$(n) 1 3 0.4 0.1

Table 3 Clustering accuracy comparison (ρ_c) expressed in percentage (%)

Dataset	K-means	FCM	T2-FCM	IPFCM	IP-T2-FCM	KM-CF	FCM-CF	T2-FCM-CF	IPFCM-CF	IP-T2-FCM-CF
Soybean	72.23	72.04	72.09	72.34	72.34	75.32	82.77	85.11	85.32	**88.72**
Robot	39.73	50.04	49.55	46.91	**51.53**	35.52	46.01	46.81	42.76	42.94
Wine	68.08	70.79	71.13	69.21	73.03	70.22	70.22	71.18	**86.35**	72.47
Heart	59.04	60.74	60.74	60.60	61.11	67.93	68.41	68.52	63.48	**69.74**
SPECT	42.49	43.06	43.06	**60.30**	56.55	48.54	45.96	48.31	50.30	51.35
WDBC	85.41	86.64	87.35	88.86	88.22	87.47	88.63	89.88	**89.96**	89.37
ImgSeg	52.04	54.38	55.94	56.71	**57.37**	50.15	47.49	50.32	52.24	53.74
Madelon	52.23	53.28	53.28	54.83	50.72	54.99	55.47	55.51	**56.16**	55.32

Table 4 Clustering quality comparison (ρ_r) expressed in percentage (%)

Dataset	K-means	FCM	T2-FCM	IPFCM	IP-T2-FCM	KM-CF	FCM-CF	T2-FCM-CF	IPFCM-CF	IP-T2-FCM-CF
Soybean	83.10	83.07	83.07	83.16	83.16	87.10	92.03	93.67	93.71	**95.25**
Robot	54.81	65.50	66.06	71.61	74.27	63.86	70.48	70.44	64.73	**70.90**
Wine	71.41	72.04	72.92	72.80	73.47	71.87	71.87	72.95	**83.65**	73.24
Heart	51.46	52.13	52.13	52.11	52.29	56.43	56.77	57.06	53.68	**57.65**
SPECT	50.64	52.19	52.19	**64.21**	61.17	54.97	52.93	54.15	56.00	57.02
WDBC	75.04	76.81	77.86	80.17	79.19	78.08	79.83	81.82	**82.01**	81.10
ImgSeg	82.26	83.85	83.99	**84.42**	84.57	80.27	80.39	81.68	82.35	82.82
Madelon	50.30	50.39	50.39	50.59	50.00	50.55	50.61	50.64	**50.79**	50.70

4.3 Results' Discussion

In the aforementioned results, two main category of clustering algorithms are presented, to know, the monoclustering (basic algorithms) and the ensemble clustering (cluster forests). We begin by comparing the results of the same category and then between the two categories.

According to Table 3, for the monoclustering algorithms, best results are achieved by IP-T2-FCM. For SPECT, WDBC, and Madelon, IPFCM slightly outperforms IP-T2- FCM with a difference of 3.75%, 0.64%, and 4.11%, respectively. Both algorithms perform the same result for the case of soybean dataset handling 72.34%. The same situation is figured for the second validity measure as illustrated in Table 4. Therefore, we expected that the combination of possibilistic and fuzzy type-2 set theory could better enhance the clustering results when including it to the cluster forests. For instance, the highest improvement is of 10.49% rate for robot dataset as given in Table 3. The same results 70.22% of ρ_c and 71.87% ρ_r are obtained for wine dataset. This confirms that changing K-means by FCM is beneficial. Therefore, we also take into consideration the variation of the base algorithm by IPFCM, T2-FCM, and IP-T2-FCM. When we compare the three variants basing on ρ_c metric, T2-FCM-CF is the best in one case that is robot dataset with a rate of 46.81%. In contrast, Table 4 shows that T2-FCM-CF does not reach a best value for any of the tested datasets. More specifically, the best results of ρ_r were raised by the hybrid variant IP-T2-FCM-CF in the case of Soybean, Robot, Heart, SPECT, and ImgSeg which are as follows: 95.25, 70.90, 57.65, 57.02, and 82.82%. Introducing the type-2 fuzzy set to IPFCM-CF has improved the results.

The first category, i.e., the monoclustering algorithms, outperforms the cluster forests-based algorithms in some situations. According to Table 3, a difference of 4.72% for robot dataset is with IP-T2-FCM with respect to T2-FCM-CF. Similarly, for ρ_r metric, a difference of 3.37% is achieved with IP-T2-FCM in comparison with IPT2-FCM-CF. For ImgSeg dataset, IP-T2-FCM outperforms IP-T2-FCM-CF to reach 3.9% for ρ_c metric and 1.75% for ρ_r metric. Note that, IPFCM algorithm in the case of SPECT dataset is able to reach 60.30 and 64.21% which outperforms IP-T2-FCM-CF with 51.35 and 57.02% for ρ_c and ρ_r metrics.

Through these findings, it is confirmed that FCM is better than K-means algorithm due to the advantage of fuzzy clustering to separate overlapping clusters. Thanks to the type-2 fuzzy set, clustering results for T2-FCM, IPFCM, and IP-T2-FCM algorithms are improved due to the robustness of type-2 FCM for noisy data. In addition, the improved possibilistic FCM with the typicality membership instead of relative membership is less sensitive to outliers. The ensemble clustering algorithms-based cluster forest has enhanced results with noise and outliers' removal due to the unsupervised feature selection step. Only strong features are kept while removing redundant and irrelevant features. In general, the proposed methods T2-FCM-CF, IPFCM-CF, and IP-T2-FCM-CF have better clustering performance than other clustering algorithms for 80% of datasets. This is explained by the benefit of ensemble clustering, feature selection, fuzzy and possibilistic clustering. Despite a feature selection step in the ensemble clustering method is used dimensionality reduction, large datasets like ImgSeg and Madelon still require high computational time. In this context, we recommend the use of high parallel solutions like Hadoop for large datasets and especially for video clustering applications as evoked in the relative study of Ayed et al. in [30].

5 Conclusion

A new possibilistic type-2 fuzzy ensemble clustering method is proposed in this work making data clustering more robust especially with the complexity of the real-world data. IP-T2-FCM-CF method contains three stages which are feature competition, base clustering algorithm (IPFCM, T2-FCM, or T2-IPFCM), and then spectral clustering.

First, feature competition is used in order to select the best features and discard noisy ones. Then, we apply one of our base clustering algorithms (IPFCM, T2-FCM, or IP-T2-FCM) which have the advantage to converge to a good solution for clean data as well as for noisy and uncertain data. Finally, the effective spectral clustering combines the partial clustering results and produces the final decision. Results show that the proposed algorithms give better results than state-of-the-art classical clustering and ensemble clustering algorithms on well-established pattern benchmarks. Our cluster forests based on IP-T2-FCM allows to deal better with outliers and to have better convergence to more adequate cluster centers. However, our method suffers of a relatively high computation time. As a future direction, we will explore the multi-objective optimization problem to consider both complexity and robustness to choose the suitable clustering method for data.

Acknowledgements The research leading to these results has received funding from the Ministry of Higher Education and Scientific Research of Tunisia under the grant agreement number LR11ES48.

References

1. Yang MS, Nataliani Y (2017) Robust-learning fuzzy c-means clustering algorithm with unknown number of clusters. Pattern Recogn 71:45–59
2. Ayed AB, Halima MB, Alimi AM (2014) Survey on clustering methods: Towards fuzzy clustering for big data. In: 2014 6th international conference of soft computing and pattern recognition (SoCPaR). IEEE, pp 331–336
3. Zhu P, Zhu W, Hu Q, Zhang C, Zuo W (2017) Subspace clustering guided unsupervised feature selection. Pattern Recogn 66:364–374
4. Choy SK, Lam SY, Yu KW, Lee WY, Leung KT (2017) Fuzzy model-based clustering and its application in image segmentation. Pattern Recogn 68:141–157
5. Jiang Z, Li T, Min W, Qi Z, Rao Y (2017) Fuzzy c-means clustering based on weights and gene expression programming. Pattern Recogn Lett 90:1–7

6. Melin P, Castillo O (2014) A review on type-2 fuzzy logic applications in clustering, classification and pattern recognition. Appl Soft Comput 21:568–577
7. Kim EH, Oh SK, Pedrycz W (2017) Design of reinforced interval type-2 fuzzy c-means-based fuzzy classifier. IEEE Trans Fuzzy Syst 26(5):3054–3068
8. Kahali S, Sing JK, Saha PK (2019) A new entropy-based approach for fuzzy c-means clustering and its application to brain MR image segmentation. Soft Comput 23(20):10407–10414
9. Bezdek JC, Ehrlich R, Full W (1984) The fuzzy c-means clustering algorithm. Comput Geosci 10(2):191–203
10. Krishnapuram R, Keller JM (1993) A possibilistic approach to clustering. IEEE Trans Fuzzy Syst 1(2):98–110
11. Krishnapuram R, Keller JM (1996) The possibilistic c-means algorithm: insights and recommendations. IEEE Trans Fuzzy Syst 4(3):385–393
12. Rubio E, Castillo O (2014) Designing type-2 fuzzy systems using the interval type-2 fuzzy c-means algorithm. In: Recent advances on hybrid approaches for designing intelligent systems. Springer, Cham, pp 37–50
13. Breiman L (2001) Random forests. Mach Learn 45(1):5–32
14. Yan D, Chen A, Jordan MI (2013) Cluster forests. Comput Stat Data Anal 66:178–192
15. Lahmar I, Ayed AB, Halima MB, Alimi AM (2017) Cluster forest based fuzzy logic for massive data clustering. In: Ninth international conference on machine vision ICMV 2016. International Society for Optics and Photonics, p 103412J
16. Ayed AB, Benhammouda M, Halima MB, Alimi AM (2017) Random forest ensemble classification based fuzzy logic. In: Ninth international conference on machine vision ICMV 2016. International Society for Optics and Photonics, p 103412J
17. Janoušek J, Gajdoš P, Radecký M, Snášel V (2016) Application of bio-inspired methods within cluster forest algorithm. In: Proceedings of the second international afro-european conference for industrial advancement AECIA 2015. Springer, Cham, pp 237–247
18. Shi J, Malik J (2000) Normalized cuts and image segmentation. IEEE Trans Pattern Anal Mach Intell 22(8):888–905
19. Filippone M, Camastra F, Masulli F, Rovetta S (2008) A survey of kernel and spectral methods for clustering. Pattern Recogn 41(1):176–190
20. Saad MF, Alimi AM (2009) Modified fuzzy possibilistic c-means. In: Proceedings of the international multiconference of engineers and computer scientists, vol 1, pp 18–20
21. Saad MF, Alimi AM (2012) An improved fuzzy clustering method (case c-means). Doctoral dissertation, University of Sfax (2012)
22. Li J, Cheng K, Wang S, Morstatter F, Trevino RP, Tang J, Liu H (2017) Feature selection: a data perspective. ACM Comput Surv (CSUR) 50(6):1–45
23. Ng AY, Jordan MI, Weiss Y (2002) On spectral clustering: analysis and an algorithm. In: Advances in neural information processing systems, pp 849–856
24. Kanaan-Izquierdo S, Ziyatdinov A, Perera-Lluna A (2018) Multiview and multifeature spectral clustering using common eigenvectors. Pattern Recogn Lett 102:30–36
25. Cai D, Chen X (2014) Large scale spectral clustering via landmark-based sparse representation. IEEE Trans Cybern 45(8):1669–1680
26. Dua D, Graff C (2019) UCI machine learning repository, University of California, Irvine, School of Information and Computer Sciences. https://archive.ics.uci.edu/ml. Accessed 30 July 2020
27. Lei Y, Bezdek JC, Romano S, Vinh NX, Chan J, Bailey J (2017) Ground truth bias in external cluster validity indices. Pattern Recogn 65:58–70
28. Ayed AB, Halima MB, Alimi AM (2016) Cluster forests based fuzzy C-means for data clustering. In: International joint conference SOCO'16-CISIS'16-ICEUTE'16. Springer, Cham, pp 564–573
29. Ayed AB, Halima MB, Alimi AM (2017) Adaptive fuzzy exponent cluster ensemble system based feature selection and spectral clustering. In: 2017 IEEE international conference on fuzzy systems (FUZZ-IEEE). IEEE, pp 1–6
30. Ayed AB, Halima MB, Alimi AM (2015) MapReduce based text detection in big data natural scene videos. In: INNS conference on big data, pp 216–223

Chapter 10
Wind Distributed Generation with the Power Distribution Network for Power Quality Control

Ali M. Eltamaly, Yehia Sayed Mohamed, Abou-Hashema M. El-Sayed, and Amer Nasr A. Elghaffar

1 Introduction

Nowadays, the power electricity demand is growing fast due to the increase in the world population. Using the fossil fuels for the electrical generation is considered as non-economical way due to the fossil fuel is nonrenewable source, and hence, the searching to utilize the renewable sources is very important for reducing the electric generation cost and the environmental aspects. To appropriately meet the consumer requirements, electricity companies have tried to improve the power quality by using the compensation techniques to overcome the drop voltage or the system disturbance [1]. There are different definitions of power quality for instance: the electricity companies define power quality as reliability that can statistically demonstrate how reliable a network to feed the loads. In contrast, electrical equipment manufacturers define power quality as guaranteeing the performance of devices based on power supply characteristics. Utilizing renewable energy sources has arrived at more important significance as it advances sustainable living and with certain special cases (biomass combustion) does not contaminant. Renewable energy sources can be utilized in either small-scale applications from the enormous estimated generation plants or in large-scope applications in areas which the asset is abundant and large conversion systems [2]. The power quality issues have been significantly paid attention by researchers and practitioners in recent years. At present, related to the highly sensitive electrical equipment, the customers are requiring an excellent stable and

A. M. Eltamaly
Electrical Engineering Department, Mansoura University, Mansoura, Egypt

Sustainable Energy Technologies Center, King Saud University, Riyadh 11421, Saudi Arabia

Y. S. Mohamed · A.-H. M. El-Sayed · A. N. A. Elghaffar (✉)
Electrical Engineering Department, Minia University, Minia, Egypt

A. N. A. Elghaffar
Project Manager, Alfanar Engineering Service, Alfanar Company, Riyadh, Saudi Arabia

A. E. Hassanien et al. (eds.), *Enabling Machine Learning Applications in Data Science*,
Algorithms for Intelligent Systems, https://doi.org/10.1007/978-981-33-6129-4_10

reliable power system, so the electric companies searching to the ability to satisfying the consumers to assess power system quality. Renewable sources can be utilized in either small-scale applications from the large estimated generation plants or in large scale applications in areas where the resource is the bountiful and enormous electrical generation and types are utilized [3]. The power quality issues have been fundamentally focused by researchers and professionals as of late. At present, related to the increase in the sensitive and critical electrical loads especially at the manufacturing part, reactions of current equipment's, expanding request to top-notch power, maker's inclination agreeable to clients, the capacity of customers to evaluate power quality and so forth expanding the quality of conveyed power has been significant for electric companies. Power quality incorporates various branches, for example, long and short disturbance, dependability, over/under frequency, voltage sag, and voltage swell [4]. Integration and exploitation of distributed generation (DG), for example, wild renewable sources, which can expand the green energy in the utility system, build the worry of voltage and frequency strength. Moreover, voltage disturbance is likewise as often as possible experienced in powerless main grid. Due to the power electronics converters, the current will ripple also cause voltage harmonics and, as a result, the utility voltage waveforms may get contorted. The presence of distributed generation along with its points of interest in both transmission and distribution network has made it is viewed as continuous issues in the power network activity and development arranging. One of these viewpoints which ought to be considered will be viewed as DGs' effects issues. In this respect, the effects of DG on distribution systems designed a quality are one of the most significant issues which ought to be examined by arranging organizers just as researchers [5]. In this manner, dispersion arranges organizers need to join DGs' effect into their grid arranging to the required target. A few sorts of research have been acted in the region of dependability appraisal in conveyance systems which are furnished with DGs. In the past works, the situation of distributed resources has been examined; in which improvement of voltage profile, diminishing the power misfortune and expanding the framework dependability have been for the most part considered [6]. So, unwavering quality assessment has been directed when conveyance systems are considered as a commercial center [5, 6]. As it is conceivable to introduce DGs in different pieces of a feeder, it is required to increase an investigative methodology for assessing the unwavering quality of a feeder contains DG in a few pieces of the feeder [7]. The renewable energy source (RES) incorporated at the circulation level is named as DG. The utility is worried because of the high infiltration level of irregular RES in dispersion frameworks as it might represent a danger to organize as far as steadiness, voltage guideline, and power quality (PQ) issues [8]. In this context, this study presents a complete investigation of DG penetration level influence on system technical aspects such as voltage profile and power losses. The simulation system in this paper has been analyzed and simulated with a sample IEE-12 busses distribution network using the power world simulator software to study the impact of wind DG on the distribution network for power quality enhancement.

2 DG Installation Based on Voltage Stability

There are more researcher focusses on the renewable energy especially solar generation and wind generation, that considered friendlier with the environment. Because of significant costs, the DGs must be assigned reasonably with ideal size to improve the system execution, for example, to diminish the system misfortune, improve the voltage profile while keeping up the system security [9]. The issue of DG arranging has as of late got a lot of consideration by power system researchers. Choosing the best places for introducing DG units and their best sizes in enormous dispersion systems is a complex combinatorial enhancement issue [10]. Various details have been utilized dependent on analytics-based techniques, search-based strategies, and blends of different methodologies [11], for example, gradient and second-order calculations [12], Hereford Ranch calculation [13], heuristic iterative search method [14], analytical method [15], hybrid fuzzy—genetic algorithm (GA) method [16]. The using of different types of DG in the correct allocation with the distribution network can enhance the operation and stability system, improving the voltage profile, which directly decrease the losses power in the system. Hence, the attention must be paid not only to decide the location for DG placement but the types of DG technologies need to be considered. In this work, voltage stability enhancement is considered to be the major criteria for the DG placement to ensure the stable operation of the system with acceptable voltage levels at the consumer nodes [17, 18]. The procedure adopted to find out the optimal locations for DG placement along with selection of different types of DG technologies in a given test system (see Fig. 1).

2.1 *Analyzing of Active and Reactive Power Flow*

The power flow analysis is very important step during the design of the power system to control the direction of power flow depending on the loads areas and the generation parts. Hence, the known of the quantities of the power system is the basic data to analyze the active and reactive power entering to the busbars. The nodal analysis method can to use to find the equations to find the driven the power flow, with considering the busbars by the node part [19]. Equation (1) is the matrix for N—busbars power system.

$$\begin{bmatrix} I_1 \\ \dots \\ I_N \end{bmatrix} = \begin{bmatrix} Y_{11} & \dots & Y_{1N} \\ \dots & \dots & \dots \\ Y_{N1} & \dots & Y_{NN} \end{bmatrix} \begin{bmatrix} V_1 \\ \dots \\ V_N \end{bmatrix} \tag{1}$$

I_i Currents value at each node
Y_{ij} Elements of the busbar admittance matrix
V_i Busses voltages

Fig. 1 Flowchart for selection of DG type and location for voltage stability

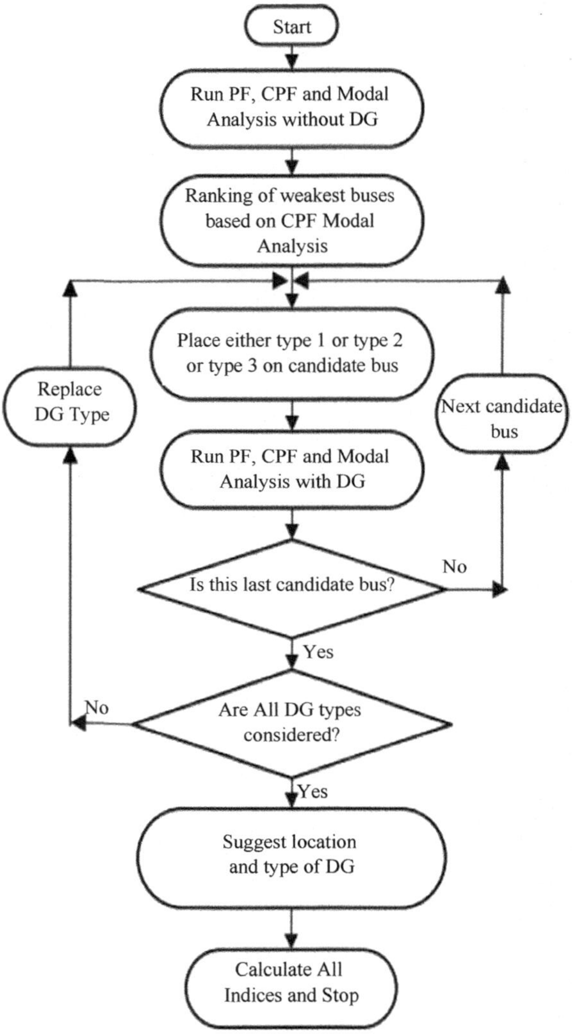

So, Eq. (2) follows to node at busbar *i*.

$$I_i = \sum_{j=1}^{n} Y_{ij} V_j \tag{2}$$

Per-unit value at Busbar *i* for active and reactive power and current injected into the system at that bus:

$$S_i = V_i I_i^* = P_i + J Q_i \tag{3}$$

V_i	Per-unit voltage at the bus
I_i*	Complex conjugate of the per-unit current injected at the bus
P_i and Q_i	Per-unit real and reactive powers.

$$I_i^* = \frac{(P_i + JQ_i)}{V_i}$$

$$I_i = \frac{(P_i - JQ_i)}{V_i^*}$$

$$(P_i - JQ_i) = V_i^* \sum_{j=1}^{n} Y_{ij} V_j = \sum_{j=1}^{n} Y_{ij} V_j V_i^* \tag{4}$$

Can be simulated as:
$Y_{ij} = |Y_{ij}|\angle\theta_{ij}$, and $V_i = |V_i|\angle\delta_i$

$$(P_i - JQ_i) = \sum_{j=1}^{n} |Y_{ij}||V_i||V_j|\angle(\theta_{ij} + \delta_j - \delta_i) \tag{5}$$

$$P_i = \sum_{j=1}^{n} |Y_{ij}||V_i||V_j| \cos(\theta_{ij} + \delta_j - \delta_i) \tag{6}$$

$$Q_i = \sum_{j=1}^{n} |Y_{ij}||V_i||V_j| \sin(\theta_{ij} + \delta_j - \delta_i) \tag{7}$$

Finally, there are four variant components to can uses to find the power flow parameters *P*, *Q*, *V*.

3 Wind Energy Conversion with the Power System

The fundamental parts of a wind turbine system are delineated (see Fig. 2), including a turbine rotor, a gearbox, a generator, a power electronic system, and a transformer for matrix association. Wind turbines catch the power from wind by methods for turbine cutting edges and convert it to mechanical power [20]. During higher wind speeds, it is imperative to add the options to allow for control and breaking point to save the generation power in a stable condition. The power constraint might be done either by slow down control, dynamic slow down, or throw control whose power curveballs are appeared (see Fig. 3). With a note that, the power might be easily restricted by turning the cutting edges either by pitch or dynamic slow down control while the power from a slowdown-controlled turbine shows a small overshoot and a lower power yield for

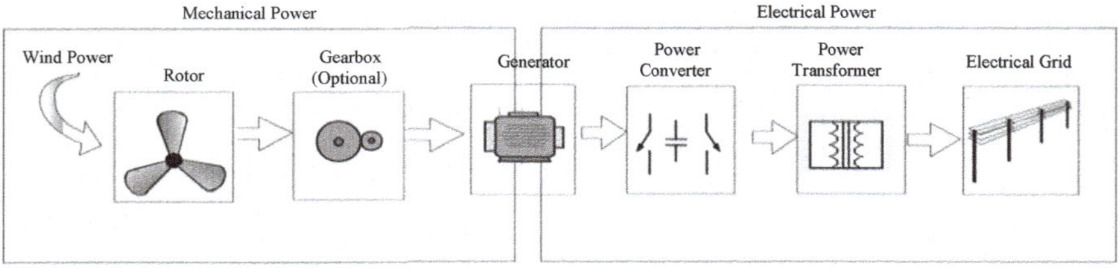

Fig. 2 Main components of a wind turbine system

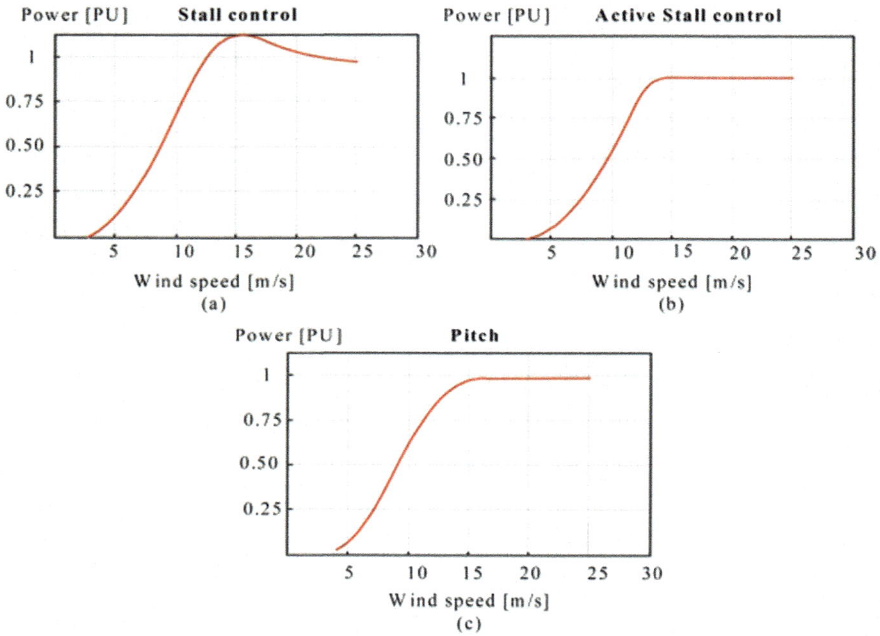

Fig. 3 Power characteristics of fixed speed wind turbines. **a** Stall control. **b** Active stall control. **c** Pitch control

higher wind speed [21–23]. The basic method to change over the low-speed, high-force mechanical power to electrical power is utilizing a gearbox and a generator with the standard speed. The gearbox adjusts the low speed of the turbine rotor to the rapid of the generator; however, the gearbox may not be essential for multipole generator systems. The generator changes over the mechanical power into electrical power, which is taken care of into a matrix perhaps through power electronic converters, and a transformer with circuit breakers and power meters. The two most normal kinds of electrical machines utilized in wind turbines are induction generators and synchronous generators. Induction generators with confine rotor can be utilized in the fixed speed wind turbines because of the damping impact. The receptive power important to empower the attractive circuits must be provided from the system or equal capacitor banks at the machine terminal that may have the risk of self-excitation when the association with the system is lost. A wound-rotor induction machine has a rotor with copper windings, which can be associated with an outer resistor or to air-conditioning systems by means of power electronic systems. Such a system furnishes an incomplete variable speed activity with a small power electronic converter and along these lines expanding energy catch and diminished mechanical burden to the system. This kind of system is a practical method to flexibly receptive power and acquires variable speed for expanded energy yield at wind speeds underneath the evaluated speed. Synchronous generators are energized by a remotely applied DC or by changeless magnets (PMs). There is extensive interest in the use of the various post synchronous generators (either with PM excitation or with an electromagnet) driven by a wind turbine rotor without a gearbox or with a low proportion gearbox. Synchronous machines powered by wind turbines may not be legitimately associated with the air conditioner framework on account of the prerequisite for critical damping in the drive train. The utilization of a synchronous generator prompts the prerequisite for a full appraised power electronic transformation system to decouple the generator from the system. While the greater part of the turbines is these days associated with the medium-voltage system, enormous seaward wind homesteads might be associated with the high-voltage and extra-high-voltage systems. The transformer is typically found near the wind turbines to maintain a strategic distance from high current streaming in long low-voltage links. The electrical assurance system of a wind turbine system ensures the wind turbine just as secures the sheltered activity of the system [21–23]. As the power yield of the wind turbine differs in a nonlinear relationship with respect to the wind speed, the point of maximum power point tracking (MPPT) control is to persistently modify the wind turbine rotor speed in such a manner to remove maximum power from the wind resource for each wind speed, it very well can be seen from Fig. 2 (see Fig. 4). Thus, in this control plot, the wind speed is taken as a boundary, while the rotor speed of the turbine or generator is a variable [24].

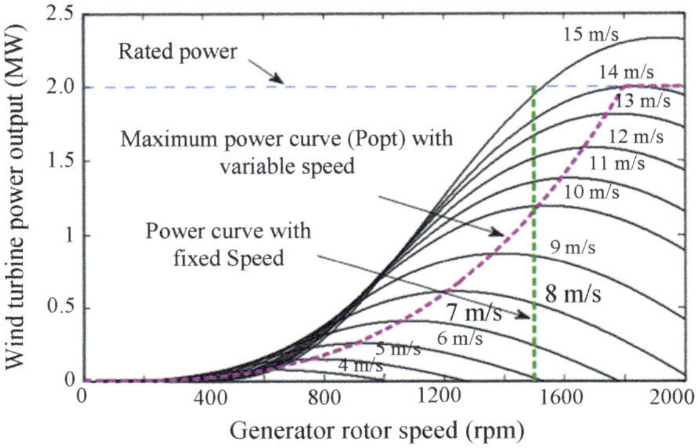

Fig. 4 Fixed and variable speed wind DGs output power

3.1 Output Power and Compensation of Wind Generation

The wind generator is the synchronous machine without the own excitation; consequently, it relies upon the input receptive power compensation by the electric network. In this way, the wind generation not considered *PQ* or *PV* in the power stream contemplates [25]. The asynchronous generator identical circuit is presented in this study (see Fig. 5), where, *U* is the yield machine voltage, *R* is the rotor resistance, X_1 is the stator reactance, X_2 is the rotor reactance, and X_m is the excitation reactance. With disregarding the stator resistance.

The total reactance rotor and stator is shown as:

$$X_\sigma = X_1 + X_2 \tag{8}$$

Th active output power can be calculated as:

$$P = \frac{SRU^2}{S^2X_\sigma^2 + R^2} \tag{9}$$

where *S* is the generator slip, which can be defined as shown in Eq. (10). And by using Eqs. (8) and (9) can obtain the reactive power as shown in Eq. (11).

$$S = R\left(U^2 - \sqrt{U^4 - 4X_\sigma^2 P^2}\right)/2PX_\sigma^2 \tag{10}$$

$$Q = \left[R^2 + X_\sigma(X_m + X_\sigma)S^2\right]/SRX_m \tag{11}$$

In the wind system, it is imperative to remunerate the responsive power by introducing the receptive compensation hardware, that additionally, to lessen the system misfortunes. Additionally, the power factor compensation can spare the steady power stream, which utilizes the equal shunt capacitors with the wind power circuit [26]. The power factor compensation condition can be appeared as:

$$\text{Cos}(\emptyset) = P/\sqrt{P^2(Q_C - Q)^2} \tag{12}$$

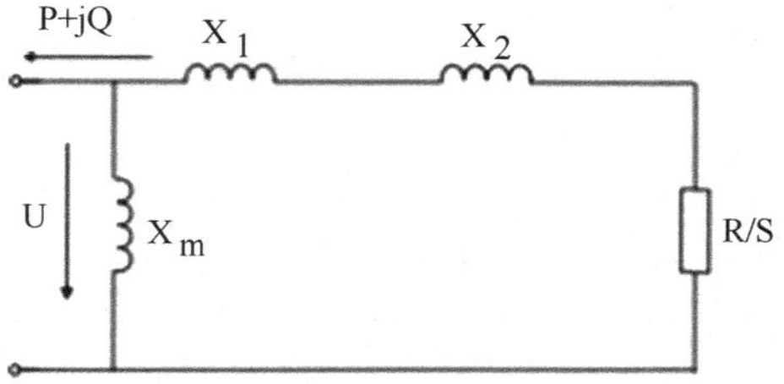

Fig. 5 A simplified wind turbine equivalent circuit

With considering the parallel capacitor group simulated by Q_C which can be demonstrate from Eq. (13).

$$Q = P\left[\sqrt{\frac{1}{(\mathrm{Cos}(\varnothing 1))^2} - 1} - \sqrt{\frac{1}{(\mathrm{Cos}(\varnothing 2))^2} - 1}\right] \tag{13}$$

At assuming the actual capacitor investment group is [n], and the reactive power capacitor compensation capacity is $Q_{N\text{-Unit}}$, that simulates at the rated voltage.

$$[n] = Q_C / Q_{N\text{-}Unit} \tag{14}$$

And, the wind generation reactive power is shown as:

$$Q' = Q_C - Q \tag{15}$$

3.2 *Wind Generation Control*

Using wind speed is considered as one of the principle difficulties to structure the wind dissemination system. By control the MPPT with the wind speeds, Fig. 4 shows an example of the output power with the variable wind speed [24].

The proportion between the wind speed and the rotor speed can be characterized as the tip-speed (λ) as shown in Eq. (16):

$$\lambda = \frac{R_p w}{v_m} \tag{16}$$

where w is the wind turbine rotational speed and v_w is the wind speed [27]. The mechanical power that extracts by the wind at constant pitch can be defined as:

$$P_m = \frac{1}{2} \cdot \rho A_r v_w^3 \cdot C_p(\lambda) \tag{17}$$

where ρ is the air density, A_r is the area swept by the blades and C_p is the turbine power coefficient. To depict the wind speed power, Fig. 6 shows the Wind turbine power control bend, which mimics the wind speed as $V1$ and $V2$. Where, every speed has one maximum power catch point at the turbine works at the ideal power coefficient (Cz_p max) [28]. There is an individual ideal power can be identified with the genera-peak speed by Eq. (18).

$$P_{\mathrm{opt}} = k_{\mathrm{opt}} w_T^3 \tag{18}$$

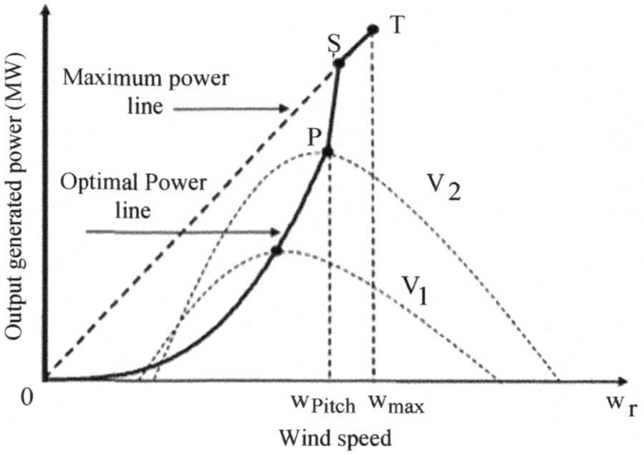

Fig. 6 Wind turbine power control curve

where k_{opt} is the unique parameter optimal power and w_t is the rotational speed.

4 Simulation and Discussion

Integration and exploitation of distributed generation (DG) systems, such as uncontrollable renewable sources, which can maximize green energy penetration in the utility network, increases the concern of voltage and frequency stability. In addition, voltage distortions and fluctuations are also frequently encountered in weak utility network systems [29]. Figure 7 shows the 12-busses sample distribution system, and it has been used for voltage stability study. This system comprises five generators including one slack busbar and 11 load busses as well as 17 transmission lines, and the system full data has been illustrated in Tables 1 and 2. Modal analysis method has applied to the 12 busses system to evaluate the voltage stability and the losses reduction of the above-mentioned system. All generators values are calculated in order to identify the weakest busbar in the system. This study has been implanted based on power world simulator software. Table 3 shows the active and reactive power losses in the branches. After adding the wind generation DG with busbar-8, it found the voltage improved as shown in Table 4. The comparison between the bus's voltage without DG and with DG is illustrated (see Fig. 8). Also, the branches losses have been decreased as shown in Table 5. Figures 9 and 10 show the comparison between active power and reactive power losses, respectively, without adding the DG and with adding the wind generation DG. Finally, after adding the DG with the distribution network, the voltage will improve and the losses will reduce.

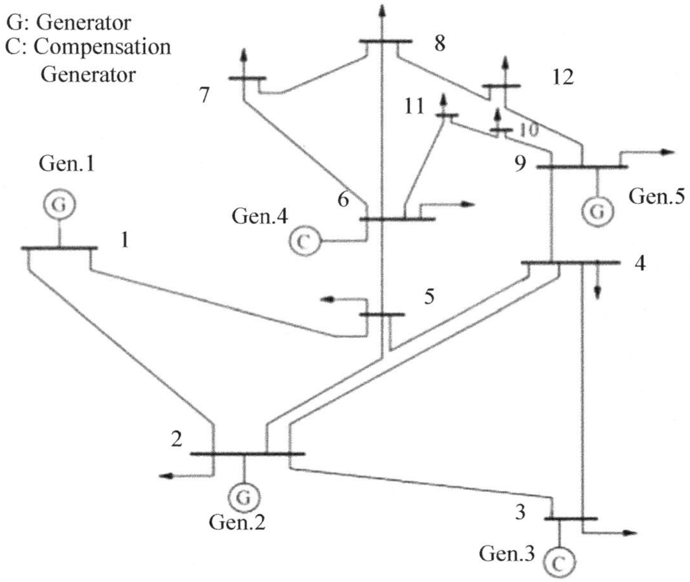

Fig. 7 IEEE-12 busses system distribution network

Table 1 Operation system data without DG with distribution network

Busbar No.	Nom. voltage	Voltage (kV)	Nom Kv PU	Angle (Deg)	Load MW	Load Mvar	Gen MW	Gen Mvar
1	22	22.000	1.00000	−18.16	50.30	18.50	75.16	44.03
2	22	22.000	1.00000	−15.26	18.00	12.58	31.01	77.20
3	22	22.000	1.00000	0.36	37.00	11.00	274.61	21.55
4	22	21.227	0.96485	−19.72	25.00	10.00	–	–
5	22	21.038	0.95625	−24.52	33.25	11.00	–	–
6	22	21.522	0.97826	−37.74	30.31	19.42	20.00	100.0
7	22	20.897	0.94989	−39.01	48.24	26.79	–	–
8	22	20.607	0.93670	−39.74	33.58	19.21	–	–
9	22	22.000	1.00001	−31.23	18.04	5.00	50.00	93.81
10	22	21.006	0.95480	−40.68	59.95	10.00	–	–
11	22	20.894	0.94973	−41.90	44.85	11.59	–	–
12	22	20.747	0.94306	−37.83	35.18	19.76	–	–

5 Conclusion

The integration of embedded power generation systems to existing power systems influences the power quality and causes voltage quality, over-voltage, reactive power,

Table 2 Initial data for the branches to link the system

Link No.	Branch from bus	Branch to bus	Branch R	Branch X	Branch Lim MVA
1	1	2	0.00000	0.18000	120.0
2	1	5	0.00000	0.20000	120.0
3	2	3	0.05000	0.21000	180.0
4	2	4	0.00000	0.20000	120.0
5	5	2	0.08000	0.24000	120.0
6	3	4	0.02000	0.30000	190.0
7	5	4	0.01000	0.30000	120.0
8	9	4	0.01000	0.20000	150.0
9	5	6	0.00000	0.20000	150.0
10	7	6	0.02000	0.05000	120.0
11	8	6	0.02000	0.30000	120.0
12	6	11	0.02000	0.23000	120.0
13	8	7	0.02000	0.22000	120.0
14	8	12	0.01100	0.18000	120.0
15	10	9	0.00200	0.21000	120.0
16	9	12	0.00100	0.21000	120.0
17	11	10	0.00100	0.13000	120.0

and safety issues. The widely popular generation resources are the wind and photovoltaic systems. Due to the penetration of renewable energy, the poor power quality arises which creates problems on electric systems. DGs are considered as a small generator that can operate stand-alone or in connection with the distribution networks and can be installed at or near the loads, unlike large central power plants. Renewable distribution generation (DG), wind turbine (WT) and photovoltaic (PV) systems present a cleaner power creation. This study introduced the using of the renewable wind DG with the distribution power network for power quality control. This works out in a good way in close relationship with the general public's normal development, where responses to a lot of pickles are respected by minute scale arrangements and nearness. In this paper, a conversation has been proposed for examining effects of the DG which centers around the wind generation on appropriation organize dependability. From the examination, it found the voltage has been improved to arrive at 1 pu in busbar 6 and busbar 8 also it is improved in another busbars. Likewise, the absolute dynamic power misfortunes have been diminished to arrive at 10.25 MW which is 17.11 MW before including the DG and the receptive power has been diminished to arrive at 92.92 MVAR which is 161.73 MVAR. At long last, from this paper, it is imperative to prescribe to the power system planner to consider the wind DGs for upgrading the power system quality.

Table 3 Active and reactive power losses through the 17 branches system without DG

Link No.	Branch from bus	Branch to bus	Branch MW loss	Branch Mvar loss	Branch % of MVA limit (max)
1	1	2	0.00	1.42	23.4
2	1	5	0.00	6.85	48.8
3	2	3	7.93	33.29	35.0
4	2	4	0.00	3.54	69.9
5	5	2	3.36	5.29	60.3
6	3	4	3.21	37.40	21.7
7	5	4	0.17	2.09	72.9
8	9	4	1.00	19.98	54.0
9	5	6	0.00	25.01	26.7
10	7	6	0.96	1.99	53.8
11	8	6	0.06	0.85	13.9
12	6	11	0.21	2.45	6.4
13	8	7	0.01	0.05	14.4
14	8	12	0.03	0.57	66.6
15	10	9	0.13	13.22	49.8
16	9	12	0.04	7.49	66.3
17	11	10	0.0	0.24	12.8

Table 4 IEEE-12 busbars operation system data after adding the wind DG with busbar 8

Busbar No.	Nom. voltage (kv)	Voltage (kV)	Nom Kv PU	Angle (Deg)	Load MW	Load Mvar	Gen MW	Gen Mvar
1	22	22.000	1.000	−13.35	50.30	18.50	75.16	34.29
2	22	22.000	1.000	−11.31	18.00	12.58	31.01	51.60
3	22	22.000	1.000	0.36	37.00	11.00	217.79	11.17
4	22	21.489	0.9767	−14.99	25.00	10.00	–	–
5	22	21.411	0.9732	−18.61	33.25	11.00	–	–
6	22	22.000	1.0000	−28.07	30.31	19.42	20.00	65.85
7	22	21.663	0.9846	−28.95	48.24	26.79	–	–
8	22	22.277	1.0125	−27.58	33.58	19.21	50.00	50.00
9	22	22.000	1.000	−23.48	18.04	5.00	50.00	54.89
10	22	21.210	0.9640	−32.12	59.95	10.00	–	–
11	22	21.188	0.963	−32.85	44.85	11.59	–	–
12	22	21.690	0.9859	−27.62	35.18	19.76	–	–

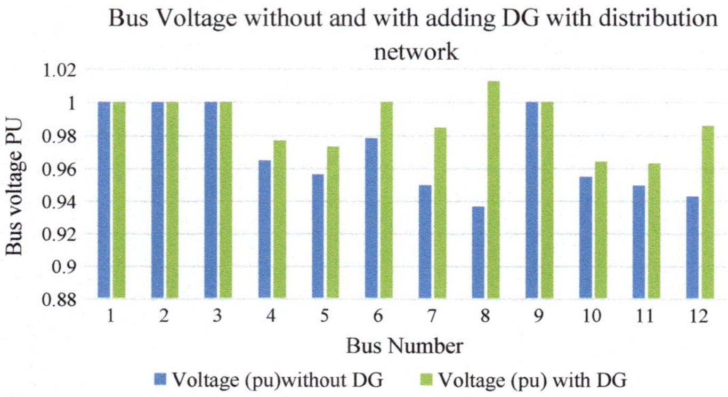

Fig. 8 Busbar voltage without and with adding DG with distribution network

Table 5 Branch losses after adding DG with the distribution network

Link No.	Branch from bus	Branch to bus	Branch MW loss	Branch Mvar loss	Branch % of MVA limit (max)
1	1	2	0.00	0.70	16.5
2	1	5	0.00	4.46	39.4
3	2	3	0.00	2.29	28.2
4	2	4	4.43	18.62	52.3
5	5	2	2.15	21.37	46.5
6	3	4	0.14	1.17	16.8
7	5	4	0.00	13.60	55.0
8	9	4	2.06	1.32	42.5
9	5	6	0.30	3.46	32.4
10	7	6	0.42	0.61	33.5
11	8	6	0.02	0.39	12.5
12	6	11	0.06	0.51	14.0
13	8	7	0.01	-0.02	4.3
14	8	12	0.55	10.95	49.3
15	10	9	0.01	2.54	29.0
16	9	12	0.10	10.93	60.4
17	11	10	0.00	0.02	7.6

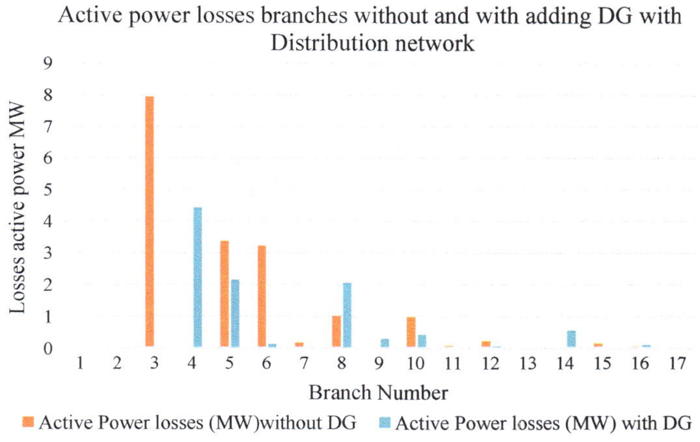

Fig. 9 Active power losses without and with adding DG with distribution network

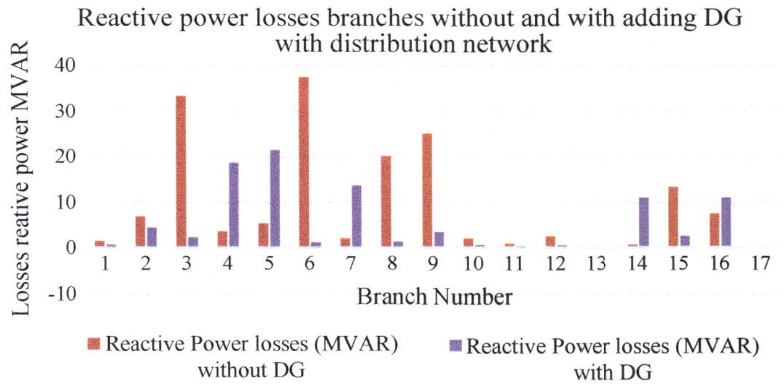

Fig. 10 Reactive power losses branches without and with adding DG with distribution network

Acknowledgments The authors wish to acknowledge Alfanar Company, Saudi Arabia for supporting to complete this research, especially thanks to Mr. Amer Abdullah Alajmi (Vice President, Sales & Marketing, Alfanar Company, Saudi Arabia) and Mr. Osama Morsy (General Manager, Alfanar Engineering Service, Alfanar Company, Saudi Arabia).

References

1. Eltamaly AM, Sayed Y, El-Sayed AHM, Elghaffar AA (2020) Adaptive static synchronous compensation techniques with the transmission system for optimum voltage control. Ain Shams Eng J. https://doi.org/10.1016/j.asej.2019.06.002
2. Eltamaly A, Elghaffar ANA (2017) Techno-economical study of using nuclear power plants for supporting electrical grid in Arabian Gulf. Technol Econ Smart Grids Sustain Energy J.

https://doi.org/10.1007/s40866-017-0031-8
3. Eltamaly A, El-Sayed AH, Yehia S, Elghaffar AA (2018) Mitigation voltage sag using DVR with power distribution networks for enhancing the power system quality. IJEEAS J ISSN 1(2):2600–7495
4. Eltamaly A, Yehia S., El-Sayed AH, Elghaffar AA (2018) Multi-control module static VAR compensation techniques for enhancement of power system quality. Ann Fac Eng J ISSN: 2601–2332
5. Fotuhi-Fi A, Rajabi A (2005) An analytical method to consider dg impacts on distribution system reliability. In: 2005 IEEE/PES transmission and distribution conference & exhibition: Asia and Pacific, Dalian, China
6. Kumawat M et al (2017) Optimally allocation of distributed generators in three-phase unbalanced distribution network. Energy Proc 142:749–754. 9th international conference on applied energy, ICAE2017, Cardiff, UK
7. Eltamaly A, Sayed Y, El-Sayed AHM, Elghaffar AA (2018) Enhancement of power system quality using static synchronous compensation (STATCOM). IJMEC 8(30):3966–3974. EISSN: 2305-0543
8. Eltamaly AM, Farh HM, Othman MF (2018) A novel evaluation index for the photovoltaic maximum power point tracker techniques. Solar Energy J https://doi.org/10.1016/j.solener.2018.09.060
9. Sudabattula SK, Kowsalya M (2016) Optimal allocation of solar based distributed generators in distribution system using Bat algorithm. Perspect Sci 8:270—272.https://doi.org/10.1016/j.pisc.2016.04.048
10. Eltamaly A, Sayed Y, El-Sayed AHM, Elghaffar AA (2019) Reliability/security of distribution system network under supporting by distributed generation. Insight Energy Sci J 2(1). https://doi.org/10.18282/i-es.v7i1
11. Laksmi Kumari RVS, Nagesh Kumar GV, Siva Nagaraju S, Babita M (2017) Optimal sizing of distributed generation using particle swarm optimization. In: 2017 international conference on intelligent computing, instrumentation and control technologies (ICICICT)
12. Rau NS, Wan Y (1994) Optimal location of resources in distributed planning. IEEE Trans Power Syst 9(4):2014–2020. https://doi.org/10.1109/59.331463
13. Eltamaly AM, Mohamed YS, El-Sayed AHM, Elghaffar ANA, Abo-Khalil AG (2021) D-STATCOM for Distribution Network Compensation Linked with Wind Generation. In: Eltamaly AM, Abdelaziz AY, Abo-Khalil AG (eds) Control and Operation of Grid-Connected Wind Energy Systems. Green Energy and Technology. Springer, Cham. https://doi.org/10.1007/978-3-030-64336-2_5
14. Griffin T, Tomsovic K, Secrest D et al (2000) Placement of dispersed generation systems for reduced losses. In: Proceedings of 33rd international conference on system sciences, Hawaii, pp 1446–1454
15. Willis HL (2000) Analytical methods and rules of thumb for modeling DG distribution interaction. In: Proceedings of IEEE power engineering society summer meeting, Seattle, USA, pp 1643–1644
16. Nara K, Hayashi Y, Ikeda K et al (2001) Application of tabu search to optimal placement of distributed generators. In: Proceedings of IEEE power engineering society winter meeting, Columbus, USA, pp 918–923
17. Kim K-H, Lee Y-J, Rhee S-B, Lee S-K, You S-K, (2002) Dispersed generator placement using fuzzy-GA in distribution systems. In: Proceedings of IEEE power engineering society summer meeting, vol 3, Chicago, IL, USA, pp 1148–1153
18. Teng J-H, Luor T-S, Liu Y-H (2002) Strategic distributed generator placements for service reliability improvement. In: Proceedings of IEEE power engineering society summer meeting, Chicago, USA, pp 719–724
19. Eltamaly AM, Elghaffar ANA (2017) Load flow analysis by gauss-seidel method; a survey. Int J Mech Electr Comput Technol (IJMEC), PISSN: 2411-6173, EISSN: 2305-0543
20. Eltamaly AM, Sayed Y, Elsayed A-HM, Elghaffar AA (2019) Analyzing of wind distributed generation configuration in active distribution network. In: The 8th international conference on modeling, simulation and applied optimization (ICMSAO'2019), Bahrain

21. Eltamaly AM, Mohamed YS, El-Sayed, AHM et al (2020) Power quality and reliability considerations of photovoltaic distributed generation. Technol Econ Smart Grids Sustain Energy 5(25). https://doi.org/10.1007/s40866-020-00096-2
22. Blaabjerg F, Chen Z, Kjaer SB (2004) Power electronics as efficient interface in dispersed power generation systems. IEEE Trans Power Electron 19(5):1184–1194
23. Budischak C, Sewell D-A, Thomson H et al (2013) Cost-minimized combinations of wind power, solar power and electrochemical storage, powering the grid up to 99.9% of the time. J Power Sources, 225, pp 60–74. https://doi.org/10.1016/j.jpowsour.2012.09.054
24. Bansal R (2017) Handbook of distributed generation electric power technologies, economics and environmental impacts. ISBN 978-3-319-51343-0 (eBook). https://doi.org/10.1007/978-3-319-51343-0
25. Tripathy M, Samal RK (2019) A new perspective on wind integrated optimal power flow considering turbine characteristics, wind correlation and generator reactive limits. Electr Power Syst Res 170, pp 101–115. https://doi.org/10.1016/j.epsr.2019.01.018.
26. Raya-Armenta JM, Lozano-Garcia et al (2018) B-spline neural network for real and reactive power control of a wind turbine. Electr Eng 100, 2799–2813. https://doi.org/10.1007/s00202-018-0749-x
27. Li L, Ren Y, Alsumiri M, Brindley J, Jiang L (2015) Maximum power point tracking of wind turbine based on optimal power curve detection under variable wind speed. In: International Conference on Renewable Power Generation (RPG 2015), Beijing, 2015, pp 1–6. https://doi.org/10.1049/cp.2015.0492
28. Cardenas R, Pena R, Perez M, Clare J, Asher G, Wheeler P (2005) Control of a switched reluctance generator for variable-speed wind energy applications. IEEE Transa Energy Conv 20 (4), pp 781–791, Dec. 2005. https://doi.org/10.1109/TEC.2005.853733
29. Eltamaly A, Sayed Y, Elsayed A-HM, Elghaffar AA (2019) Impact of distributed generation (DG) on the distribution system network. Ann Fac Eng Hunedoara Int J Eng Tome XVII

Prof. Dr. Ali M. Eltamaly (Ph.D.-2000) is Full Professor at Mansoura University, Egypt, and King Saud University, Saudi Arabia. He received B.Sc. and M.Sc. degrees in Electrical Engineering from Al-Minia University, Egypt, in 1992 and 1996, respectively. He received his Ph.D. degree in Electrical Engineering from Texas A&M University in 2000. His current research interests include renewable energy, smart grid, power electronics, motor drives, power quality, artificial intelligence, evolutionary and heuristic optimization techniques, and distributed generation. He has published 20 books and book chapters, and he has authored or coauthored more than 200 refereed journals and conference papers. He has published several patents in the USA patent office. He has supervised several M.S. and Ph.D. theses and has worked on several national/international technical projects. He has obtained distinguish professor award for scientific excellence from Egyptian Supreme Council of Universities, Egypt, in June 2017, and he has been awarded with many prizes by different universities in Egypt and Saudi Arabia. He is participating as an editor and associate editor in many international journals and chaired many international conferences' sessions. He is Chair Professor of Saudi Electricity Company Chair in power system reliability and security, King Saud University, Riyadh, Saudi Arabia.

Prof. Dr. Yehia Sayed Mohamed was born in Egypt. He received B.Sc. and M.Sc. degrees in Electrical Engineering from Asuit University, Egypt, and Ph.D. degree from Minia University, Egypt. Currently, he is Professor and Chairman of Electrical Engineering Department, Faculty of Engineering, Minia University, Egypt. He has been with the Department of Electrical Engineering, Faculty of Engineering, Minia University, as Teaching Assistant, Lecturer Assistant, Assistant Professor, Associate Professor, and Professor. He is Distinguished Professor from the Egyptian Supreme Council of Universities, Minia University, Egypt. His current research interests include electrical machines, power electronics, energy management system, distribution automation system, power system quality, and renewable energy. He has supervised a number of M.Sc. and Ph.D. theses and has supervised and published multi-books, many book chapters, and a lot of technical papers.

Prof. Dr. Abou-Hashema M. El-Sayed received his B.Sc. and M.Sc. in Electrical Engineering from Minia University, Minia, Egypt, in 1994 and 1998, respectively. He was a Ph.D. student in the Institute of Electrical Power Systems and Protection, Faculty of Electrical Engineering, Dresden University of Technology, Dresden, Germany, from 2000 to 2002. He received his Ph.D. in Electrical Power from the Faculty of Engineering, Minia University, Egypt, in 2002, according to a channel system program, which means a Scientific Co-operation between the Dresden University of Technology, Germany, and Minia University, Egypt. Since 1994, he has been with the Department of Electrical Engineering, Faculty of Engineering, Minia University, as Teaching Assistant, Lecturer Assistant, Assistant Professor, Associate Professor, and Professor. He is Distinguished Professor from the Egyptian Supreme Council of Universities, Minia University, Egypt. He was Visiting Researcher at Kyushu University, Japan, from 2008 to 2009. He is Head of Mechatronics and Industrial Robotics Program, Faculty of Engineering, Minia University, from 2011 till now. His research interests include protection systems, renewable energy, and power system.

Dr. Amer Nasr A. Elghaffar received his B.Sc., M.Sc., and PhD degrees in Electrical Engineering from Minia University, Egypt, followed with Power and Machine in 2009, 2016, and 2021, respectively. Currently, he is Project Manager in Alfanar Engineering Service (AES), Alfanar Company, Saudi Arabia. In parallel, he is Researcher in Electrical Engineering Department, Minia University, Minia, Egypt. His current research interests include renewable energy, power system protection and control, microgrid, power quality, and high voltage. In 2013, he joined Alfanar company, Saudi Arabia, as Testing and Commissioning Engineer. He has the enough experience about the electrical high-voltage substations testing and commissioning field. He has published a lot of technical papers and attended a lot of conferences in the electrical power system field. He has the licenses and certifications on Agile in The Program Management Office (PMI), Exam Tips: Project Management Professional (PMP), Project Management Foundations: Risk, Managing Project Stakeholders, Leading Productive Meetings (PMI and CPE), and Goal Setting.

Chapter 11
Implementation of Hybrid Algorithm for the UAV Images Preprocessing Based on Embedded Heterogeneous System: The Case of Precision Agriculture

Rachid Latif, Laila Jamad, and Amine Saddik

1 Introduction

Precision agriculture has been recently a very interesting domain of research, especially with the interesting approaches that appeared in different contexts for different aims. A lot of researches were elaborated in the field of classifying and weed detection, [1–4], fruits counting [5–8], or other applications, based on image processing, deep learning, and Internet of Things. Regarding the embedded systems domains, it is usually preferred to work with image processing, thanks to its low energy consumption and space. In 2019, an image processing approach was elaborated in vegetation index computing based on UAV images [9], this approach is using an images database token by UAV, then used in computing NDVI and NDWI indexes. Yet, the application of such a approach in reality will certainly face some challenges that we aim to solve in our article. Therefore, the work we present is in service of precision agriculture where we treat a major concern of image processing. In our research, we aimed to concretize a real case application of a precision agriculture algorithm, based on image processing using UAV images thanks to their high resolution. In this application, the database of images used will certainly contain blurred images because of camera motion or camera shake, which will lead to false results if we use it, or at least will decrease the quality of precision algorithm. This is exactly why we propose in this work an additional pretreatment block to be used for precision agriculture algorithms that are based on UAV images processing. This pretreatment block will contain a deblurring algorithm that aims to apply the filtering operation in only blurred images before being used in treatment, so that the results will be improved.

This article is devised to seven parts, where every part will be explained and detailed as it should. Firstly, we present an introduction, and then, an overview in precision agriculture is to explore the different techniques used in this field, after that

R. Latif · L. Jamad (✉) · A. Saddik
LISTI, ENSA Ibn Zohr University Agadir, 80000 Agadir, Morocco

A. E. Hassanien et al. (eds.), *Enabling Machine Learning Applications in Data Science*, Algorithms for Intelligent Systems, https://doi.org/10.1007/978-981-33-6129-4_11

the main problem of data collected using UAV, to arrive to our proposed approach and its results and the two final parts.

2 Precision Agriculture: Application and Algorithms

Precision farming was a concept that emerged in the USA in the 1980s [10]. The use of fertilizers, phytosanitary products, and water is increasing sharply, and it is becoming important to optimize their use in order to protect people and the environment [10].

In precision agriculture, a lot of algorithms were elaborated, in different contexts such as weed detection, plants counting, or vegetation index computing. In every different context, many approaches were presented.

In this way, the work of Pudelko et al. [11] presents a study evaluating the possibility to monitor agricultural soils using a remote-controlled flight model. The work is divided in two parts. The first part aims to show the advantages and disadvantages of flight with motor glider as a platform. The second part is based on the acquisition and development of photos taken with this type of platform. The work also presents an example of detection of areas where differences are visible inside the structure of the plant cover. The example shows the presence of weeds and domestic diseases [11].

Sa et al. [12] present a classification approach based on multispectral images collected by an aerial micro-vehicle to detect weeds. They use neural networks to obtain a set of data. They also used a field with different levels of herbicides, which gives plots that contain only crops or weeds. Subsequently, they used the NDVI vegetation index to characterize the plots. The work is divided into three parts, the first part is the construction of the database, and the second part is related to the preprocessing of this database, and the last is for the dense semantic segmentation framework (DSSF) algorithm [12].

In like manner, counting crops is a difficult task for agriculture today. Moreover, the demand for food supplies creates a need for the performance of farming activities more efficiently and accurately. Use of remote sensing images can help better control the crop population and predict yields, future profits, and disasters. The prediction of yields of agricultural soils is very important for high yields. Several methods have been developed, for example, we find in the work of [13] a numerical counting method for corn soils to see the density of the plants.

In the same way, other approaches in vegetation index computing were presented, one of the recent works is [9], where authors R. Latif et al. worked on studding two famous vegetation indexes, normalized difference vegetation index (NDVI) and normalized difference water index (NDWI), using image processing in order to find out if the plants need water or have an issue, so that farmers can take their decisions to improve the plots. This work was experimented on a heterogeneous embedded Odroid-XU4 architecture based on two CPUs and a GPU. Figure 1 shows steps of the algorithm in [9].

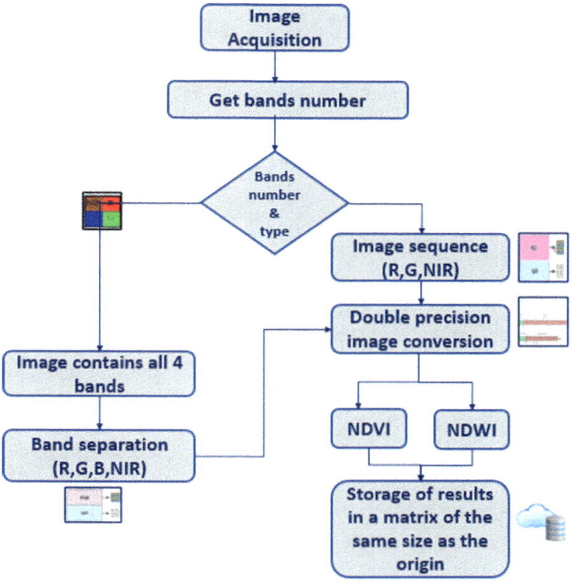

Fig. 1 Algorithm overview proposed in [9]

Figure 1 presented gives an overview of the based algorithm used in [9], standing on image processing to compute vegetation indexes NDVI and NDWI in order to study the state of the crop. Precision agriculture algorithms can be developed using different tools, for instance Internet of Thing, convolutional neural network (CNN), deep neural network (DNN), image processing, or other [14–19]. In embedded systems domain, it is usually preferred to work with a method that has low consumption of memory and power, which is in our case: image processing.

3 Image Processing in Precision Agriculture

To set up a precision agriculture algorithm, it is usually known that it should start with an image acquisition block, which should be based on tool equipped with a camera, such as drones, UAVs or others, in order to collect the necessary data. Then comes the treatment block specified based on the treatment that we want to apply. Figure 2 presents a general precision agriculture algorithm based on image processing.

3.1 Challenge and Proposed Method

Using an UAV for data collection will attend a major concern, and this concern is data. The data collected will certainly contain unblurred and clean images, but due

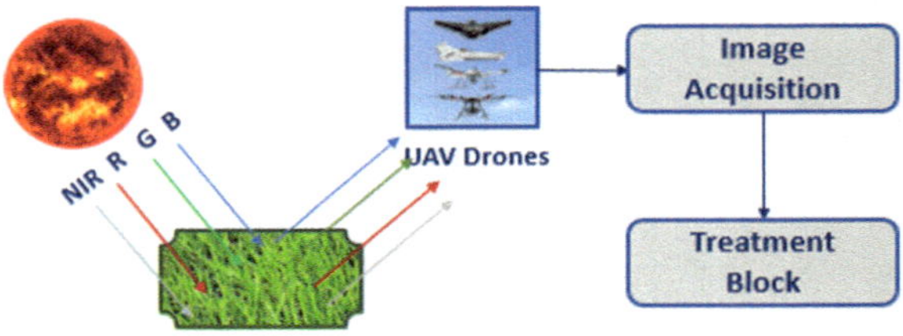

Fig. 2 General precision agriculture algorithm based on image processing

to camera shake and UAV motion, some images will contain a blur. These blurred images if used will lead to false results, because the information that they contain was changed because of the motion happened. Figure 3 shows the containing of images database collected by UAVs.

Classifying images. With this option, the system will use an additional algorithm to classify images into two categories, such as the algorithm elaborated in [20], blurred and unblurred ones, then use only unblurred ones. Here, another problem is facing us: Data loss. Data loss means that the volume of data that we collected will go low, and this is not a very preferred option, especially when we look for good results.

Preprocess the data. This option gives another proposed architecture for the general algorithm, and it proposes an additional bloc to the algorithm for preprocessing images before being used in the treatment, for instance, the work presented in [21]. This additional bloc is dedicated to eliminate the blur or noise that the image contains before being used in treatment. This solution gives us the opportunity to have the same volume of data collected by the UAV camera, with clear images, all with correcting what the shake of camera did to the image. The global general used algorithm becomes as shown in Fig. 4.

Figure 4 comes with the global precision agriculture algorithm proposed for image processing with the additional preprocessing block, dedicated to filtering and preparing data for use. In our case, this additional bloc will contain a deblurring algorithm, to increase the results of the camera shake.

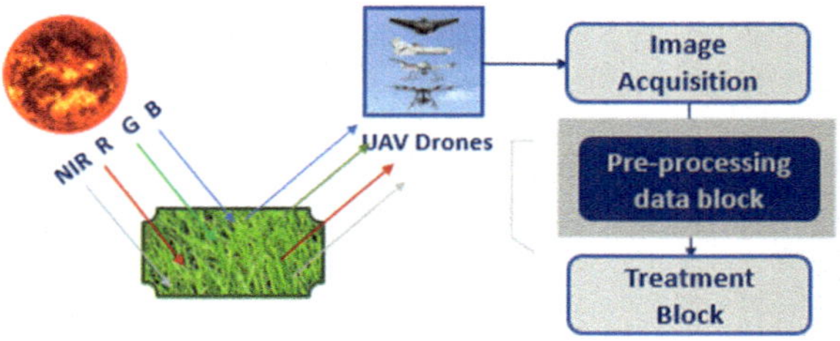

Fig. 3 Images data of a moving camera

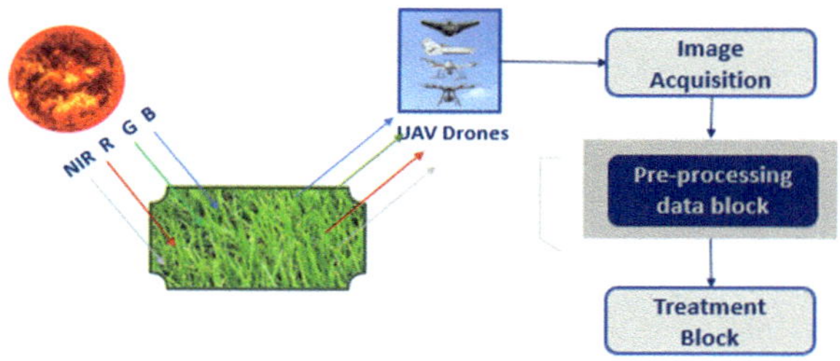

Fig. 4 Global precision agriculture algorithm based on image processing with the additional block

4 Deblurring Algorithms

A blur is the state of that which has no clear outlines, with a certain lack of clarity, precision. This can be happening in a capture of a scene, due to a camera shake or camera motion. In a camera, moving captures, the object motion blur, and camera shake blur are the more possible happening types of blur, due to camera motion or camera shake. It is the conclusion of the relative movement between the camera and the area we want to capture during the assimilation time of the image [22].

To illustrate the system, we should start by the illustration and representation of the images taking system considered as the convolution of a kernel of blur that we may call in our case C, and A is the random array with the possibility distribution resolute standing on the perfect capture and the kernel K [23]. Below is the global used equation:

$$A = K \otimes I + N \tag{1}$$

K Blur kernel
A Random array
I Ideal sharp image
N Noise

Generally, deconvolution or deblurring algorithms are divided into two categories based on informations that we have. If we have no information about the kernel blur, then we proceed a blind deconvolution approach, yet if we do have it, we proceed a non-blind approach.

4.1 *Blind Approach*

Blind deconvolution approaches are used when we have no information about the blur kernel or the noise. In blind convolution approach, the image and the filter are unidentified, which leads to a lack of data [24]. The principle idea of this deconvolution is to overhaul the primitive capture without using blur kernel [25]:

$$y = \text{Nnk} \otimes xi \tag{2}$$

xi Visually possible sharp image
Nnk Non-negative blur kernel

4.2 Non-blind Approach

In this non-blind deconvolution approach, we have a trace track to start. We generally have here information's about the spread point function to estimate the blur filter. In non-blind approach, *K* (kernel) is considered to be identified [24]. We obtain a sharp image *I* when applying the deblurring deconvolution on capture.

$$A = K \otimes I + N \tag{3}$$

K Blur kernel
A Random array
I Ideal sharp image
N Noise

4.3 Deblurring Algorithms

A lot of algorithms were elaborated and experimented for the two kinds of deblurring approaches, in different fields, agriculture, or others. For the blind approach, we can mention the work of Jia et al. [25] in 2007 where they propose a method basing on transparency to deblur images without having any idea about the kernel which is a blind approach, passing by a lot of papers, arriving to recent ones, such as in [26] appeared in 2018 which comes with a blind normalized deconvolution approach. In the other hand, non-blind approaches also took a very large amount of interest, for cases where the kernel is known, such as in [27, 28] based on different methods, such as deep learning and others.

In our case, images token or captured are not all the same, the kernel convolution cannot be known, so the best approach that we can work on is the blind deconvolution approach. One of the most known approaches is the direct Fourier transform (DFT) [29], this approach will be the base of our proposed algorithm, but the challenge faced is the estimation of variance in order to know if the image is blurred or not.

5 Proposed Work

In our proposed work, we need to remedy the data loss problem by using a preprocessing block of pretreatment.

The pretreatment block in our case will take in lead a deblurring algorithm that we opt to create as a hybrid approach based on the discrete Fourier transform algorithm that generally comes as follow:

Figure 5, is a global based discrete Fourier transform algorithm, used in deblurring algorithms standing on DFT. This algorithm can be adapted to the previous schema proposed, so that the DFT algorithm takes the place in the pretreatment block. This algorithm starts with resizing the image preprocesses to the optimal size before proceeding deblurring step and then comes the step of applying DFT algorithm on the input capture to next transform the two parts of the complex result into magnitude before converting it into logarithmic scale and normalizing step. After applying this preprocessed deblurring approach, the capture should be going to the treatment block before being stored. The application of algorithm on images was not a very good idea, because in a real case concretization, the database of images will not be all blurred. The algorithm should contain a blur measurement to know if the image has a blur or not, if yes, the image will be deblurred then treated, if not it will be directly treated.

The challenge as we mentioned before is that applying this algorithm on a database will give false results, that is simply because not all captures are blurred, but the application of the algorithm will deblur all the captures, even the clean ones. So, we should make sure first if the image is blurred or not. This is why it is important to measure the blur that the image contain, to know if we need a preprocessing application or not, and to do this, we opt the use of the noise estimation approach for the measure. The blur value should be compared to a threshold, and in our case, it was 20, then we take the decision to apply deblurring or nit. We applied the hybridization

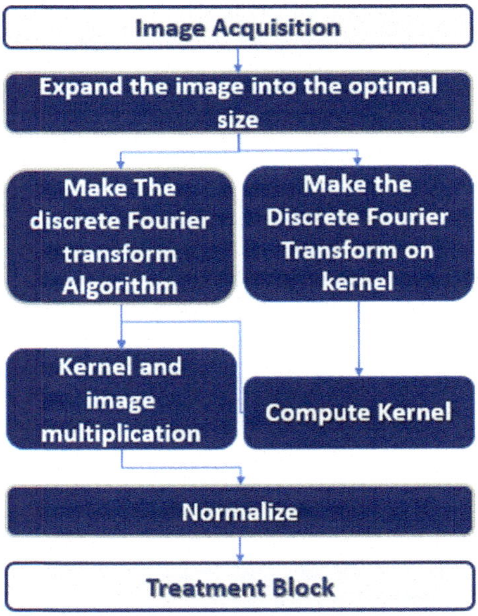

Fig. 5 Global based DFT algorithm

of the two algorithms, and then we got the following algorithm as a final proposed approach. Figure 9 illustrates our proposed approach (Figs. 6, 7, and 8).

Figure 6 is an illustration of our proposed hybrid algorithm, based on DFT deblurring approach and the noise estimation approach. This global proposed architecture is dedicated to process the input image and estimate its noise, and then decide if the image is blurred or not after comparing the noise to a threshold. If the image is blurred, it will be filtered by the deblurring process then sent to the treatment block, if not it will be directly sent to the treatment block before being stored. The algorithm was applied on real case images. Images used in our case were taken with good quality and no blur. We added a motion blur to the clean images before applying the algorithm, then we explored the results of deblurring this noised created captures to

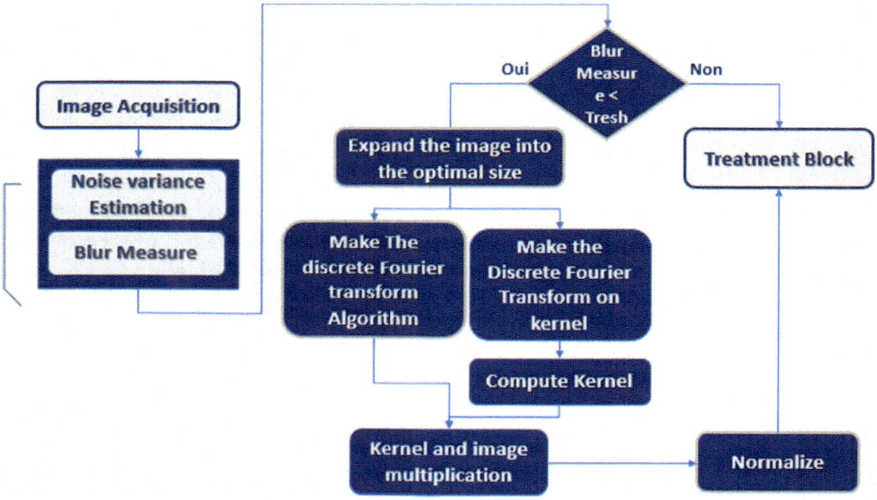

Fig. 6 Our proposed algorithm

Fig. 7 Our used images

Fig. 8 Blurred images

Fig. 9 OpenCL architecture for the proposed algorithm

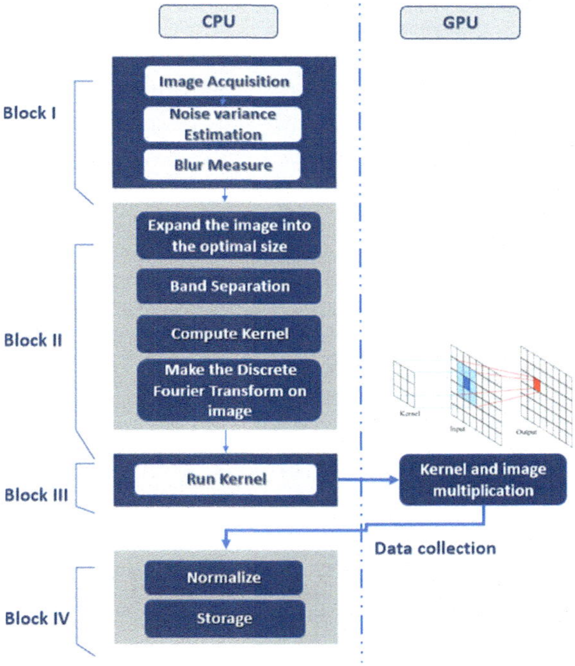

see how much the algorithm can give satisfying results. Images that we used in our case are the following.

6 Results and Discussion

The images used and mentioned before in Fig. 9 are clean with no blur, and the idea of the experience is to add a motion blur before treatment to compare results. The following captures are the result of adding motion blur with specified characteristics. A motion blur has two basic parameters, which are the following:

- Angle of blur: 50°
- Distance of blur: 10 pixels.

Figure 8 shows the blurred images resulted after adding a motion blur to the clean original images. The approach proposed in our work was experimented on a computer desktop and an embedded architecture based on two CPUs and a GPU. The desktop is equipped with a GeForce GPU MX130 which is a graphic card used on notebooks. In the other hand, we have the two CPU and one GPU based architecture which is the ODROID-XU4 architecture equipped with Samsung exynos5422 with the big ARM cortex A15 @2 GHz and the little cortex A7 @1.4Ghz, with a GPU Mali-T628 MP6 with 2Gbyte LPDDR3 RAM.

Our algorithm of deblurring UAV images for a precision agriculture exploration is written in C/C++, and it actually gave good results with an acceptable processing time. To improve the processing time, we converted the code into the famous coding language titled OpenCL in order to explore the use of GPU and guarantee a parallelism processing of blocks. Besides, OpenCL or Open Computing Language is the combination of an API and a programming language derived from C and as its name suggests "open" it is an open standard, open source. This computer language is designed to program heterogeneous parallel systems comprising a multi-core CPU and a GPU, and it was created in June 2008 [41]. In our case, the multiplication between kernel's DFT and image's DFT takes a little bit more time than other blocks, so we decided to send this part to GPU, while all other block will be treated in CPU. This multiplication should be done for every band of bands in images, which will take a little bit more time because of the back and forth from CPU to GPU. So, the solution was to calculate the kernel and DFT of each band all in CPU and then pass the lead to GPU in order to make the computing of multiplication of the three bands in once. In Fig. 9, we present the schema of the proposed algorithm converted to OpenCL, becoming as follow:

After converting the algorithm to OpenCL, Fig. 9 gives an overview of the new architecture, standing on tasks that both CPU and GPU will do. The proposed parallelized approach has the same steps as the C++ algorithm, yet all the units work together in order to better the performance of the code. The CPU here will be released from the heaviest part which is the computing of multiplication of the kernel and the image, while the GPU will take the lead to do this task, to guarantee a specific parallelism between tasks. After applying our parallelized algorithm on the blurred images in Fig. 10.

The step of embedding is the last step proceeded, where we got the results of our algorithm. The implementation of the algorithm was done on a desktop computer

Fig. 10 Resulted images

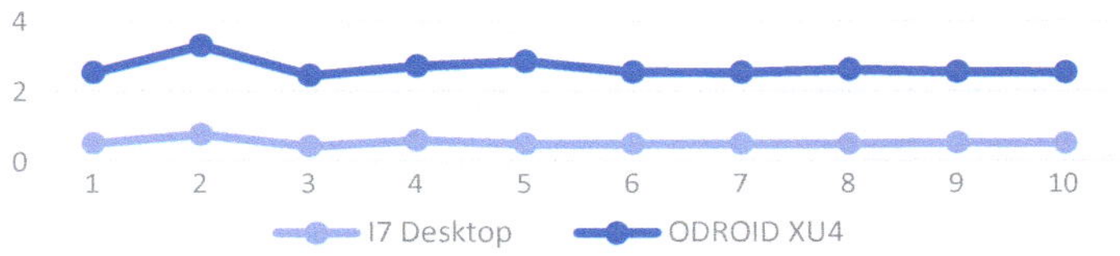

Fig. 11 Processing time in an i7 core computer and an ODROID-XU4 architecture

as mentioned before, and on an ODROID-XU4 embedded architecture. On the XU4 board, it gives a 0.02 s as a processing time for each image. This gave a 50 frames per second. While on a desktop, it gave 0.006 s for each image, on the average of around 100 images per second.

After testing the algorithm in two different architectures, we elaborated Fig. 11 to illustrate the comparison between the processing times of each architecture.

7 Conclusion and Perspectives

Developing agriculture field becomes a necessity for a good sustainable development, that is why precision agriculture is a very interesting field of research. A good database is the key of success of any good algorithm, without data, and the algorithm can never be viable. In like manner, preprocessing of database before use is a very huge and important step to avoid data loss and bad results, especially for a real case concretization, so that the host of the system gets satisfying results and be able to take the best decisions for the field he studies and improve the precision of the system. The preprocessing step is a step that comes after taking on consideration a lot of environmental challenges. Our proposed approach is an additional block, not only for the case of our application, but for any image processing algorithm explored in precision agriculture using images captured by UAV, taking the lead of deblurring and preparing data for a good use in the main treatment block, and it has given good results embedded on i7 Desktop and ODROID-XU4 board.

Besides, our proposed algorithm was parallelized to give better results in a real-time embedding, and the experiment showed that the algorithm can preprocess over 50 images per second which is a good score regarding agriculture domain. As perspectives, we are looking forward to use an environment reference for a real case study to take in consideration all the environment's constraints. Also, we look forward to embed our algorithm in different embedded architectures in order to compare the performance of the approach in different environments [30–39].

References

1. López-Granados F (2011) Weed detection for site-specific weed management: mapping and real-time approaches. Weed Res 51:1–11. https://doi.org/10.1111/j.1365-3180.2010.00829.x
2. Perez-Jimenez A, López F, Benlloch-Dualde J-V, Christensen S (2000) Colour and shape analysis techniques for weed detection in cereal fields. Comput Electron Agric 25:197–212. https://doi.org/10.1016/S0168-1699(99)00068-X
3. El-Faki MS, Zhang N, Peterson DE (2000) Weed detection using color machine vision. Trans ASABE (Am Soc Agric Biol Eng) 43:1969–1978. https://doi.org/10.13031/2013.3103
4. Thompson JF, Stafford J, Miller P (1991) Potential for automatic weed detection selective herbicide application. Crop Protect 10:254–259. https://doi.org/10.1016/0261-2194(91)90002-9
5. Rahnemoonfar M, Sheppard C (2017) Deep count: fruit counting based on deep simulated learning. Sensors (Basel, Switzerland) 17.https://doi.org/10.3390/s17040905
6. Maldonado Jr W, Barbosa JC (2016) Automatic green fruit counting in orange trees using digital images. Comput Electron Agric 127:572–581. https://doi.org/10.1016/j.compag.2016.07.023
7. Chen SW, Skandan S, Dcunha S, Das J, Okon E, Qu C, Taylor C, Kumar V (2017) Counting apples and oranges with deep learning: a data driven approach. IEEE Robot Autom Lett 1–1. https://doi.org/10.1109/LRA.2017.2651944.
8. Song Y, Glasbey CA, Horgan GW, Polder G, Dieleman J, van der Heijden G (2014) Automatic fruit recognition and counting from multiple images. Biosys Eng 118:203–215. https://doi.org/10.1016/j.biosystemseng.2013.12.008
9. Latif R, Saddik A, Elouardi A (2019) Evaluation of agricultural precision algorithms on UAV images 1–4. https://doi.org/10.1109/ICCSRE.2019.8807604
10. Ternoclic Homepage. https://ternoclic.com/infos/explication-agriculture-de-precision/
11. Rafał P, Stuczynski T, Borzecka M (2012) The suitability of an unmanned aerial vehicle (UAV) for the evaluation of experimental fields and crops. Zemdirbyste 990014:431–436
12. Sa I, Chen Z, Popovic M, Khanna R, Liebisch F, Nieto J, Siegwart R (2017) WeedNet: dense semantic weed classification using multispectral images and MAV for smart farming. IEEE Robot Autom Lett. https://doi.org/10.1109/LRA.2017.2774979
13. Gnädinger F, Schmidhalter U (2017) Digital Counts of maize plants by unmanned aerial vehicles (UAVs). Remote Sens 9:544. https://doi.org/10.3390/rs9060544
14. Khanna A, Kaur S (2019) Evolution of internet of things (IoT) and its significant impact in the field of precision agriculture. Comput Electron Agric 157:218–231. https://doi.org/10.1016/j.compag.2018.12.039
15. Popovic T, Latinović N, Pesic A, Zecevic Z, Krstajic B, Đukanović S (2017) Architecting an IoT-enabled platform for precision agriculture and ecological monitoring: a case study. Comput Electron Agric 2017:255–265. https://doi.org/10.1016/j.compag.2017.06.008

16. Shkanaev AY, Sholomov DL, Nikolaev DP (2020) Unsupervised domain adaptation for DNN-based automated harvesting 112. https://doi.org/10.1117/12.2559514
17. Aggelopoulou K, Bochtis D, Fountas S, Swain K, Gemtos T, Nanos G (2011) Yield prediction in apple orchards based on image processing. Precis Agric 12:448–456. https://doi.org/10.1007/s11119-010-9187-0
18. Honkavaara E, Saari H, Kaivosoja J, Pölönen I, Hakala T, Litkey P, Mäkynen J, Pesonen L (2013) Processing and assessment of spectrometric, stereoscopic imagery collected using a lightweight UAV spectral camera for precision agriculture. Remote Sens 5:5006–5039. https://doi.org/10.3390/rs5105006
19. Primicerio J, Di Gennaro SF, Fiorillo E, Genesio L, Lugato E, Matese A, Vaccari FP (2012) A flexible unmanned aerial vehicle for precision agriculture. Precis Agric. https://doi.org/10.1007/s11119-012-9257-6
20. Liu R, Li Z, Jia J (2008) Image partial blur detection and classification. IEEE Int Conf Comput Vis Pattern Recogn 1–8. https://doi.org/10.1109/CVPR.2008.4587465
21. Luo G, Chen G, Tian L, Qin Ke, Qian S-E (2016) Minimum noise fraction versus principal component analysis as a preprocessing step for hyperspectral imagery denoising. Can J Remote Sens 42:00–00. https://doi.org/10.1080/07038992.2016.1160772
22. Ben-Ezra M, Nayar S (2004) Motion-based motion deblurring. IEEE Trans Pattern Anal Mach Intell 26:689–698. https://doi.org/10.1109/TPAMI.2004.1
23. Agarwal S, Singh OP, Nagaria D (2017) Deblurring of MRI image using blind and non-blind deconvolution methods. Biomed Pharmacol J 10:1409–1413. https://doi.org/10.13005/bpj/1246
24. Almeida M, Figueiredo M (2013) Parameter estimation for blind and non-blind deblurring using residual whiteness measures. IEEE Trans Image Process Publ IEEE Signal Process Soc 22.https://doi.org/10.1109/TIP.2013.2257810
25. Jia J (2007) Single image motion deblurring using transparency. CVPR 1–8.https://doi.org/10.1109/CVPR.2007.383029
26. Jin M, Roth S, Favaro P (2018) Normalized blind deconvolution: 15th European conference, Munich, Germany, proceedings, part VII.https://doi.org/10.1007/978-3-030-01234-2_41
27. Bar L, Kiryati N, Sochen N (2006) Image deblurring in the presence of impulsive noise. Int J Comput Vis. https://doi.org/10.1007/s11263-006-6468-1
28. Schuler CJ, Christian Burger H, Harmeling S, Scholkopf B (2013) A machine learning approach for non-blind image deconvolution. In: Proceedings/CVPR, IEEE computer society conference on computer vision and pattern recognition, pp 1067–1074. https://doi.org/10.1109/CVPR.2013.142.
29. Machine Version Study guide: https://faculty.salina.k-state.edu/tim/mVision/freq-domain/image_DFT.html
30. Menash J (2019) Sustainable development: meaning, history, principles, pillars, and implications for human action: literature review. Cogent Soc Sci 5. https://doi.org/10.1080/23311886.2019.1653531
31. Kusnandar K, Brazier FM, Kooten O (2019) Empowering change for sustainable agriculture: the need for participation. Int J Agric Sustain 1–16. https://doi.org/10.1080/14735903.2019.1633899
32. Dicoagroecologie Homepage. https://dicoagroecologie.fr/encyclopedie/agriculture-de-precision/. last accessed 21 Nov 2016
33. Be Api Homepage. https://beapi.coop/
34. Miles C (2019) The combine will tell the truth: on precision agriculture and algorithmic rationality. Big Data Soc 6:205395171984944. https://doi.org/10.1177/2053951719849444
35. Soubry I, Patias P, Tsioukas V (2017) Monitoring vineyards with UAV and multi-sensors for the assessment of water stress and grape maturity. J Unmanned Veh Syst 5.https://doi.org/10.1139/juvs-2016-0024
36. Chen J (2004) A simple method for reconstructing a high-quality NDVI time-series data set based on the Savitzky? Golay filter. Remote Sens Environ 91.https://doi.org/10.1016/S0034-4257(04)00080-X

37. Torres-Sánchez J, López-Granados F, Serrano N, Arquero O, Peña-Barragán JM (2015) High-Throughput 3-D monitoring of agricultural-Tree plantations with unmanned aerial vehicle (UAV) technology. PLoS ONE 10:E0130479. https://doi.org/10.1371/journal.pone.0130479
38. Adão T, Hruška J, Pádua L, Bessa J, Peres E, Morais R, Sousa J (2017) Hyperspectral imaging: a review on UAV-based sensors, data processing and applications for agriculture and forestry. Remote Sens 2017:1110. https://doi.org/10.3390/rs9111110
39. Gago J, Douthe C, Coopman R, Gallego P, Ribas-Carbo M, Flexas J, Escalona J, Medrano H (2015) UAVs challenge to assess water stress for sustainable agriculture. Agric Water Manag 153.https://doi.org/10.1016/j.agwat.2015.01.020

Chapter 12
SLAM Algorithm: Overview and Evaluation in a Heterogeneous System

Rachid Latif, Kaoutar Dahmane, and Amine Saddik

1 Introduction

The interdisciplinary field of robotics has seen great progress, especially the research area-based mobile robots. In the past, mobile robots were controlled by heavy, bulky, and expensive computer systems, and this made transportation difficulties because it had to be linked using cable or wireless devices. Nowadays, we can find mobile robots with a variety of actuators and sensors that are controlled by inexpensive, small, and light embedded systems that are carried on-board the robot. The development of embedded systems that can be found inside the robot and connects the different subsystems can solve the problem [1]. The evolution of the computer vision has led to an increase in the use of cameras as the only source of information or with other sensors such as odometry or laser in mobile robots [2]. The choice of sensors is made taking into account the improvement of the autonomy and the precision of mobile robots, in order to build robust models of the environment and ensure the autonomous navigation function of the robot which includes the phases: mapping, localization, and planification. The robot has to estimate the pose called localization and obtain a model map of the environment named mapping at the same time without prior knowledge of the environment [3]. To perform the localization task, we have many positioning methods include the Global Positioning System (GPS) [4], inertial measurement unit (IMU), and wireless signal. However, GPS-based satellite positioning system can only work outdoors, and the signal is transmitted by satellite which can be received and processed, but the accuracy of the GPS estimation depends on the number of satellites detected by the sensors. The IMU represents an indoor system capable to integrate the movement of mobile but it suffers from sensors errors and nonlinearity error, the cumulative error affects the measurement of the robot. In wireless technology, there is no need for wires, we can connect our devices to the

R. Latif · K. Dahmane (✉) · A. Saddik
LISTI, ENSA Ibn Zohr University Agadir, 80000 Agadir, Morocco

A. E. Hassanien et al. (eds.), *Enabling Machine Learning Applications in Data Science*, Algorithms for Intelligent Systems, https://doi.org/10.1007/978-981-33-6129-4_12

network, but this solution generates a number of problems in connection with localization; among this problem we find that a number of measurements have to be done for determining the position of a mobile node; also, the choice of the localization technique depends on the given environmental conditions [5]. Giving the limits of the previous methods, we pushed to use different techniques that allow the robot to explore the environment and map while staying localized. The SLAM algorithm can be used in indoor and outdoor applications will the mobile robot collect the data from the environment which is contained landmarks with own sensors then interpret them and after the construction of navigation map, determine its location in that map.

Recently in [6], the authors proposed two main algorithms of localization to be analyzed linear, Kalman filter (KF), and extended Kalman filter (EKF). Their essential contribution involved one-dimensional SLAM using a linear KF which consists of five phases, such as a motionless robot with absolute measurement, moving vehicle with absolute measurement, a motionless robot with relative measurement, moving vehicle with relative measurement, and moving vehicle with relative measurement while the robot location is not detected. Moreover, the authors dissected the localization performance of SLAM with EKF. The proposed SLAM-based algorithms present good accuracy compared to each other and also with other algorithms regarding SLAM [7]. In another study, Sualeh et al. 2019 present a state-of-the-art-based SLAM and discuss the insights of existing methods. Starting with a classical definition of SLAM, a brief conceptual overview, and the formulation of a standard SLAM system is provided. The work presents a taxonomy of newly developed SLAM algorithms with a detailed comparison metrics, in order to survey the currently available techniques, to classify and identify the future directions [8]. Nguyen et al. [9] developed an algorithm named Hessian ORB Overlapped FREAK (HOOFR) based on the modification of the ORB (Oriented FAST and rotated BRIEF) detector and the bio-inspired descriptor: Fast Retina Key-point (FREAK). This new algorithm has a better compromise between precision and processing times. The HOOFR-SLAM algorithm uses images taken by a stereo camera to perform simultaneous localization and mapping, and the authors also proposed an implementation of the algorithm on a heterogeneous architecture CPU-GPU using CUDA and OpenCL language-based CPU-FPGA with adopting the embedded platforms such as JETSON Tegra X1 equipped with 4-Core ARM and A57 4-Core ARM A53 @ 1.3–1.9 GHz and Intel core i7 laptop @ 3.40 GHz [9]. More recently, we can find in [10] a proposition of a high-level implementation on an SoC-FPGA of a feature extraction algorithm dedicated to SLAM applications called the HOOFR extraction algorithm with the integration of a bucketing detection method to have a distribution of key points in the image that provides robust performance but requires a lot of computation on the embedded CPU. The heterogeneous FPGA-based architecture where the design has been validated is an Intel Arria 10 SoC-FPGA with 54 fps throughput at 1226 $\times$ 370 pixels or 14 frames per second at 1920 $\times$ 1080 pixels. using the OpenCL programming language, also the authors performed a performance evaluation of the FPGA-based implementation compared to the embedded GPU-based implementation using a public data set [10].

The aim of this paper is to realize an implementation of a bioinspired SLAM algorithm in a heterogeneous system CPU-GPU for the mobile robot, to assure an autonomous navigation. Using a laptop that comprises the NVIDIA GeForce 940MX, with boost clock: 1241 MHz, Memory Clock 4000 MHz effective, and Intel Core i7 @ 2.70 GHz the evaluation of the algorithm is based on the New College dataset recorded by a stereo camera. For the result, we find the processing time on the robot operating system (ROS) equals to 160.43 ms using CUDA language.

The work presented is composed of four parts: the first part for the introduction, the second part is devoted to present an overview on SLAM algorithm embedded for mobile robotic, the third section present the evaluation and result obtained by the use of embedded CPU-GPU systems, and the last section is consecrate to the conclusion.

2 SLAM: Overview and Approach

Over the last decade, the issue of simultaneous localization and mapping (SLAM) has obtained more attention from researchers and remains an influential topic in robotics. Currently, various algorithms of the mobile robot SLAM have been reviewed [6]. The robot used in the SLAM topic is equipped with functions of perception, decision, and action. This machine is described as autonomous if it is able to decide its actions to achieve its goal, so the aim here is to perform correctly its task of localization and mapping even if in the case where they will be faced with unforeseen situations without human implication. Navigation is the set of techniques and allows us to define the robot's coordinates while referring to a defined fixed point. In the absence of sufficient information on the desired environment and on its positioning and the algorithms: generation of trajectory, avoidance of obstacles, the robot cannot function properly. In addition, location and mapping, the two tasks that the robot must perform are linked. We do not have an environmental map or location data, but in this situation, the robot must be able to find this information simultaneously. The accomplishment of tasks can be ensured using SLAM algorithms: Simultaneous Localization and Mapping, and that is why, the SLAM algorithms occupy an important place between the different algorithms applied in the field of autonomous navigation. Figure 1 shows the several stages of SLAM.

The first step is the acquisition of data from various sensors: called perception, and this step consists of connecting the robot to the input from the sensors and extracting from them reference points that designate a characteristic or position in the environment that can easily be re-observed and tracked to give the robot the identification of one's position in relation to one's environment. After that we have the association and fusion of data operation from the different sensors by filtering and analyzing. The extracted landmarks are compared with previous sets of detecting landmarks while trying to match the different landmarks. As the robot moves, there is a succession of movements and an update of the position, which allows the SLAM to locate the machine, since the relative change of the marker position is used to calculate the change of position of the robot. After this, we have the step of prediction of the next position and orientation of the robot that means we estimate the state, after the association of the new reference points with the preceding reference points, the new

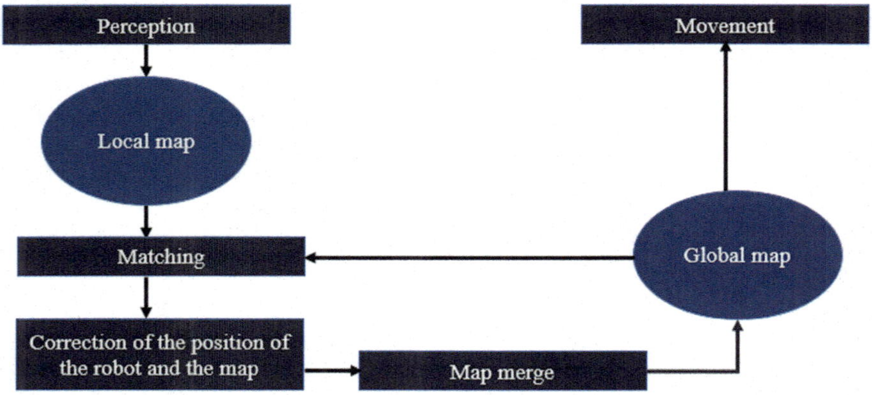

Fig. 1 SLAM process

state of the robot can be calculated from the difference in relative position of each reference point from the previous position to the new position. Then we evaluate the validity of the conformity of the map taking into account the sensor data and the prediction, so in this step we update the global map and the position of the robot, in order to allow the map, the improvement. The new landmarks have been associated with newly retrieved landmarks, so that the most accurate map of the environment can be reproduced. Previously retrieved landmarks are recalculated with the information so that the most accurate map of the environment can be reproduced [11].

The probabilistic approaches are used in SLAM problems to model and reduce the uncertainty and noise generated by the motion of the robot that increases the uncertainty of the position and map estimations and adds errors and noise to the observations [12]. A probabilistic algorithm is based on the idea of representing information through probability densities. Almost all the probabilistic techniques are based on the Bayes rule:

$$p(x/d)p(d) = p(d/x)p(x) \tag{1}$$

Equation 1 informs us that if want to get a quantity x that represents here a robot position based on measurement data d represents the odometry, we can do that by multiplying the two terms *p(d/x)* and *p(x)*. The term *p (d/x)* defines the way a sensor measurement *d* of a real-world value *x* is generated. It is the probability of observing the measurement d under the hypothesis *x*. *p(x)* is called the prior, it specifies the probability that *x* is the value in the real world before the arrival of any data. *p(d)* can be viewed as a normalization constant [12]. Of the main techniques used by regular SLAM algorithms that were proven to be successful at reducing the amount of uncertainty and noise inherent to the robot motion and to the sensors measurements we find the technique called Kalman filter [12]. This filter gives a recursive solution to the filtering of linear data. It allows the position of the robot to be measured successively after each measurement acquisition. When the global map of the environment is formed as the robot explores, this filter also makes it possible to estimate the position of each geometric element introduced into the map. The Kalman filter cycle consists of two basic stages, the prediction step that can estimate the state of the system at time t using the time series $(t - 1)$. The correction step allows us the estimation of the state of the system made at time t using the information received by the different sensor at time t. On the other hand, we can found also the particulate filter technique [13]. A particulate filter is a recursive filter-based EKF that enables one to estimate the posterior state using a set of particles. Unlike parametric filters like the Kalman filter, a particulate filter represents a distribution by a set of samples created from this distribution and also is able to process strongly nonlinear systems with non-Gaussian noise. The use of particulate has a disadvantage, and it suffers from the difficulties encountered when defining the number of particles. Indeed, the quality of the estimate is strongly correlated with the discretization of the search space, but it is difficult to find an optimal number of particles [14].

The most known approaches such as the EKF, FastSLAM, GraphSLAM, visual slam and ORBSLAM, the EKF SLAM approach of solving the SLAM are based on Kalman filter (KF) and particle filters (PF). The EKF SLAM approach of solving the SLAM problem presents an extension of the Kalman filter and is by using sensor data accumulated from the movement of the robot. Besides, it is also necessary to gather information on the environment. Employing this data, the algorithm can define where the robot is supposed to be positioned on a map, as well as specific benchmarks observed [15]. The robustness of the EKF algorithm is low [16]. Due to the large computational load, the EKF SLAM algorithm is more proper for smaller maps containing fewer landmarks. In terms of efficiency, the EKF SLAM algorithm is very costly.

Another solution for the SLAM problem using the FastSLAM algorithm, which makes use of the Rao-Blackwellized particle filter. The pose of the robot is probabilistically dependent on its previous pose, whereas the landmark locations are probabilistically dependent on the position of the robot. For EKF's SLAM algorithm, we do not have this possibility; at each update, it has to recalculate its covariance matrix, which leads to bad scaling in large maps. The fact that the data association from measurements to landmarks is particle-based which reduces the impact of a wrong data association, we can assume that the FastSLAM algorithm is significantly more robust than the EKF SLAM algorithm. Nevertheless, the FastSLAM has an advantage that he reduces algorithmic complexity compared to EKF SLAM. An updated one has been developed, named FastSLAM 2.0. FastSLAM 2.0 is similar to its predecessor, with the difference that when estimating the motion, it also considers the range sensors. This creates better guesses in general and will result in more particles having higher importance factors. In this context, the authors in [17] provided the study of the portability of SLAM algorithms on heterogeneous embedded architectures. This study was based on the hardware/software co-design approach in order to propose an optimal implementation based on the chosen algorithms. The algorithms selected in this work are: FastSLAM2.0, ORB SLAM, RatSLAM, and Linear SLAM. These algorithms are evaluated on embedded architectures in order to study the possibility of their portability. Additionally, the authors presented a case study of the FastSLAM 2.0 algorithm dedicated to large-scale environments implemented in different embedded architectures such as the Tegra X1 system-on-chip (SoC) which integrates processor 4 × ARM Cortex A57 and 4 × ARM Cortex A53 CPUs @ 1.9 GHz, and also SLAM algorithm implemented on a high-performance desktop Core 2 Quad Q6600, @ 2.40 GHz and on the T4300 dual-core, @ 2.10 GHz laptop computer another implementation are proposed on the ODROID-XU4 which uses a Quad-Core ARM Cortex15, and on The Panda board, ES includes an ARM Dual-Core Cortex A9 processor @ 1 GHz [17].

GraphSLAM is another way of solving the SLAM problem, based on graph optimization. The purpose of GraphSLAM is to solve the SLAM problem by using a graph. The graph comprises nodes that represent the poses of the robot $x_0, \ldots, x_t$ as well as landmarks in the map which are denoted as $m_0, \ldots, m_t$ as there are also constraints between the position so this is not enough to address the SLAM problem $x_t, x_{t-1}, x_{t-2}, \ldots, x_{t-n}$ of the robot, and the landmarks $m_0, \ldots, m_t$. These constraints represent the distance between adjacent locations such as x_{t-1}, x_t as well as the distance between the locations to the landmarks $m_0, \ldots, m_t$. These constraints can be constructed due to information from the odometry source u_t. Graph SLAM is considered to be an offline SLAM approach because it does not run and update while the robot is moving and collecting data. The fact that older data associations can be reexamined if a wrong data association has been performed, the robustness of the GraphSLAM algorithm is higher than the EKF algorithm [17].

Another approach to solving the SLAM problem is the visual-based SLAM also known as V-SLAM. In this way, we can found a diversity of solutions using different visual sensors including monocular [18], stereo [19], omnidirectional [20], and combined color and depth (RGB-D) cameras [21]. Davison et al. proposed one of the first V-SLAM solutions. They employed a single monocular camera and formed a map by extracting sparse features of the environment and matching new features to those already observed using a normalized sum-of-squared difference correlation [18]. Henry et al. present the first implementation of an RGB-D mapping approach that employed an RGB-D camera, and this information is used to obtain a dense 3D reconstructed environment [21, 22]. Oriented FAST and Rotated BRIEF (ORB) is also a solution to the SLAM issue, and it is an algorithm that combines the FAST and BRIEF algorithms [23]. The algorithm starts by using FAST algorithm to obtain the corner data from an image but because this does not produce multi-scale features, so a scale pyramid of the image is used where the FAST algorithm is applied to each level of the pyramid. The corner features extracted are then oriented by applying an intensity centroid on each feature. This method assumes that a corner's intensity is offset from the center, which is represented by a vector used as the orientation of the feature [24]. Recently, Stefano et al., present a method to modify an open-source SLAM algorithm version ORB-SLAM2 in order to run this algorithm in real time on an embedded NVIDIA Jetson TX2 board and they obtain a representation of the original algorithm as a graph, which allows subdividing efficiently the computational load between CPU and GPU, which results in a processing speed of about 30 frames per second, in order to have a possibility to run this algorithm in real-time performance on four different sequences of the KITTI dataset while maintaining good accuracy [25]. Latif et al., present a study on some proposed methods to solve the SLAM problem. The extended Kalman filter (EKF) has the advantage of giving the uncertainty on the position of the robot also the landmarks in the course of time. Concerning their disadvantage, this solution suffers from problems of complexity, also a consistency problem, such as the FastSLAM approach. However, the FastSLAM has an advantage that can reduce algorithmic complexity compared to EKF SLAM. The strong point of graph slam is that allows avoiding the propagation of linearization errors, despite GraphSLAM gives more precise results [26]. Because of these drawbacks of the probabilistic algorithm, a bioinspired slam has been developed, the bio-inspired techniques were proven to be able to solve the SLAM problem as well using methods that are robust, flexible, and well-integrated into the robot. The bio-inspired algorithms are inspired by biological systems, mainly the human and rodent's brain [12]. Rodents and insects show the ability to store and organize visual cues and can update the pose estimate and locate himself. Rodents, in particular, are better than humans in facing navigation problems, without external, only by using estimates of self-motion or path integration. Bio-inspired SLAM based on minimal information about the environment, this system use odometry and observation sources collected by vision systems, for example, the lasers or cameras to update and review the map estimated [27].

RatSLAM represents a biologically inspired approach that offers a solution to the problem of large-scale mapping and localization based on the visual SLAM algorithm systems provided by Milford et al. [28]. It uses a simplified computational model of the rodent hippocampus to build an online map in real time. Visual ambiguity and data association problems are managed by maintaining multiple competing robots pose estimates. RatSLAM corrects cumulative errors in odometry by a map correction algorithm. The biologically inspired part in RatSLAM is represented by the pose cells that can imitate the behavior of place cells and head direction cells. The self-motion cues that can update the activity in the pose cells are used to drive path integration, the kernel of the system is a three-dimensional continuous attractor network, (CAN), which is used to maintain an estimate of the system's current pose. RatSLAM can map a complex road network using a single webcam positioned on a car. RatSLAM generated a coherent map of the entire environment at real-time speed, and these advantages have not been done by using the probabilistic SLAM algorithms [12]. Further, we have the bio-inspired algorithm named FREAK, it was introduced by considering the topology of the human retina and the observations in neuroscience. The human retina deduces information from the visual field using the gaussian comparison of gaussian differences of different sizes and coding these differences in binary mode as a neural network.

3 Evaluation and Result

In this paper, we use a biological system called RatSLAM that represents an implementation of the hippocampal model, which can perform competitive SLAM results in real-world environments with a camera sensor and optionally sensors that gather odometrical data. It presents a rough computational model of the hippocampus part of the rodent. RatSLAM uses techniques of landmark detecting with odometrical information in order to build a Competitive Attractor Network (CAN) which forms a topological representation of adjacent world locations. RatSLAM consists of different processing units which are presented here in short, in the order RatSLAM consists of several processing units which are presented here in short in the order: local view, pose cell, the experience map block, and odometry, data that can be extracted from the bag file. We evaluated this algorithm on The New College Dataset [29] the dataset is available online as ROS bag file which is the method of standardized for storing ROS message data, it contains a variety of tools to record play analyze and visualize the message data in ROS bag file, recorded by a robot using a stereo camera with resolution of 512 × 382 pixels gathered while traversing 2.2 km through a college's grounds and adjoining parks, the stereo imagery captured at 20 Hz, and 5 view omni-directional images at 5 Hz.

The Open-Source version of the RatSLAM named Open RatSLAM algorithm has provided by Ball et al., in 2019 [30] using ROS (Robot Operating System) a meta-operating system it assumes that there is an underlying operating system that will help it carry out its tasks. It is open source and it provides the services expected from an operating system, including controlling low-level devices, implementing commonly used features, passing messages between processes, and managing packages. The architecture of ROS consists of five elements: a ROS master, nodes, publishers, subscribers, and topics. The ROS master allows the location and communication between the nodes of the robotic system, and generally, initiates the node's communication. The node allows communication between another node. A message transmitted by a node or a subject in a ROS system is called a publisher, and the message that is received by a node or subject in a ROS system is called a subscriber. Publishing and subscribing a message of a specific name type is known as a subject. Open RatSLAM is a modular version of RATSLAM integrated with ROS. Open RatSLAM is made up of connected ROS nodes representing cells. Open RatSLAM takes mono images and odometry as standard ROS messages. Figure 2 shows, the blocks that comprise the Open RatSLAM.

The first block named Local View Cells which means an expandable network of units, each representing a distinct visual scene in the environment. When a new visual scene is observed, a new local view cell is generated and associated with the raw pixel data and an excitatory link β is learned between this local view cell and the centroid of the dominant activity packet in the pose cells at this time. When this view is seen again by the robot, the local view cell is activated and injects activity into the pose cells using this exciting link. The Pose Cell network is the heart of RatSLAM and it forms three-dimensional localization and orientation (x', y', θ') for placing the robot in the real environment (x, y, θ) which presents the dimensions of the cell array correspond to the three-dimensional pose of a ground-based robot [31], the coordinates x and y represent the displacement environment, and the angle θ is the rotation. RatSLAM algorithm is focused on iterations of the network based

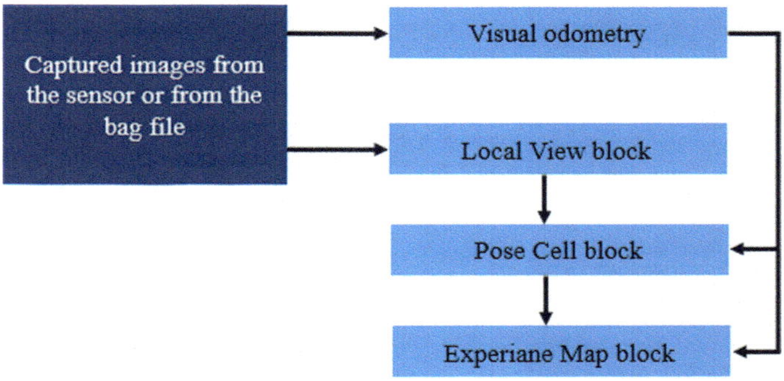

Fig. 2 Open RatSLAM blocks

on competitive pose cell attractors. The dynamics of competitive attractors guarantees that the total activity in the pose cells remains constant. The activity packages located close to each other come together and reinforce each other, merging similar poses representations. Separate activity packages representing multiple installation hypotheses compete. Then the representation of space is provided to the experience map block by the pose cells which corresponds well to the metric layout of the environment in which a robot moves. Nevertheless, as the odometer error accumulates and loop closure events happen, the space represented by the pose cells becomes discontinuous adjacent cells. The network can represent physical places separated by great distances. The experience card is a graphics card that offers a unique estimate of the robot's pose by combining the information from the pose cells and the local view cells.

Each of these three main blocks represents a process executing simultaneously. After a study that took place on the block of local view, we evaluated the algorithm using the C/C++ language in order to estimate the processing time and to interpret it. The first local view block is converted into CUDA language with the heterogeneous architecture CPU-GPU based on Nvidia desktop. The use of the CUDA language running on the GPU part of the system used will allow us to be able to complement the CPU architecture by giving capacity to achieve repetitive calculation involving massive amounts of data. The graphics processor is a multicore architecture with a computational power doubles every ten months, surpassing Moore's approach. Heterogeneous system enables employing more than one kind of processor, to work efficiently and cooperatively, and it has demonstrated their ability to become potential candidates for system-on-chip (SoC) in a hardware–software co-design approach. The parallel programming language CUDA used on GPUs and CPUs is supported by the heterogeneous system. Compute unified device architecture (CUDA) based on the standard C/C++ language represents a parallel programming paradigm allowing to use of GPU resources. For the parallelizable part of the computation, CUDA enables developers to speed up compute-intensive applications by harnessing the power of GPUs. CUDA allows an explicit hierarchization of memory areas private, local, global, allowing to finely organize transfers between them [32]. Grouping of

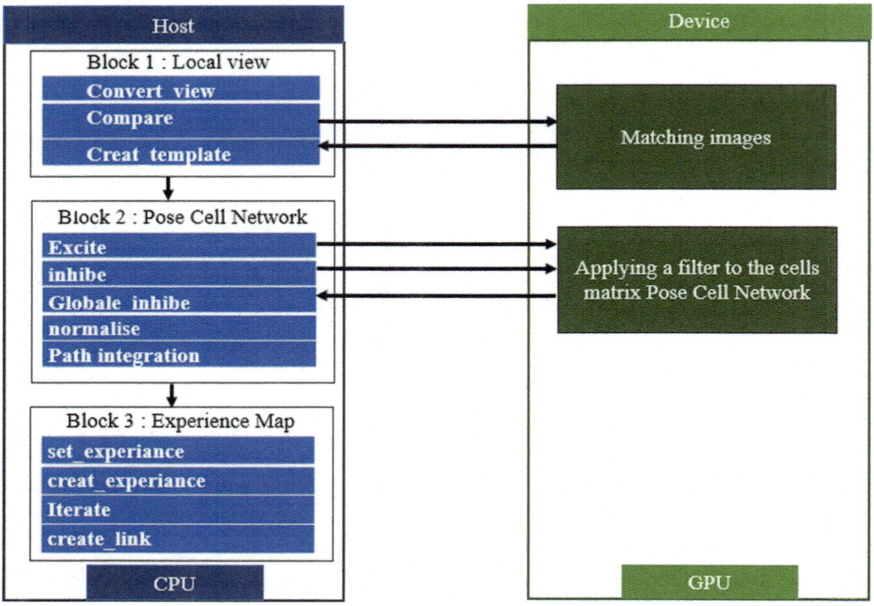

Fig. 3 Block architecture diagram

threads in grids: 1D, 2D, or 3D grid. The local grids are ordered in the global grid allowing access to the global memory, of the advantages of using CUDA we will have relatively standardized access to computing power. Figure 3 shows our proposed CPU-GPU architecture of the Open RatSLAM algorithm.

When each new image is collected, the algorithm checks if the current view is identical or similar to the stored images models, model by model, so that it decides whether or not to add a new model to the vector of the different models, then the size of the visual template increases and the similarity calculation time also increases. In this work, the calculation of the matching operation is performed in parallel so that the similarity is calculated for the images collected with all the templates stored at the same time. In the second block, we apply a filter to the cells of the pose cells matrix to excite these cells, the multiplication of the matrix by the pose cell matrix is carried out box by box, each box represents a pose cell, then we have a large number of iterations for this operation of 204 image sequences; there are 129,925 iterations, also the inhibit function responsible for inhibiting the activation of cells by the use of convolution matrix. The multiplication of the two cell-to-cell matrices, then for both methods, we can apply the 3D multiplication (x, y, θ) in parallel by the GPU. At the end, the Experience Map block creates the map based on the observations of the robot. The map is represented by a graph in which each node is an experiment. Table 1 shows the processing time, for the blocks on homogeneous architecture CPU and heterogeneous CPU-GPU, for the block 1 called local view the use of C/C++ gave us a time of 170.611 ms and the use of the CUDA language gave us a time

Table 1 Total execution time (ms)

Tools		Laptop, CPU (C/C++)	Laptop, CPU-GPU Nvidia GeForce 340MX (CUDA)
	Time of local view block (ms)	170.611	160.43
Total time (ms)		304.611	292.73

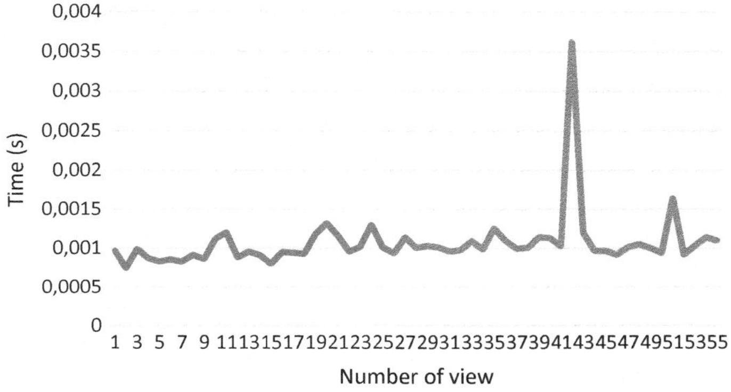

Fig. 4 Evaluation in GeForce 940Mx of 55 views of the function compare of the Bloc local view with CUDA

of 160.43 ms, the block pose cell has a time of 79.7 ms with the C/C ++ language which is already small and we have a time of 78.0 ms using the CUDA, for the block named experience map we have a processing time equal to 54.3 ms using the C /C++ language. We find that the average execution time of the global code of an image in CPU: 82.742032 ms and using the CUDA language we find 69.23 ms. Figure 4 presents the evolution of times of the two architectures.

In Fig. 4, the execution time of the local view block compares function-based CUDA language in the Nvidia GeForce 940MX card of the laptop @ 1241 MHz, using also the Intel Core i7 CPU @ 2.70 GHz, the time variations are between 0.75 ms and 3.62 ms and which gives 1.07 ms in average.

4 Conclusion

This work presents the implementation of the bioinspired SLAM algorithm RatSLAM, on CPU-GPU architecture with using CUDA language, then we obtained as a result of the evaluation an execution time of 170,611 ms in the processing of the algorithm used, resulting in a processing time of 5 frames per second, which allows us to benefit from the technological progress and allows us to respond more to the constraint of real time. The use of conventional systems such as desktop does not imply the processing of algorithms on embedded systems in real time; for this reason as future work, we aim to use heterogeneous embedded systems such as the Xavier board proposed by Nvidia. This implementation will be based on a combination between the H/S Co-Design approach and the equilibration approaches, which will allow us to have a real-time implementation by exploiting all the resources of the architecture used.

References

1. Braunl T (2006) Embedded robotics: mobile robot design and applications with embedded systems: Second edition, pp 1–455. https://doi.org/10.1007/3-540-34319-9
2. Payá L, Gil A, Reinoso O (2017) A state-of-the-art review on mapping and localization of mobile robots using omnidirectional vision sensors. J Sens 2017:1–20. https://doi.org/10.1155/2017/3497650
3. Mahrami M, Islam MN, Karimi R (2013) Simultaneous localization and mapping: issues and approaches. Int J Comput Sci Telecommun 4(7):1–7. ISSN 2047-3338
4. Hoffman-Wellenhof B, Lichtenegger H, Collins J (2001) Global positioning system: theory and practice. Springer, New York, p 382
5. Varga AK (2013) Localization techniques in wireless sensor networks. Product Syst Inf Eng 6:81–90
6. Ullah I, Su X, Zhang X, Cha D (2020) Simultaneous localization and mapping based on Kalman Filter and extended Kalman Filter, Hindawi, Wirel Commun Mobile Comput 2020(12). Article ID 2138643. https://doi.org/10.1155/2020/2138643.
7. Kurt-Yavuz Z, Yavuz S (2012) A comparison of EKF, UKF, FastSLAM2.0, and UKF-based FastSLAM algorithms. In: Proceedings of the IEEE 16th international conference on intelligent engineering systems (INES' 12), pp 37–43
8. Sualeh M, Kim G-W (2019) Simultaneous localization and mapping in the epoch of semantics: a survey. Int J Control Autom Syst 17(3):729–742
9. Nguyen D-D (2018) A vision system based real-time SLAM applications. Hardware Architecture [cs.AR]. Université Paris-Saclay, 2018. English. NNT: 2018SACLS518ff. tel-02398765
10. Nguyen D-D, El Ouardi A, Rodriguez S, Bouaziz S (2020) FPGA implementation of HOOFR bucketing extractor-based real-time embedded SLAM applications. J Real-Time Image Proc. https://doi.org/10.1007/s11554-020-00986-9
11. Stevenson N An Introduction to SLAM How a robot understands its surroundings https://www.doc.ic.ac.uk/~ns5517/topicWebsite/slamOverview.html
12. Nafouki C, Conradt J (2015) Spatial navigation algorithms for autonomous robotics
13. Bouraine S (2007) Contribution à la localisation dynamique du robot mobile d'intérieur b21r en utilisant la plateforme multi sensorielle. Master's thesis,Université de Saad Dahleb de Blida, Faculté des sciences d'ingénieur
14. Bougouffa A, Hocine A (2017) Contribution à La Localisation et la Cartographie Simultanées (SLAM) dans un Environnement Urbain Inconnu
15. EKF SLAM for dummies (2005) https://ocw.mit.edu/courses/aeronautics-and-astronautics/16-412j-cognitive-robotics-spring-2005/projects/1aslam_blas_repo.pdf. Accessed 24 Mar 2017. Cited on pages 6, 15, 17, and 18.
16. Thrun S, Burgard W, Fox D (2005) Probabillistic robotics, 1 edn. MIT Press, Cambridge, Mass
17. Abouzahir M, Elouardi A, Latif R, Bouaziz S, Tajer A (2017). Embedding slam algorithms: Has it come of age? Robot Auton Syst
18. Davison A (2003) Real-time simultaneous localisation and mapping with a single camera. In: Ninth IEEE international conference on computer vision, Proceedings, pp 1403–1410

19. Mahon I, Williams S, Pizarro O, Johnson-Roberson M (2008) Efficient view-based slam using visual loop closures. IEEE Trans Robot 24(5):1002–1014. https://doi.org/10.1109/TRO.2008.2004888
20. Kim S, Oh S (2008) Slam in indoor environments using omnidirectional vertical and horizontal line features. J Intell Robot Syst 51(1):31–43
21. Henry P, Krainin M, Herbst E, Ren X, Fox D (2012) RGB-D mapping: using kinect-style depth cameras for dense 3D modeling of indoor environments. Int J Robot Res 31(5):647–663
22. Yousif K, Bab-Hadiashar A, Hoseinnezhad R An overview to visual odometry and visual SLAM: applications to mobile robotics. Intell Indus Syst 1(4):289–311
23. Rublee E, Rabaud V, Konolige K, Bradski G (2011) ORB: an efficient alternative to SIFT or SURF. In: 2011 international conference on computer vision, pp 2564–2571
24. Tsunemichi Tubman R (2016) Fusion-SLAM by combining RGB-D SLAM and Rat SLAM
25. Aldegheri S, Bombieri N, Daniele Bloisi D, Farinelli A (2019) Data flow ORB-SLAM for real-time performance on embedded GPU boards. In: 2019 IEEE/RSJ international conference on intelligent robots and systems (IROS), pp 1–6. https://doi.org/10.1109/IROS40897.2019.8967814
26. Latif R, Saddik A (2019) SLAM algorithms implementation in a UAV, based on a heterogeneous system: a survey. In: 2019 4th world conference on complex systems (WCCS), Ouarzazate, Morocco, pp 1–6, https://doi.org/10.1109/ICoCS.2019.8930783
27. Müller S, Weber C, Wermter S (2014) RatSLAM on humanoids—a bio-inspired SLAM model adapted to a humanoid robot. Artificial neural networks and machine learning—ICANN 2014, pp 789–796. https://doi.org/10.1007/978-3-319-11179-7_99
28. Milford MJ, Wyeth GF, Prasser D (2004) Ratslam: a hippocampal model for simultaneous localization and mapping. In. 2004 IEEE international conference on robotics and automation, 2004 proceedings. ICRA'04, vol 1. IEEE, pp 403–408
29. Smith M, Baldwin I, Churchill W, Paul R, Newman P (2009) The new college vision and laser data set. Int J Robot Res 28(5):595–599
30. David BALL. Open RatSLAM from Internet: https://github.com/davidmball/ratslam.2019
31. Ball D, Heath S, Wiles J, Wyeth G, Corke P, Milford M (2013) OpenRatSLAM: an open source brain-based SLAM system, Autonomous Robots
32. Wikipedia, Compute Unified Device Architecture. https://fr.wikipedia.org/wiki/Compute_Unified_Device_Architecture#Avantages

Chapter 13
Implementing big OLAP Cubes using a NoSQL-Based approach: Cube models and aggregation operators

Abdelhak Khalil and Mustapha Belaissaoui

1 Introduction

Organizations have realized that it is in their interest to consider "Data" as an asset in the same way as their other strategic assets. In fact, data is no longer just the result of the organization's processes, but it comes from multiple sources from its different ecosystems (customers, suppliers, etc.) and from a multitude of capture means (big data, connected objects, social networks, open data ...). The computer science engineering field interested in transforming data into valuable insights is called business intelligence. In business intelligence, an OLAP cube is a method of storing data in multidimensional form, typically for reporting purposes. It can be seen that as the result of a set of group by queries performed against data warehouse, each cell of the cube holds a meaningful value called measure. There are typically two implementations of OLAP cube: The first one called ROLAP technique [2] performs complex structured query language (SQL) group by queries over relational star schema. As the result is not stored, at each consultation, the query must be executed. The second implementation called MOLAP, in which all cuboids are pre-calculated and stored in a multidimensional data structure persisted on disk, then consulted using multidimensional expression language (MDX), this approach has the advantage of faster response time as the query runs directly on the cube. With the emergence of big data phenomenon, storing and processing a huge amount of data with eventually a large number of dimensions become a very challenging concern and exponentially cost in memory and time; this is the primary motivation of our work. Great deal of research in OLAP focuses on OLAP computation technique under ROLAP and MOLAP technologies rather than exploring other new modern approaches. Over the

A. Khalil (✉) · M. Belaissaoui
LEFCG-SIAD Laboratory, Hassan First University of Settat, Settat, Morocco
e-mail: a.khalil@uhp.ac.ma

M. Belaissaoui
e-mail: mustapha.belaissaoui@uhp.ac.ma

A. E. Hassanien et al. (eds.), *Enabling Machine Learning Applications in Data Science*, Algorithms for Intelligent Systems, https://doi.org/10.1007/978-981-33-6129-4_13

last few years, NoSQL databases have gained a lot of popularity in OLTP systems and become the big trend of database management systems (DBMS); currently, this technology has reached a high level of maturity.

The aim of this paper is the to propose a new way of modeling the OLAP cube and building OLAP engines based on key value databases, and to answer how to extend basic OLAP operations to NoSQL key value model. For this purpose, we study two data models and we provide transformation process to NoSQL logical model.

The remaining of this paper is organized as follow. In the next section, we review the state of the art. Then, in Sect. 3, we present our contribution and data modeling for the proposed approach. In Sect. 4, we conduct an experiment to validate our proposal, and finally, we conclude this paper and discuss future work.

2 State of the Art

In the last decade, much publications have been devoted to NoSQL systems [9], particularly in the scope of business intelligence [8, 10] and more specifically in OLAP cube building under NoSQL databases [5]. Previous researches had focused on OLAP cube computation. In [11], the author has proposed and analyzed a new approach to map a relational database into sorted sets on a key value store in order to create OLAP-like reporting systems. In [12], the author presents a new approach to build MOLAP engines on key value stores for large-scale data analysis and discuss how to match the star schema with cube model databases. In [1], an OLAP data structure based on NoSQL graph databases is presented by the author using OLAP queries based on the cipher declarative language, without providing any complete experimental campaign. In [4], the author studied the multidimensional data models for data warehousing and the computation of pre-aggregate OLAP cubes with document-oriented NoSQL. Later, the same author proposed an extended version of OLAP cuboids under document-oriented NoSQL model that are not possible with relational databases using nesting and arrays called nested and detailed cuboid[3]. Within the same vein of ideas, in [7], the author proposed an aggregation operator called Columnar NoSQL CUBE (CN-CUBE), for column-oriented NoSQL database management systems. This operator allows OLAP cubes to be computed using column-oriented NoSQL data warehouses.

3 Contribution

In this section, we present our contribution for modeling cubes within NoSQL key value stores. The use case bellow serves as a running example.

Example 1. We consider a data warehouse to observe the sales of a company's products. The dimensions are: *Customer* described by *CustomerKey, Supplier, Part,* and the *Date*. The fact table is: *LineItem* identified by *lineItemKey*. Measures are: *Quantity, price, and status.*

3.1 Background

Definition 1 (OLAP Cube): An OLAP cube is a multidimensional dataset which encapsulates quantitative information (measures) and its characteristics (dimensions) to analyze data for decisions making purpose. If the number of dimensions is greater than 3, we call it hypercube.

Definition 2 (OLAP Schema): An OLAP schema is multidimensional logical data model that defines a multidimensional structure for storing one or more cubes in a database. It defines two essential concepts, dimensions which contain the qualitative label that identify the cube's data and fact which holds the quantitative values called measures.

For example, in Fig. 1, the fact we want to observe is product's sales for a given store represented by *LineOrder* table; according to many axe of analysis, the quantity is the measure which represent the unit count to evaluate good sales.

Definition 3 (Key Value store): Key values store is a type of database which stores data as a group of key value pairs. The key must unique to identify one single value. The value contains data referenced by a given key, generally there is no limitation about data being stored, and it could be a numerical value, a document or event another key value pair depending on restrictions imposed by the database management system. Some database software has the feature to specify a data type for the value like Oracle NoSQL.

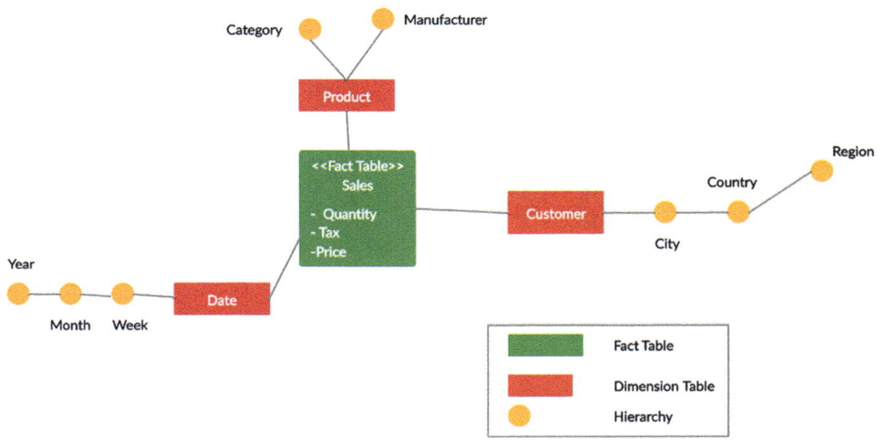

Fig. 1 OLAP schema

Oracle NoSQL is a shared-nothing distributed key value store which provides features like horizontal scalability and high availability. Data is partitioned across shards; a single key value pair is always associated with a unique shared in the system. At the API level, this NoSQL-type database provides a table API and key value paradigm. Oracle NoSQL database supports in its recent versions feature like parent–child join and aggregation functions.

3.2 Approach Overview

There are several possibilities regarding representing data cube under NoSQL database like JSON format, tree structure, nested table… We investigated two data models; each one differs in term of structure and complexity. The first one uses nested records and table hierarchies to store aggregated measures and dimensions. The second one uses the simple key value structure. As there are no assumptions regarding data being stored within the key and the value, we profit from this schema flexibility to represent cube cells in the value and dimensions in the key.

3.3 First Approach

This approach uses the tabular structure layered on the top of the key value API which allows implementing a physical model for the OLAP schema under key value database. We propose two logical models to store fact and associated dimensions (Fig. 2), and we provide examples of aggregation queries to build OLAP cubes:

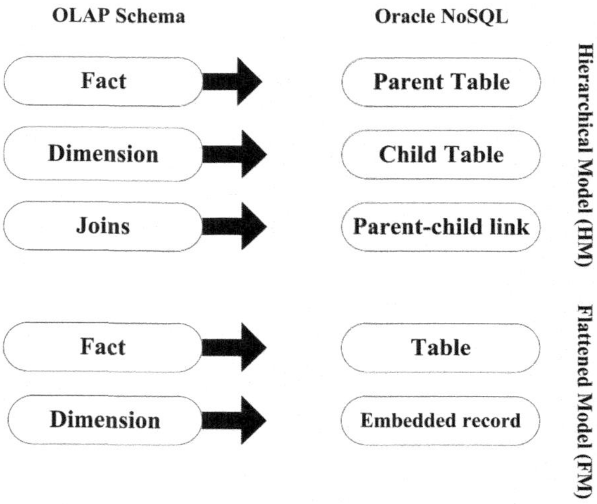

Fig. 2 Transformation rules from OLAP schema to Oracle NoSQL

3.3.1 Flattened Model (FM)

In this model, fact and associated dimensions are stored jointly in the same table at the same level using nested record, that is, where the flattened name comes from. Let us consider an OLAP schema defined by a triplet $(Name^{MS}, F, \langle D_1, ...D_n \rangle)$, where:

- $Name^{MS}$ is the name of the multidimensional schema.
- F is its fact containing n measures $\{m_1, ...m_n\}$.
- D_i is a dimension among the n dimension of MS with m parameters $\{p_1, ..., p_m\}$.

The SQL-like instruction to create this schema is:

```
CREATE TABLE IF NOT EXISTS FACT_TABLE (
id_fact Type^id,
m1  Type^m1,
...
mn  Type^mn,
PRODUCT_DIM record (
p1  Type^p1,
...
pm  Type^pm,
   )
PRIMARY KEY (id_fact)
)
```

As the entire OLAP schema is stored in one single table, there is no join operation needed. The query to build the OLAP cube has the following form:

```
SELECT PRODUCT_DIM.p1,…,PRODUCT_DIM.pm,AGGREGATION
FROM FACT_TABLE
WHERE RESTRICTIONS
GROUP BY PRODUCT_DIM.p1,…,PRODUCT_DIM.pm
```

3.3.2 Hierarchical Model (HM)

In this model, fact and dimensions are stored in separate tables and are linked with a parent–child relationship. Unfortunately, NoSQL systems do not have the concept of foreign key; thus, join operations are not supported, and tracking record relations is delegated to the client application layer; furthermore, generally, they do not provide aggregation operators. Oracle NoSQL overcomes this issue by providing two interesting features:

- Join operations are supported among table belonging to the same hierarchy.
- Support of basic aggregate functions like sum, count, avg...

Adopting the same use case and formalism the SQL-like instruction to create the schema has the following form:

```
CREATE TABLE IF NOT EXISTS FACT_TABLE (
```

```
id_fact Type^id,
m1  Type^m1,
....
mn  Type^mn,
PRIMARY KEY (id_fact)
)
CREATE TABLE IF NOT EXISTS FACT_TABLE.PRODUCT_DIM (
Id_prod Type^id,
m1  Type^m1,
....
mn  Type^mn,
PRIMARY KEY (id_prod)
)
```

PRODUCT_DIM inherits implicitly its parent table's primary key; as a result, the dimension table's primary key is composed of two field id_fact and id_prod.

We can build OLAP cube with the following way:

```
SELECT PRODUCT_DIM.p1,…,PRODUCT_DIM.pj,AGGREGATION
FROM NESTED TABLE (FACT_TABLE.PRODUCT_DIM  ANCESTORS(FACT_TABLE))
WHERE RESTRICTIONS
GROUP BY PRODUCT_DIM.p1,…,PRODUCT_DIM.pj
```

3.4 Second Approach

This approach is based on a principle; all DBMS extend key value database, and they can be represented as key value pair using hashing technique. We propose two data models to store OLAP cube within key value physical model: In the first model, we propose a formal representation for storing all possible cuboids, and the second one is based on an aggregation operator called *KV-operator*. It is noteworthy that key value database provides basic CRUD operation (Create, Read, Update, and Delete). Hence, complex data querying is delegated to the client application level.

3.4.1 From Relational Data Warehouse to Key Value Cube

This model allows storing aggregated measures according to dimensions. Assuming that the cube is a multidimensional structure built on the star schema, it has two main components: the aggregated value and axes of analysis (dimensions). To represent a dimension parameter we need to specify three components: a prefix which describe the dimension name, a suffix which define the parameter name and identifier which uniquely identify the dimension tuple. A dimension attribute can be modeled in key value structure as follow:

$\langle Name^D \rangle : \langle Key^D \rangle : \langle Param^D \rangle \Rightarrow Value^{Param}$, where:

- $Name^D$ is the dimension name.

- Key^D is the dimension key.
- $Param^D$ is a dimension attribute called parameter.
- $Value^{Param}$ is the parameter value.

Example:
Product: 12: name => "Xiaomi pro 12"
Product: 12: reference => "AF4589"
An OLAP cuboid can be represented by the following schema:
$\langle Name^C \rangle : \langle Key^{D_i} : ... : Key^{D_n} \rangle : \langle Name^M \rangle \Rightarrow Value^M$, where:

- $\langle Name^C \rangle$ is the name of the cuboid.
- $\langle Key^{D_i} : ... : Key^{D_n} \rangle$ is a set of dimension keys.
- $\langle Name^M \rangle$ is name of the aggregated value called measure.
- $Value^M$ is value of the measure.

Example: Sales: 12:458:56: Sum(quantity):8745.

3.4.2 KV-Operator for Key Value Store

The aim of this subsection is to propose an aggregation operator for key value-oriented DBMS. This operator allows to calculate OLAP cubes from data warehouses implemented under key value store in three steps.

Step 1. It consists in defining a view on the attributes (dimension and measures) necessary for the calculation of the OLAP cube. Only the values which satisfy the predicates are taken into account. The result obtained is ,therefore, a relation R composed of the keys which represent the axes of analysis and the value representing the measure to be aggregated. This relationship is an intermediate result to constitute all the parts of the OLAP cube. This strategy avoids the need to return to data warehouse for the calculation of aggregates. At this stage, the relation R already makes it possible to obtain the total aggregation and that according to all the columns representing the dimensions (Fig. 3).

Step 2. In this phase, each key dimension of the relation *R* is hashed with the values which compose it to obtain the list of the positions of these values. The values of these lists are binary, and they can correspond to "1" or "0"; the "1" indicates that the hash value exists at this position and "0" otherwise. These lists allow you to have the aggregates of each dimension separately (Fig. 4).

Step 3. In this phase, each key dimension of the relation R is hashed with the values which compose it to obtain the list of the positions of these values. The values of these lists are binary, they can correspond to "1" or, "0" and the "1" indicates that the hash value exists at this position and "0" otherwise. These lists allow you to have the aggregates of each dimension separately.

Key	Value
1 : 4 : 2 : 'quantity'	48
1 : 4 : 2 : 'price'	154.23
2 : 2: 4 : 'price'	125.00
3 : 3: 4 : 'price'	120.12
3 : 2: 4 : 'price'	155.00
3 : 2: 4 : 'quantity'	49

Extracting tuples that meets the query predicates →

Key	Value
1 : 4 : 2 : 'price'	154.23
2 : 2: 4 : 'price'	125.00
3 : 3: 4 : 'price'	120.12
3 : 2: 4 : 'price'	155.00

Fig. 3 Data extraction from data warehouse

Key	Value
1 : 4 : 2 : 'price'	154.23
2 : 2 : 4 : 'price'	125.00
3 : 3 : 4 : 'price'	120.12
3 : 2 : 4 : 'price'	155.00

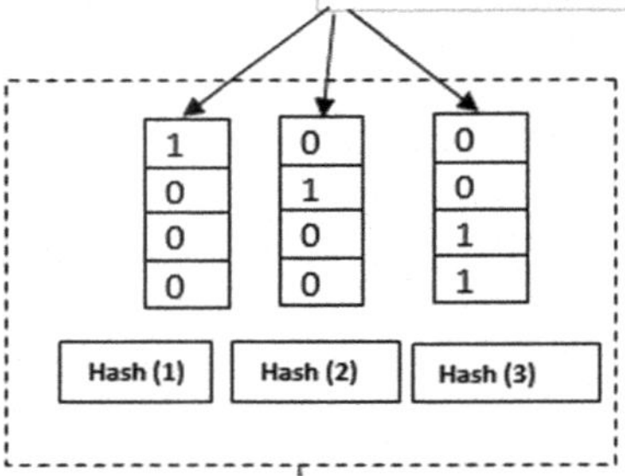

Key	Value
1 : all : all : 'price'	154.23
2 : all : all : 'price'	125.00
3 : all : all : 'price'	275.12

Fig. 4 Hashing positions and performing aggregations

4 Implementation

4.1 Experiments Setting

In what follow, we will give an overview about tools and technical concepts that is relevant to our implementation.

4.1.1 Data Generation

For data generation, we used an extended version of a decision support benchmark called the TCP benchmark [6] which is implemented in the form of Java 1.8 application and support different logical models (flat, star, snowflake, and flat) and multiple formats compatible with relational storage systems and NoSQL.

It is noteworthy that our objective is not to measure the performance of the database management system product used in this experiment, but to evaluate the performance of our modeling approach under key value store. For this, we refrain from using the ad hoc query load and we create custom queries allowing to perform basic OLAP operations.

4.1.2 Software/Hardware Architecture

In the experiment, we use Docker swarm to deploy oracle NoSQL as a container in two different configurations: single node and a cluster of 3 Docker images, each node having one copy of data. Docker container pulls the latest release of Oracle NoSQL database community edition. Our physical machine has an Intel Core i7 CPU with 16 GB of RAM running with Linux Debian distribution. The software setup is described in Fig. 5.

We implement our second approach using LevelDB which is an open source fast key value store written by Google. Staying true to DevOps approach, we use the existing image in Docker Hub to deploy an instance of LevelDB.

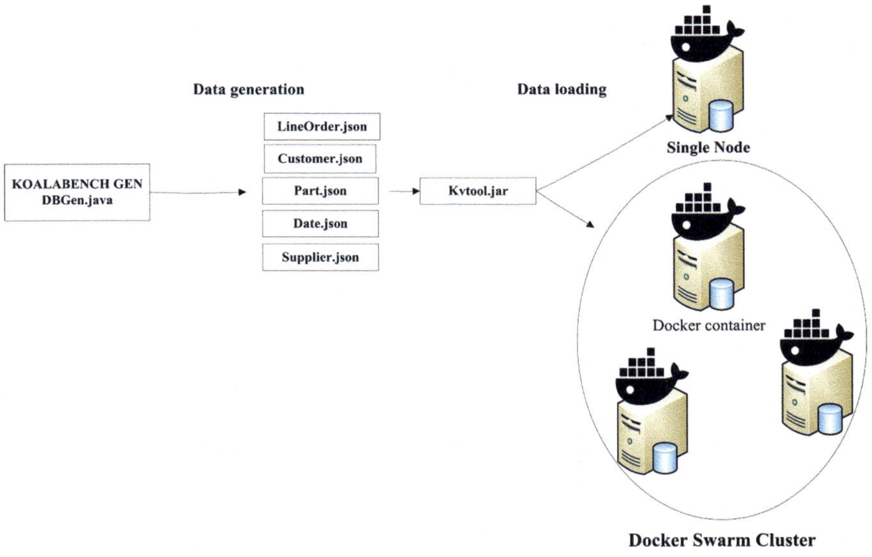

Fig. 5 Software architecture

4.1.3 Basic OLAP Cube Operations Using LevelDB.

In this subsection, we give an overview of basic OLAP operation that we can perform against a data cube stored within key value store, and we take the example of dicing and slicing operations.

A slice is a specialized filter that helps to visualize information according to a particular value in a dimension. Let us assume that we want to know the total revenue by product with name "DEVICE" sold across all market; with SQL, we would perform the following query:

SELECT article, SUM(revenue) FROM sales.

WHERE article = "*DEVICE*" GROUP BY article;

When using LevelDB, we have to implement the following algorithm to perform this operation:

1. *Integer get Total Revenue By Dimension (dimensionAttributeKey String)*
2. *Declare variable totalRevenue = 0*
3. *Get a keyIterator from levelDBStore and moves the iterator to the keys starting with "sales".*
4. *While iterator has next element do*
5. *Get and split the key with ':' separator into an array*
6. *If an array element is equal to dimensionAttributeKey do*
7. *totalRevenue = totalRevenue + value*
8. *end if*
9. *end while*
10. *return totalRevenue*
11. *end of function body*

Dicing is quite similar to slicing; slice operation focuses on a particular dimension attribute, while dicing select a subset over multiple dimensions. In this case, the algorithm we defined receives multiple input parameters representing dimensions' attributes concerned by the dice operation.

4.2 Result and Evaluation

Experiment 1

The objective of this experiment is to measure storage space metric consumed by each model in the first approach (flattened model FM and hierarchical model HM). For that, we generate an increasing volume of data using the TCP-H generator with three scale factor, respectively, sf = 1, sf = 5, and sf = 10. To describe the metadata model, we use Oracle NoSQL Avro schema which defines fields and their data type as shown below.

The result we obtained is shown in Fig. 7

Figure 7 depicts the storage space required by each model; we observe that the two model show a linear behavior when scaling up and a little difference in disk

```
kv-> show table -name lineItemHM
{
  "json_version" : 1,
  "type" : "table",
  "name" : "lineItemHM",
  "shardKey" : [ "lineNumber" ],
  "primaryKey" : [ "lineNumber" ],
  "fields" : [ {
    "name" : "lineNumber",
    "type" : "LONG",
    "default" : null
  }, {
    "name" : "quantity",
    "type" : "INTEGER",
    "default" : null
  }, {
    "name" : "extendedPrice",
    "type" : "DOUBLE",
    "default" : null
  }, {
    "name" : "discount",
    "type" : "DOUBLE",
    "default" : null
  }, {...}
  ]
}
```

```
kv-> show table -name lineItemFM
{
  "json_version" : 1,
  "type" : "table",
  "name" : "lineItemFM",
  "shardKey" : [ "lineNumber" ],
  "primaryKey" : [ "lineNumber" ],
  "fields" : [ {
    "name" : "lineNumber",
    "type" : "LONG",
    "nullable" : false,
    "default" : null
  }, {
    "name" : "quantity",
    "type" : "INTEGER",
    "nullable" : true,
    "default" : null
  },{
    "name" : "supplier_dim",
    "type" : "RECORD",
    "fields" : [ {
      "name" : "supplierKey",
      "type" : "LONG",
      "nullable" : true,
      "default" : null
    }, {
```

Fig. 6 Avro schema for fact table (HM and FM)

space between the two implementations; EM requires more space than HM, which is easily explained by height data redundancy in flattened model. It is noteworthy that data redundancy is very important in both models which is permitted and even encouraged in NoSQL architecture.

Experiment 2

The aim of this experiment is to evaluate query performance in term of elapsed time to process a query request; for that, we populate our database with a random dataset equivalent to sf = 10, and we gradually increase the number of dimensions involved in query criteria. The experiment is made with two setups: a single node and a cluster of three nodes. The result we obtained is shown in Fig. 8.

Figure 8 depicts the processing time for multidimensional queries performed against NoSQL data warehouse to build OLAP cubes. We observe that flattened model outperform in comparison with hierarchical model, the difference in query response time varied from 20 to 100% as the number of dimension increases. This observation is justified by the number of joins between tables belonging to the same hierarchy in HM.

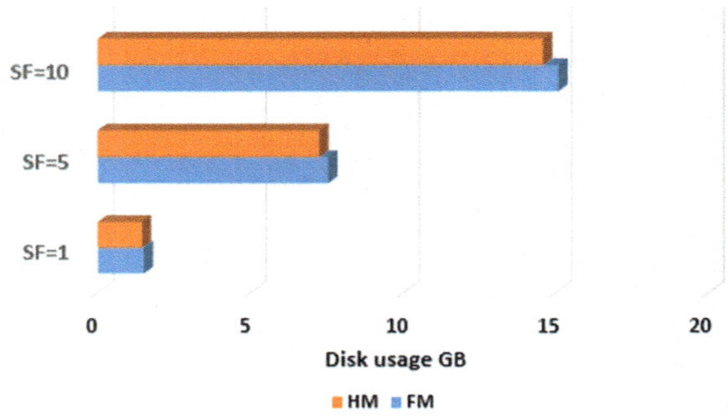

Fig. 7 Disk space by model and scale factor

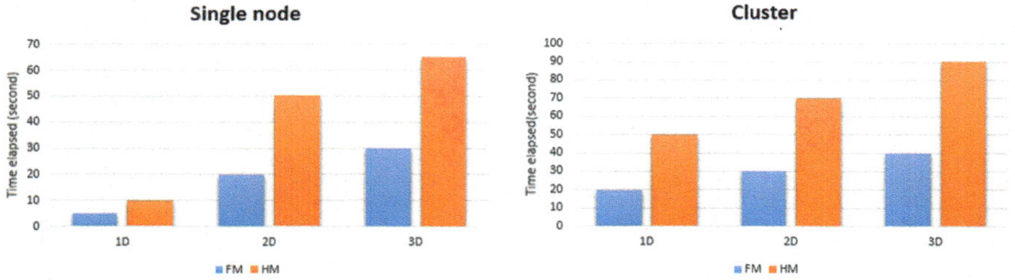

Fig. 8 Query processing elapsed time

5 Conclusion

In this work, we address the concern of using NoSQL databases in OLAP systems. Thus, we analyze two approaches in order to implement OLAP cubes under key value store. The first one is based on a tabular data structure layered on the top of key value schema, and the second one uses key value API. In the first approach, we propose two logical models called flattened model and hierarchical model. Experiences show that FM offers better query performance but requires more disk space. In the other hand, in the second approach, we propose transformation rules from relational data warehouse to NoSQL OLAP cube and implement an algorithm to perform basic OLAP operations against the cube. Finally, we present our aggregation operator to compute OLAP cube from data warehouse implemented under key value store.

This paper offers interesting perspectives for implementing OLAP systems using NoSQL technologies. Thus, future work will provide an extended experimental campaign to measure the performance of our proposed KV-operator in comparison with existing solutions. We look ahead also to use MapReduce framework and Spark for OLAP cube computation.

References

1. Castelltort A, Laurent A (2014) NoSQL graph-based OLAP analysis. In: KDIR 2014—Proceedings of the international conference on knowledge discovery and information retrieval. pp 217–224. https://doi.org/10.5220/0005072902170224
2. Chaudhuri S, Dayal U, Ganti V (2001) Database technology for decision support systems. Computer 34(12). https://doi.org/10.1109/2.970575.
3. Chavalier M et al (2016) Document-oriented data warehouses: models and extended cuboids, extended cuboids in oriented document. In: Proceedings—International conference on research challenges in information science. 2016-August. https://doi.org/10.1109/RCIS.2016.7549351
4. Chevalier M et al (2009) Implementation of multidimensional databases in column-oriented NoSQL systems. To cite this version: HAL Id: hal-01363342'
5. Chevalier M et al (2015) Benchmark for OLAP on NoSQL technologies comparing NoSQL multidimensional data warehousing solutions. In: Proceedings—International conference on research challenges in information science. pp 480–485. https://doi.org/10.1109/RCIS.2015.7128909
6. Chevalier M et al (2017) Un benchmark enrichi pour l'évaluation des entrepôts de données noSQL volumineuses et variables. Eda
7. Dehdouh K et al (2014) Columnar NoSQL cube: agregation operator for columnar NoSQL data warehouse. In: Conference Proceedings—IEEE international conference on systems, man and cybernetics. pp 3828–3833. https://doi.org/10.1109/SMC.2014.6974527
8. Duda J (2012) Business intelligence and NoSQL databases. Inf Syst Manage 1(1):25–37. https://doi.org/10.1080/07399018408963028
9. Han J et al (2011) Survey on NoSQL database. In: Proceedings—2011 6th international conference on pervasive computing and applications, ICPCA 2011. pp 363–366. https://doi.org/10.1109/ICPCA.2011.6106531
10. Kurpanik J (2017) Nosql databases as a data warehouse for decision support systems. J Sci Gen Tadeusz Kosciuszko Military Acad Land Forces 185(3):124–131. https://doi.org/10.5604/01.3001.0010.5128
11. Loyola L, Wong F, Pereira D (2012) Building OLAP data analytics by storing path-enumeration keys into sorted sets of key-value store databases. In: Data analytics 2012, the first international conference on data analytics, pp 61–70
12. Rabl T et al (2014) Advancing big data benchmarks: proceedings of the 2013 workshop series on big data benchmarking WBDB.cn, Xi'an, China, July 16–17, 2013 and WBDB.us, San José, CA, USA October 9–10, 2013 Revised Selected Papers', Lecture Notes in Computer Science (including subseries Lecture Notes in Artificial Intelligence and Lecture Notes in Bioinformatics), 8585, pp 155–170. doi: https://doi.org/10.1007/978-3-319-10596-3.

Chapter 14
Multi-objective Quantum Moth Flame Optimization for Clustering

Yassmine Soussi, Nizar Rokbani, Ali Wali, and Adel M. Alimi

1 Introduction

Clustering consists in dividing a dataset into k clusters, and this is based on the similarities of objects belonging to each cluster [15]. The clustering problem can be defined as a mono-objective optimization problem with one objective function presented by one cluster validity criteria.

Taking into account that the classification problems have become more complex and different cluster validity criteria have appeared, it has therefore become necessary to consider the clustering as a multi-objective optimization problem [7, 22] with different objective functions presented by different cluster validity criteria.

Various multi-objective clustering techniques in the literature have been proposed to solve the multi-objective clustering optimization problem including: The multi-objective clustering with automatic determination of k (k is the number of clusters), **MOCK**, proposed by [20]; the variable string length point symmetry-based, **VGAPS**, proposed by [5]; the two multi-objective simulated annealing methods,

Y. Soussi (✉)
University of Sousse, ISITCom, 4011 Sousse, Tunisia
e-mail: yassmine.soussi@ieee.org

Y. Soussi · N. Rokbani · A. Wali · A. M. Alimi
University of Sfax, National Engineering School of Sfax (ENIS), REGIM-Lab.: REsearch Groups in Intelligent Machines, LR11ES48, 3038 Sfax, Tunisia
e-mail: nizar.rokbani@ieee.org

A. Wali
e-mail: ali.wali@ieee.org

A. M. Alimi
e-mail: adel.alimi@ieee.org

N. Rokbani
University of Sousse, High Institute of Applied Science and Technology of Sousse, Sousse, Tunisia

A. E. Hassanien et al. (eds.), *Enabling Machine Learning Applications in Data Science*, Algorithms for Intelligent Systems, https://doi.org/10.1007/978-981-33-6129-4_14

GenClustMOO and **GenClustPESA2**, proposed by [25]; the multi-objective clustering algorithm based on artificial bee optimization, **cOptBees-MO**, proposed by [7]; the multi-objective particle swarm optimization based on simulated annealing algorithm, **MOPSOSA**, proposed by [1]. These multi-objective clustering techniques are used as comparatives algorithms with the proposal.

In this study, MOQMFO algorithm used three cluster validity indices as objective functions: the **I-Index** [7, 25], the **Sym-Index** [4, 7] and the **Con-Index** [7, 24]; the optimization of these objectives functions aims to detect the correct number of clusters and find a good clustering solution for each dataset. Here, **F-measure** [12, 25] is the external validation measure used to evaluate and compare the final clustering solutions obtained by MOQMFO and its competitors.

This paper is organized as follow: In Sect. 2, MFO algorithm and QMFO algorithm were presented. In Sect. 3, the multi-objective form of QMFO (MOQMFO) was introduced. In Sect. 4, an application of MOQMFO in the multi-objective clustering optimization domain was made. Section 5 presents MOQMFO flowchart. In Sect. 6, some experiments have been carried out. And finally, in Sect. 7, a conclusion with some perspectives was performed.

2 The Quantum-Behaved MFO (Q-MFO)

2.1 Moth Flame Optimization (MFO)

MFO is a new metaheuristic population-based method, and it was proposed by [21] based on the flying mode of moths while looking for light sources (flames). In a given problem, the positions of moths and flames are considered solutions. Moths present the actual solutions (actual positions), while flames present the best solutions (best positions). The numbers of moths "nMoths" and the number of flames "nFlames" are equal (nMoths = nFlames). When moving to a flame, moth changes its position using a logarithmic-spiral-function; the position of a moth is then updated using Eq. (1).

$$S\left(\text{Moth}_i, \text{Flame}_j\right) = D_i.\text{e}^{\text{bt}}.\text{Cos}(2\pi t) + \text{Flame}_j \tag{1}$$

- S defines the spiral function used by the moth while moving;
- D_i is the Euclidian distance between the i-th moth and the j-th flame, illustrated in Eq. (2).

$$D_i = \left|F_j - M_i\right| \tag{2}$$

- b is a constant for establishing the logarithmic-spiral form;
- t is a number chosen randomly from $[-1, 1]$.

MFO parameters are simple and do not require a large adjustment that makes MFO algorithm simple, easy to implement and robust.

In the rest of this document, we combined the quantum theory [9, 17, 30, 31] into MFO and proposed a quantum-behaved MFO (QMFO) algorithm. Next section will describe the application of the quantum technique in MFO.

2.2 Quantum-Behaved MFO (QMFO)

The quantum mechanics is a new area that has been influenced by several factors namely: The atomic theory [6] proposed by Bohr in 1928; the quantum mechanics [14] discovered by Heisenberg; the wave mechanics [29] discovered by Schrödinger. According to Feynman [10], the simulation of quantum mechanical systems is continuous and fast with quantum computers than with ordinary computers. Following the evolution of quantum computing applications over the past three decades, the concept of quantum computing, once thought to be a mere theoretical alternative, has become a real alternative [16].

The methods of evolutionary computation and swarm intelligence are classifications of population-based methods known as optimization algorithms. These methods have become increasingly in demand in research communities over the past 20 years. In 1995, particle swarm optimization (PSO) technique was introduced by [18] as an optimization algorithm, and it was inspired by the social behavior of particles (a swarm is a group of particles).

Taking into consideration the creation of quantum-based optimization techniques inspired by the combination of three areas: the quantum computing, the mathematics and the computer science; several hybridization of quantum theory and optimization algorithms have been developed namely the quantum particle swarm optimization (QPSO) how is a variant of particle swarm optimization proposed in [26–28]. By referring to these previous studies, it has been proven that QPSO shows good convergence and good performance compared to PSO. In [17], authors explored the applicability of QPSO in the data clustering domain and proved that QPSO performance is better than PSO performances and that is because the global convergence behavior of QPSO.

In this paper, quantum-behaved MFO (QMFO) is a search technique inspired by QPSO. The new quantum equations introduced in QMFO are gathered to those used in QPSO clustering algorithm [17]. As QPSO, QMFO guarantees global convergence and makes moths more dynamic with wave function instead of their positions and the positions of theirs corresponding flames. A new variable is introduced called mean best position "**mbest**"; this variable has a role in improving Q-MFO functionality.

Since the flames present the best "**pbest**" positions of the moths, the best pbest of these flames, called the best overall "**gbest**", and the best average flame among these flames, called "**mbest**", are calculated from the set of flames using Eq. (3).

$$\text{mbest} = \frac{1}{\text{nFlames}} \sum_{j=1}^{\text{nFlames}} (\text{Flame}(j)) \tag{3}$$

In QMFO, the moth moves and changes its position using Eq. (5).

$$\text{Flame}_j = \phi \cdot \text{Flame}_j + (1 - \phi) \cdot \text{gbest} \tag{4}$$

$$\text{Moth}_i = D_i e^{\text{bt}} \cdot \text{Cos}(2\pi t) + \text{Flame}_j \pm \alpha \cdot |\text{mbest} - \text{Moth}_i| \cdot \ln\left(\frac{1}{u}\right) \tag{5}$$

The convergence behavior of the moths in QMFO is influenced by:

- The new Flame_j calculated in Eq. (4), who is a randomly determined point between Flame_j and gbest, is the attractor of moth_i;
- ϕ and u which are two random numbers distributed uniformly in the interval [0,1];
- α, named "Contraction–Expansion-Coefficient", who is a relative parameter to QMFO algorithm.

3 Multi-Objective QMFO (MOQMFO)

MOQMFO is the multi-objective form of QMFO algorithm, it combines the characteristics of the QMFO and the multi-objective optimization [8, 11].

MOQMFO uses the concept of "non-dominance-Pareto" to store the best solutions obtained during each generation into a "Repository": The repository is a set of non-dominated solutions called "pareto set solutions".

4 MOQMFO for Clustering

In the multi-objective clustering optimization [22] area, a moth or flame is presented by a matrix ($k*d$); k and d present, respectively, the number of clusters and the number of features related to the dataset: As showed in Eq. (6), each line-vector in this matrix presents a cluster center.

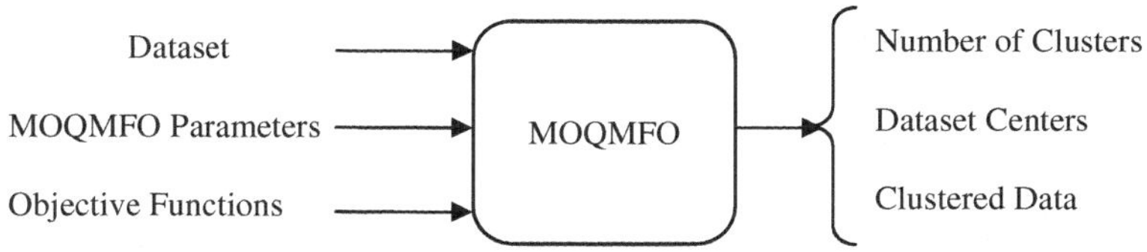

Fig. 1 MOQMFO architecture

$$\text{Moth}_i = \begin{bmatrix} C_{i1} \\ \cdot \\ \cdot \\ \cdot \\ C_{ij} \\ \cdot \\ \cdot \\ \cdot \\ C_{ik} \end{bmatrix}, i = [1, .., \text{nMoths}] \quad (6)$$

To calculate the fitness of each moth or flame, a combination of three objective functions is used. In our case, clusters validity index (the **I-Index** [7, 25], the **Sym-Index** [4, 7] and the **Con-Index** [7, 24] is used as objectives functions.

The architecture of our proposed methodology MOQMFO for clustering is presented in Fig. 1.

5 MOQMFO Flowchart

After selecting the range of values of k (number of clusters) related to the dataset using MFO-based mean-distance clustering algorithm, the MOQMFO algorithm was run n (n = number of runs) times for each value of k, and in each run, the optimal "F-Measure" as well as its corresponding number of clusters and optimal position of clusters centers "Best-Position" are selected and then stored. In the end of all runs, a set of n values of F-measure with its corresponding number of clusters and positions of clusters centers is obtained; the higher value of F-measure, with its corresponding "Best-Position" of clusters centers and its corresponding value of k, is then selected from this set.

The flowchart of the proposed method, MOQMFO for clustering, is detailed in Fig. 2. MOQMFO flowchart is based on four phases: the initialization, the optimization, the decision making1 and the decision making2.

The second and third phases are repeated n times (n equals to MaxIteration).

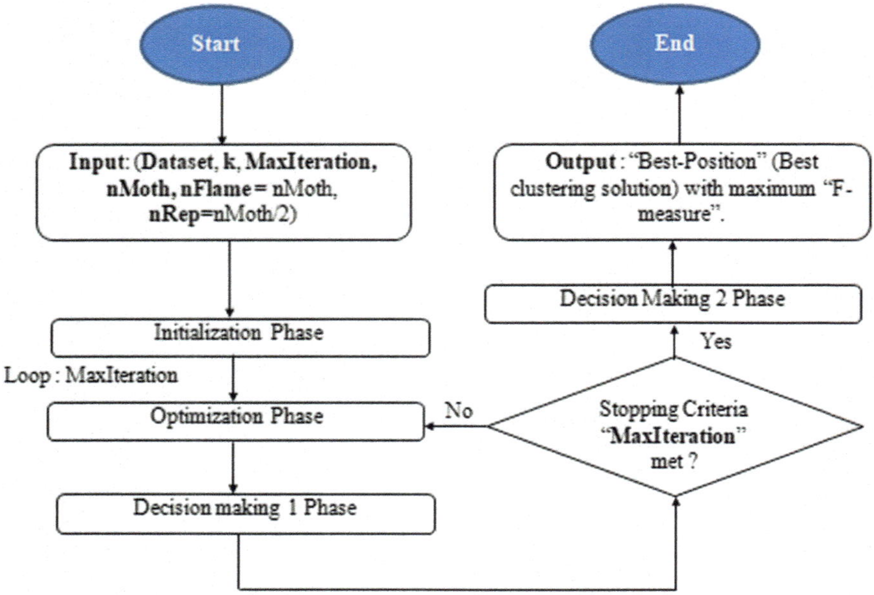

Fig. 2 MOQMFO flowchart

As input, we have the dataset, the number of clusters (k), the number of moths (nMoth), number of Flames (nFlame), the maximum number of iteration (MaxIteration) and the repository size (nRep).

The flowchart of the "**Initialization**" phase, showed in Fig. 3, consists in generating the initials moths, flames and repository members.

The "**Optimization**" phase consists in updating the moths and flames positions and fill the repository.

The **"Decision Making1"** phase (red blocs) consists in selecting the best position with the best cost, in the pareto set solutions produced in each iteration, then

Fig. 3 MOQMFO initialization phase flowchart

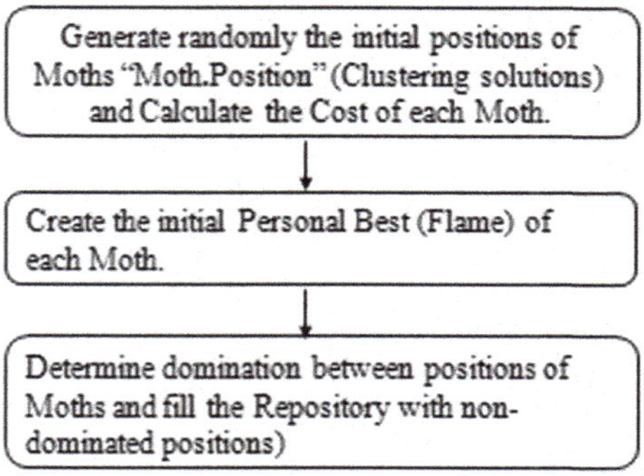

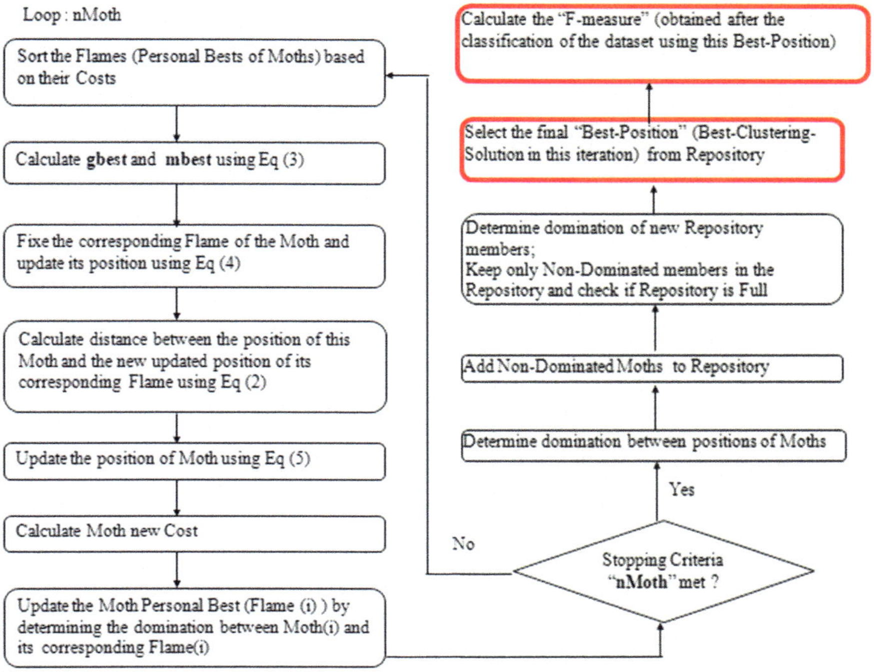

Fig. 4 Flowchart of the **"Optimization"** phase and **"Decision Making1"** phase of the MOQMFO algorithm

calculating the F-measure obtained after the classification of the dataset using this best position which represents the clusters centers positions;

Figure 4 illustrates the flowchart of the **"Optimization"** phase and **"Decision Making1"** phase of the MOQMFO algorithm.

The **"Decision making2"** phase: consist in selecting the best position with the best "F-measure" in the set of "F-measure" produced in the end of all iterations.

6 Experiments

6.1 Datasets

In the experimentations part, 20 datasets, including **Iris** [19], **Cancer** [19], **New thyroid** [19], **Wine** [19], **Liver Disorder** [19], **Glass** [19], **Sph_5_2** [2], **Sph_4_3** [2], **Sph_6_2** [2], **Sph_10_2** [3], **Sph_9_2** [3], **Pat 1** [23], **Pat 2** [23], **Long 1** [13], **Sizes 5** [13], **Spiral** [13], **Square 1** [13], **Square 4**, **Twenty** [13] and **Forty** [13], are used to evaluate the performances of MOQMFO and the competing algorithms.

Table 1 Description of datasets

	Dataset	# Points	Dimension	# Clusters
Real-life datasets	Iris	150	4	3
	Cancer	683	9	2
	New thyroid	215	5	3
	Liver disorder	345	6	2
	Glass	214	9	6
	Wine	178	13	3
Artificial datasets	Sph_5_2	250	2	5
	Sph_4_3	400	3	4
	Sph_6_2	300	2	6
	Sph_10_2	500	2	10
	Sph_9_2	900	2	9
	Pat 1	557	2	3
	Pat 2	417	2	2
	Long 1	1000	2	2
	Sizes 5	1000	2	4
	Spiral	1000	2	2
	Square 1	1000	2	4
	Square 4	1000	2	4
	Twenty	1000	2	20
	Forty	1000	2	40

These datasets are divided into two types: real-life and artificial datasets. In Table 1, there is a representation of these datasets: number of cases (#points), number of features (dimension) and number of clusters, corresponding to each one.

6.2 *Scenario*

MOQMFO was executed 30 times with the input parameters shown in Table 2.

Table 2 MOQMFO Parameters

Parameter	Value
Number-moths	50
Number-flames	50
Repository-size	25
Number-iterations	100

F-measure [12, 25] is the external validation measure used to evaluate and compare the final clustering solutions obtained by MOQMFO and the algorithms used for comparison (GenClustMOO, MOCK, VGAPS, GenClustPESA2, cOptBees-MO and MOPSOSA).

Higher values of F-measure imply better clustering, and when F-measure reaches the value 1, the clustering is optimal.

6.3 Results and Discussion

The MOQMFO algorithm was applied with six real-life datasets and 14 artificial datasets. Comparisons are based on F-measure values and number of clusters (k) obtained with MOQMFO and its competitors.

Obtained results, illustrated in Table 3, show that MOQMFO had a competitive performance when compared with the other multi-objective clustering techniques.

According to Table 3, MOQMFO algorithm provided the real number of clusters for all real-life and artificial datasets.

Compared with its competitors, MOQMFO gave the best values of F-measure with the following datasets: Iris (F-Measure = 1), Newthyroid (F-Measure = 0.96), LiverDisorder (F-Measure = 0.75), Wine (F-Measure = 0.91), Sph_4_3 (F-Measure = 1), Sph_6_2 (F-Measure = 1), Long 1 (F-Measure = 1), Sizes 5 (F-Measure = 0.99), Twenty (F-Measure = 1) and Forty (F-Measure = 1).

However, MOQMFO provided acceptable values of F-measure for the following cases: Cancer (F-Measure = 0.91), Glass (F-Measure = 0.60), Sph_5_2 (F-Measure = 0.96), Sph_10_2 (F-Measure = 0.96), Sph_9_2 (F-Measure = 0.80), Pat 1 (F-Measure = 0.77), Pat 2 (F-Measure = 0.82), Spirale (F-Measure = 0.65), Square 1 (F-Measure = 0.97) and Square 4 (F-Measure = 0.92).

The clustering results of the artificial datasets based on MOQMFO clustering algorithm are presented in Fig. 5: Sph_5_2 presented in Fig. 5a, Sph_4_3 illustrated in Fig. 5b, Sph_6_2 showed in Fig. 5c, Sph_10_2 presented in Fig. 5d, Sph_9_2 illustrated in Fig. 5e, Pat 1 depicted in Fig. 5f, Pat 2 illustrated in Fig. 5g, Long 1 showed in Fig. 5h, Sizes 5 presented in Fig. 5i, Spiral showed in Fig. 5j, Square 1 illustrated in Fig. 5k, Square 4 depicted in Fig. 5l, Twenty presented in Fig. 5m, and Forty illustrated in Fig. 5n.

7 Conclusion and Perspectives

In this paper, a hybridization of quantum theory and the moths flames (QMFO) optimization algorithm is proposed, and the multi-objective form of QMFO (MOQMFO) is then applied in multi-objective clustering optimization area.

To evaluate the performance of MOQMFO, a scenario based on three cluster validity criteria (I-index, Con-index and Sym-index), used as objective functions,

Table 3 F-measure values and number of clusters obtained with MOQMFO and the competing algorithms, best values are bold

Dataset	#Clusters	MOPSOSA		GenClustMOO		GenClustPESA2		MOCK		VGAPS		cOptBee-MO		MOQMFO	
		k	F-Measure	*k*	F-Measure	*k*	F-Measure	*k*	F-Measure	*k*	F-Measure	*k*	F-Measure	*k*	F-Measure
Iris	3	3	0.92	3	0.79	3	0.93	2	0.78	3	0.76	3	**0.86**	3	**1.00**
Cancer	2	2	**0.98**	2	0.97	2	**0.98**	2	0.82	2	0.95	3	0.94	2	0.91
New thyroid	3	3	0.89	3	0.86	9	0.69	2	0.74	5	0.66	4	0.86	3	**0.96**
Liver Disorder	2	2	0.69	2	0.67	5	0.60	2	0.67	2	0.70	2	0.67	2	**0.75**
Glass	6	6	0.57	6	0.49	5	0.53	5	0.53	5	0.53	2	**0.88**	6	0.60
Wine	3	–	–	3	0.71	13	0.44	3	0.73	6	0.62	2	0.65	3	**0.91**
Sph_5_2	5	5	**0.98**	5	0.97	5	0.94	6	0.91	5	0.55	–	–	5	0.96
Sph_4_3	4	4	**1.00**	4	**1.00**	4	**1.00**	4	**1.00**	4	**1.00**	–	–	4	**1.00**
Sph_6_2	6	6	**1.00**	6	**1.00**	6	**1.00**	6	**1.00**	6	**1.00**	–	–	6	**1.00**
Sph_10_2	10	10	**0.99**	10	**0.99**	12	0.94	6	0.72	7	0.76	–	–	10	0.96
Sph_9_2	9	9	**0.92**	9	0.69	8	0.66	9	0.73	9	0.49	–	–	9	0.80
Pat 1	3	3	**1.00**	3	0.95	3	0.95	10	0.55	4	0.42	–	–	3	0.77
Pat 2	2	2	**1.00**	2	**1.00**	2	**1.00**	11	0.55	4	0.59	–	–	2	0.82
Long 1	2	2	**1.00**	2	**1.00**	2	**1.00**	2	**1.00**	3	0.50	–	–	2	**1.00**
Sizes 5	4	4	0.98	4	0.97	3	0.88	2	0.80	5	0.82	–	–	4	**0.99**
Spiral	2	2	**1.00**	2	**1.00**	2	**1.00**	3	0.95	6	0.38	–	–	2	0.65
Square 1	4	4	**0.99**	4	**0.99**	4	**0.99**	4	**0.99**	4	**0.99**	–	–	4	0.97
Square 4	4	4	**0.94**	4	0.92	4	0.88	4	0.90	2	0.93	–	–	4	0.92
Twenty	20	20	**1.00**	20	**1.00**	24	0.95	20	**1.00**	20	0.48	–	–	20	**1.00**
Forty	40	40	**1.00**	40	**1.00**	40	0.98	40	**1.00**	2	0.10	–	–	40	**1.00**

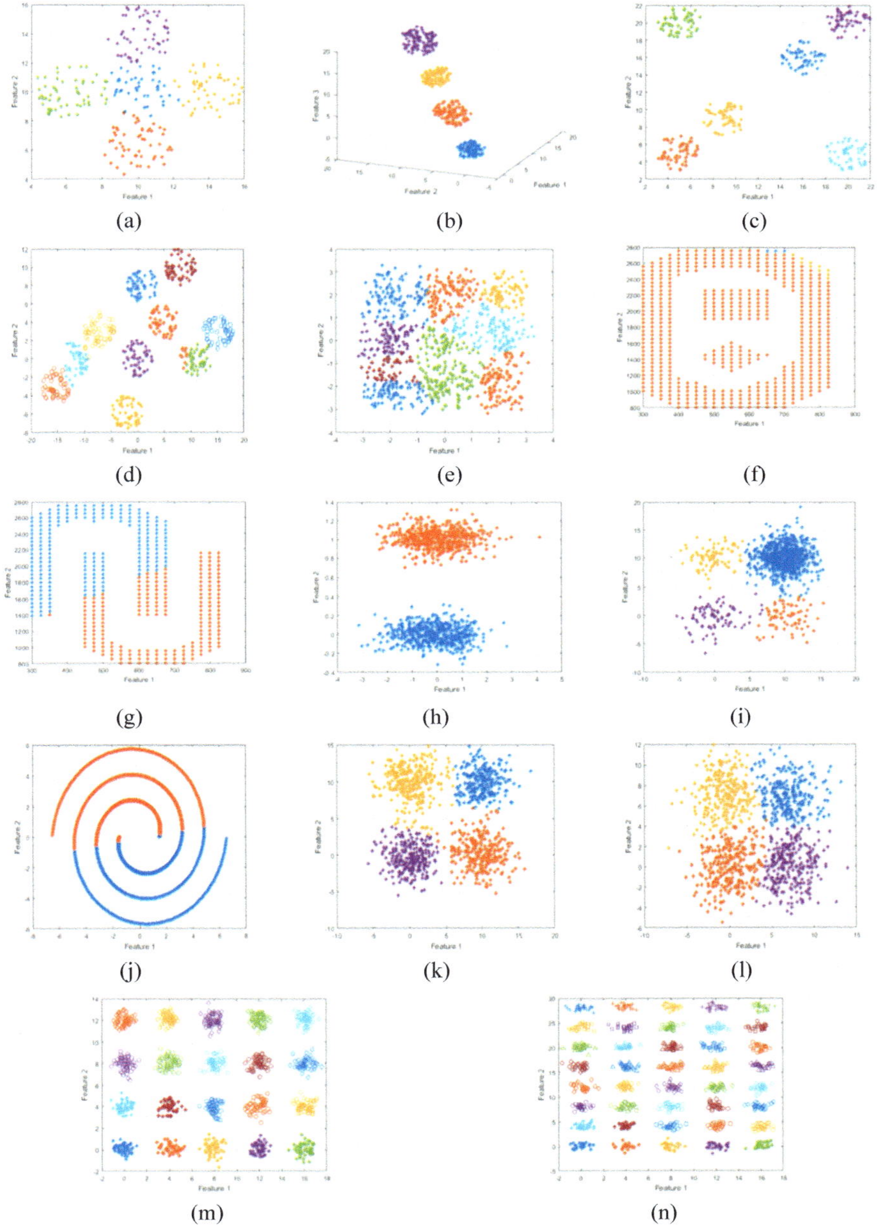

Fig. 5 Graphical representations of artificial datasets based on MOQMFO clustering technique

is carried out. Twenty datasets (including Iris, Cancer, New thyroid, Wine, Liver Disorder, Glass, Sph_5_2, Sph_4_3, Sph_6_2, Sph_10_2, Sph_9_2, Pat 1, Pat 2, Long 1, Sizes 5, Spiral, Square 1, Square 4, Twenty and Forty) and six multi-objective algorithms (GenClustMOO, MOCK, VGAPS, GenClustPESA2, cOptBees-MO and MOPSOSA) are used in this scenario. F-measure is the metric used in the evaluation of MOQMFO performances.

The results provided in the experiments part show that MOQMFO had a high capacity to provide an optimal clustering solution, with the correct number of clusters over all datasets, and a competitive F-measure value comparing with other multi-objective clustering optimization algorithms.

Best results of MOQMFO are provided in Iris, New thyroid, Wine, Liver Disorder, Sph_4_3, Sph_6_2, Long 1, Sizes 5, Twenty and Forty datasets.

Our perspectives in this subject consist in improving MOQMFO performances on Glass, Sph-9–2, Pat 1, Pat 2 and Spiral datasets, using other cluster validity criteria as objective functions and introducing other new techniques for the MFO algorithm and apply them in the multi-objective clustering optimization area.

References

1. Abubake A, Baharum A, Alrefaei M (2015) Automatic clustering using multi-objective particle swarm and simulated annealing. PloS ONE 10(7)
2. Bandyopadhyay S, Maulik U (2002) Genetic clustering for automatic evolution of clusters and application to image classification. Pattern Recogn 35(6):1197–1208
3. Bandyopadhyay S, Pal SK (2007) Classification and learning using genetic algorithms: applications in bioinformatics and web intelligence. Springer Science & Business Media
4. Bandyopadhyay S, Saha S (2007) GAPS: A clustering method using a new point symmetry-based distance measure. Pattern Recogn 40(12):3430–3451
5. Bandyopadhyay S, Saha S (2008) A point symmetry-based clustering technique for automatic evolution of clusters. IEEE Trans Knowl Data Eng 20(11):1441–1457
6. Bohr N (1928) The quantum postulate and the recent development of atomic theory1. Nature 3050(121):580–590
7. Cunha D, Cruz D, Politi A, de Castro L (2017) Bio-inspired multiobjective clustering optimization: A survey and a proposal. Artif Intell Res 6(2):10–24
8. Deb, K. (2001). Multi-objective optimization using evolutionary algorithms. Wiley, p 16
9. Dereli S, Köker R (2020) A meta-heuristic proposal for inverse kinematics solution of 7-DOF serial robotic manipulator: quantum behaved particle swarm algorithm. Artif Intell Rev 53(2):949–964
10. Feynman RP (1982) Simulating physics with computers. Int J Theor Phys 21(6–7):467–488
11. Freitas AA (2004) A critical review of multi-objective optimization in data mining: a position paper. ACM SIGKDD Explor Newsl 6(2):77–86
12. Fung BC, Wang K, Ester M (2003) Hierarchical document clustering using frequent itemsets. In: Proceedings of the 2003 SIAM international conference on data mining. Society for Industrial and Applied Mathematics, pp 59–70
13. Handl J, Knowles J (2007) An evolutionary approach to multiobjective clustering. IEEE Trans Evol Comput 11(1):56–76
14. Heisenberg PR (1929) The uncertainty principle. Phys Rev 34(1):163–164
15. Jain AK (2010) Data clustering: 50 years beyond K-means. Pattern Recogn Lett 31(8):651–666
16. Jiacun W (2012) Handbook of finite state based models and applications. CRC

17. Jun S, Wenbo X, Bin Y (2006) Quantum-behaved particle swarm optimization clustering algorithm. In: International conference on advanced data mining and applications, pp 340–347
18. Kennedy J, Eberhart R (1995) Particle swarm optimization. In: Proceedings of ICNN'95-international conference on neural networks IEEE, vol 4, pp 1942–1948
19. Lichman M (2013) UCI machine learning repository
20. Matake N, Hiroyasu T, Miki M, Senda T (2007) Multiobjective clustering with automatic k-determination for large-scale data. In: Proceedings of the 9th annual conference on genetic and evolutionary computation, pp 861–868
21. Mirjalili S (2015) Moth-flame optimization algorithm: A novel nature-inspired heuristic paradigm. Knowl-Based Syst 89:228–249
22. Mukhopadhyay A, Maulik U, Bandyopadhyay S (2015) A survey of multiobjective evolutionary clustering. ACM Comput Surv 47(4):1–46
23. Pal SK, Mitra S (1994) Fuzzy versions of Kohonen's net and MLP-based classification: performance evaluation for certain nonconvex decision regions. Inf Sci 76(3–4):297–337
24. Saha S, Bandyopadhyay S (2012) Some connectivity based cluster validity indices. Appl Soft Comput 12(5):1555–1565
25. Saha S, Bandyopadhyay S (2013) A generalized automatic clustering algorithm in a multiobjective framework. Appl Soft Comput 13(1):89–108
26. Sun J, Feng B, Xu W (2004) Particle swarm optimization with particles having quantum behavior. In: Proceedings of the 2004 congress on evolutionary computation (IEEE Cat. No. 04TH8753), vol 1, pp 325–331
27. Sun J, Xu W, Feng B (2004) A global search strategy of quantum-behaved particle swarm optimization. In: IEEE conference on cybernetics and intelligent systems, vol. 1, pp 111-116
28. Sun J, Xu W, Feng B (2005) Adaptive parameter control for quantum-behaved particle swarm optimization on individual level. In: 2005 IEEE international conference on systems, man and cybernetics, vol 4, pp 3049–3054
29. Wessels L (1979) Schrödinger's route to wave mechanics. Stud Hist Philos Sci Part A 10(4):311–340
30. Yang S, Wang M (2004) A quantum particle swarm optimization. In: Proceedings of the 2004 congress on evolutionary computation (IEEE Cat. No. 04TH8753), vol. 1, pp 320–324
31. Yu C, Heidari AA, Chen H (2020) A quantum-behaved simulated annealing enhanced moth-flame optimization method. Appl Math Model

Chapter 15
On Optimizing the Visual Quality of HASM-Based Streaming—The Study the Sensitivity of Motion Estimation Techniques for Mesh-Based Codecs in Ultra High Definition Large Format Real-Time Video Coding

Khaled Ezzat, Ahmed Tarek Mohamed, Ibrahim El-Shal, and Wael Badawy

1 Introduction

The paper purposed illustrates and compares the four most iconic three step search motion estimation algorithms and its impact in the scope of the mesh-based technique in ultra high-resolution video sequences. Motion estimation is critical in definition of moving objects and frame difference. A comparative result for a real practical experiment using raw video sequences is presented, comparing different motion estimation algorithms, from the ordinal three step search, leap three step search, grid three step search to diamond three step search, which usually applied on 8×8 blocks.

However, a larger block sizes will be investigated in conjunction with the large video formats existing today mainly for two reasons decreasing the computational power or time needed and also reducing the size of motion estimation vector, which reflected as much lower video size compared to other smaller block sizes for either saving or streaming purposes. The proposed experiments are using the adaptive

K. Ezzat (✉) · A. Tarek Mohamed · I. El-Shal
School of Information Technology and Computer Science, Center of Informatics, Nile University, 26th of July Corridor, City of Sheikh Zayed, Giza, Egypt
e-mail: Kh.Ezzat@nu.edu.eg

A. Tarek Mohamed
e-mail: Ahm.Tarek@nu.edu.eg

I. El-Shal
e-mail: i.hamed@nu.edu.eg

W. Badawy
School of Engineering and Technology, Badr University in Cairo, University road, Badr City, Cairo 11829, Egypt
e-mail: wael@waelbadawy.com

A. E. Hassanien et al. (eds.), *Enabling Machine Learning Applications in Data Science*, Algorithms for Intelligent Systems, https://doi.org/10.1007/978-981-33-6129-4_15

mechanism criteria that have been used in HASM in splitting structures as 128, 265, 512 and 1024 with adaptive criteria based on the Minimum Sum Absolute Difference, for detecting the grid, which has more motion and targeting it for further splitting, which is ideal mechanism in reducing the size of the video sequence with the same visual and statistical video quality needed, the statistical video quality is also measured by Peak Signal to Noise Ratio (PSNR) between predicted and the original frame, and compared to in encoding in H.264 and H.265. The performance results show the superiority of the HASM performance when optimized.

The rest of this paper is organized into five more sections. In Sect. 2, we will provide a summary of the related work done in this field over the past years. In Sect. 3, we will introduce comparative algorithms details, and Sect. 4 introduces the adapted video sequences as well as it introduces the experimental results and metrics. In the last section, we will provide our conclusion.

2 Literature Review

The hierarchical adaptive structure mesh "HASM" was introduced by the PI in 2000 as efficient scalable technique for video encoding [1] followed by a sequence of patents to protect the commercial right of those technologies [2–5]. HASM has been used as foundation for motion tracking and protected by consequent patents [6–9]. It has been used in different applications for leak detection, human detection on belts and protected under patents [10–13], respectively. Several architectures have been proposed [14–18]. A multiplication free algorithm has been proposed [19]. However, HASM was not efficient in modeling large video formats such as HD, 4K, and 8K. New algorithms introduced to enable HASM to model HD, 4k, and 8k video formats are developed [20, 21]

Motion estimation is an effective way to utilize the temporal dependencies between the video frames. Types of methods are used to estimate the motion among consecutive frames are block-based motion estimation and mesh-based motion estimation. The assumption is that every block in the current frame has a matching block in the next frame. This algorithm tries to find the matching block with an error metric such as the minimum sum absolute difference Min (SAD) at the other frame. Hence, we got the motion vector for this block. This way enables us to represent every block in the other image with the motion vector.

3 Mesh Generation Techniques

Mesh is a surface that is constructed out of a set of polygons that are joined by common edges or a collection of vertices, edges, and faces that defines the shape of a polyhedral object in 3D computer graphics. The hierarchical adaptive structured

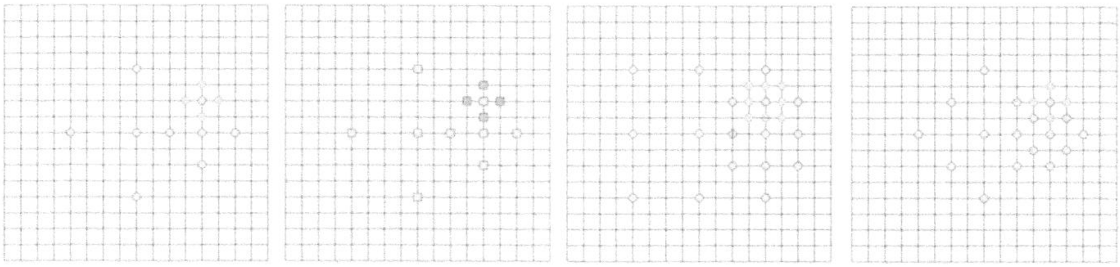

Fig. 1 Overview of different three step search algorithms

mesh (HASM) is a technique that caught the object's movements using coarse-to-fine adaptive mesh. The generated mesh at its coarse level is uniform mesh with a defined resolution. Then, each patch with more motion elements (dynamics) is split into smaller patches. So, it captures more detailed information about the motion of the patch.

Motion estimation based on blocks of pixels is known as an estimation of block matching motion. The basic idea of block matching algorithms is to divide the current frame into a matrix of non-overlapping macro blocks and to calculate motion vectors for each frame of block of the video frame, after the frame divides into pixel blocks, the block movement in the current frame is estimated and compared with all overlapping blocks in the search window. The area of search window previous frame is obtained by selecting the corresponding block, block with the same spatial location as the current block and add pixels in each direction to the search area in the previous frame.

The motion vectors in mesh nodes are estimated by generating an 8×8 block around the node and then using block- matching to generate the motion vector. The search criterion used is three step search (TSS) shown in Fig. 1. It has a very limited computation requirement. TSS gave the region of motion vectors between $[-7, 7]$ moreover four bits are enough to represent the x component of a motion vector, and four bits represent y component for compression purpose. One byte represents one motion vector, also reducing computational complexity.

Mesh for sequence frames is generated using various methods:

1. Uniform Hierarchical Mesh
2. Hierarchical Adaptive Structured Mesh (HASM)
3. Hierarchical Adaptive Merge/Split Structured Mesh (HAMSM).

However, in this paper, the uniform hierarchical mesh and the hierarchical adaptive structured mesh are adopted with various types of motion vector estimators are tested with a relatively larger block size:

1. Ordinal Three Step Search
2. Leap Three Step Search
3. Grid Three Step Search
4. Diamond Three Step Search.

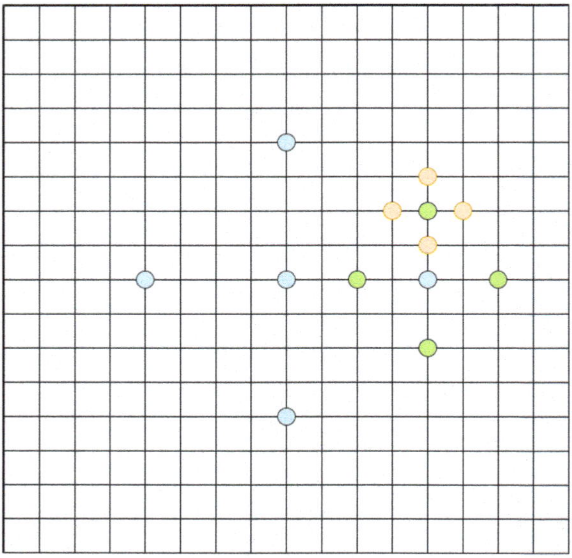

Fig. 2 Ordinal three step search

3.1 Ordinal Three Step Search

The general idea of ordinal three step search is that it starts the search from the block location as the center with the step size $S = 4$. It searches four locations around the center with step size $\pm S$ in each direction. From these eight searched locations, it selects the one having the least distortion using the Min (SAD) and makes it the new search center, then reduces step size by half $S = 2$, and repeats the search for two more iterations until the step size S is equal to 1. When it finds the location with the least SAD, it selects that block as the best match, as in Fig. 2.

3.2 Leap Three Step Search

Leap three step search algorithm works ideally the same as the ordinal three step search. In which, it assumes that either the video sequence has a low number of frames or the objects within each frame are moving rapidly, where the search area for blocks within each step is deduced by the block size, it moves as $S = 4$ * Block Size, and then for the two more iterations, it reduces the step size by half $S = /2$ * Block Size. So, each move step is shifted by block size instead of pixels, which works fine in low frame rate video sequences as in Fig. 3.

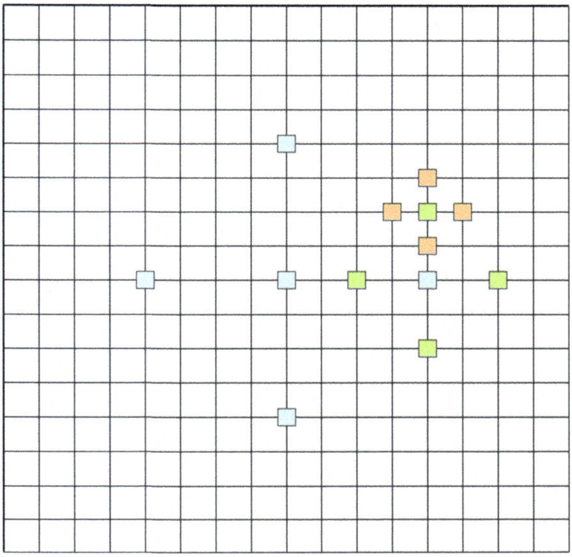

Fig. 3 Leap three step search

3.3 Grid Three Step Search

The grid three step search on the other hand initiates a grid-like ordinal but with four more search areas, one in each diagonal, as the search area is having eight search nodes instead of four with the same steps as the ordinal three step search far from the center, although it has double the computational complexity but it has a wider search region with theoretically can lead to better block matching criteria as in Fig. 4.

3.4 Diamond Three Step Search

The diamond three step search algorithm initiates a grid also of eight nodes, but unlike the grid three step search as the scattered nodes are a square-box shape, the diamond three step search has a diamond shape as in the first step the horizontal and vertical vertices have $S = 4$ in each direction as well as the diagonal nodes have $S =$ 2, in the second step the $S = /2$ in all nodes, and in the third step, there all the nodes have $S = 1$ as on Fig. 5.

*Where the S is the Stride.

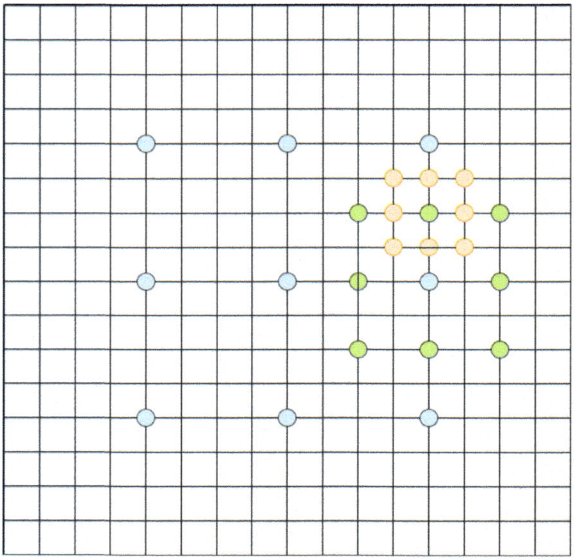

Fig. 4 Grid three step search

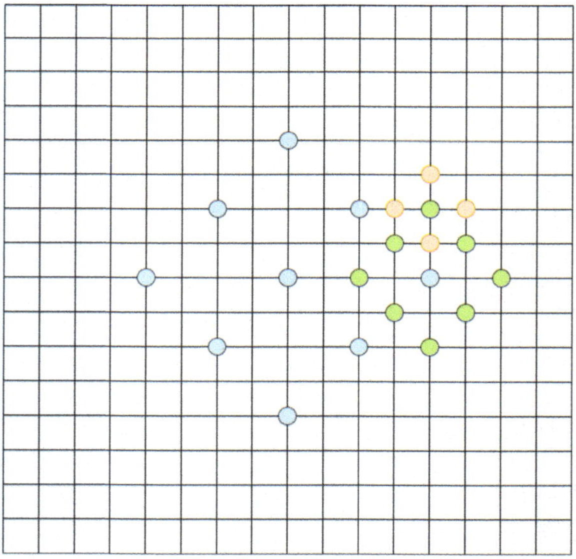

Fig. 5 Diamond three step search

4 Experimental Results and Discussion

The experiments done in this paper made under equal conditions and parameters to ensure the feasibility comparing the results. The motion vectors of mesh nodes are estimated by generating size of 8×8 block size around the node each with uniform hierarchical mesh which has a fixed block size and hierarchical adaptive structured mesh which splits if the threshold of PSNR does not meet the required block quality, for capturing further more details in the frame, and then using block- matching to generate the motion vector. The search criterion used is all the three step search (TSS) algorithms mentioned in this paper shown in Fig. 1 which has a very limited computation requirement. TSS gave the region of motion vectors between [−7, 7]; moreover, four bits are enough to represent the x component of a motion vector, and four bits represent y component for compression purpose. One byte represents one motion vector, also reducing computational complexity.

4.1 Video Sequences

Four test sequences were used for simulation. All of them Colored 4K 3840 × 2160 8 bit 120 FPS YUV RAW Video Format, Figs. 6, 8, 10 and 12 show the original f_1 sequence frames, while Figs. 7, 9, 11 and 13 illustrate the f_1 frames with the HASM mesh applied on the top.

1. Beauty Sequence:
2. ShakeNDry Sequence:
3. Jockey Sequence:
4. Bosphorus Sequence.

Fig. 6 Original f1 beauty sequence frame

Fig. 7 Mesh applied to f1 beauty sequence frame

Fig. 8 Original f1 ShakeNDry sequence frame

4.2 Experimental Results

Four types of three step search (TSS) were used with hierarchical adaptive structured mesh (HASM) to produce these results which are ordinal TSS, diamond TSS, grid TSS and leap TSS. These measures are peak signal to noise ratio (PSNR) which represents the quality of the conducted images. The block size that used was 256 × 256 with respect to the splitting mechanism which has a depth of four layers: 256 × 256, 128 × 128, 64 × 64, and 32 × 32. The experiment took place on a

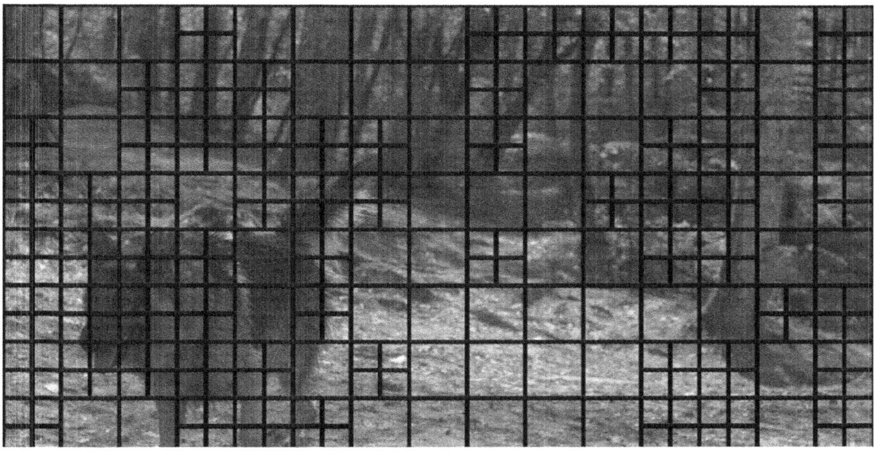

Fig. 9 Mesh applied to f1 ShakeNDry sequence frame

Fig. 10 Original f1 jockey sequence frame

PC with specification CPU Intel Core i9-9900K, 32 GB RAM, and Graphics Card Nvidia RTX 2070. Here, we show only the results of testing the algorithms on four different 4K sequences on ten frames from each sequence for easy comparisons. Table 1 shows the average PSNR results for each algorithm using HASM with 256 block size, while Table 2 shows the average matching time per frame in seconds for each algorithm using HASM with 256 block size.

Fig. 11 Mesh applied to f1 jockey sequence frame

Fig. 12 Original f1 busphorus sequence frame

4.3 Discussion

In general, in terms of speed, the ordinal TSS has majority. On the other hand, its quality is not the worst among these algorithms. The diamond and grid TSS came after ordinal both gave almost the same results in terms of speed, but they have better PSNR than ordinal. As it can be seen, leap TSS outputs the worst results in both speed and PSNR.

Fig. 13 HASM applied to f1 busphorus sequence frame

Table 1 Average PSNR results for each algorithm + HASM 256 block size

Algorithm	Beauty	ShakeNDry	Jockey	Busphorus
Ordinal	31.87	32.41	31.98	9.8
Diamond	31.88	32.55	31.96	10
Grid	31.81	32.1	31.96	10.25
Leap	30.65	29.12	30.55	15.6

Table 2 Average time/frame in seconds for each algorithm + HASM 256 block size

Algorithm	Beauty	ShakeNDry	Jockey	Busphorus
Ordinal	20.1	15.92	20	32.4
Diamond	20.75	15.78	19.84	32.5
Grid	21.46	15.71	20.11	32.48
Leap	15.3	16.86	20.5	30.22

From the above, some of the algorithms are good in speed, but they suffered in terms of quality. We believe that the grid and diamond TSS are the best choices for quality while have nearly same speed as the ordinal. Both grid and diamond TSS algorithms introduced almost the same quality. The reason for the speed degradation, compared to the ordinal and leap, is that grid and diamond search in a bigger region of interest area of the image. Lastly, we believe that the ordinal TSS gives average results in terms of speed and quality, while if the ultimate target is highest frame quality trying the diamond or the grid is the best.

5 Conclusion

In this report, different algorithms were presented and discussed. The best results obtained from diamond and grid three step search in terms of statistical quality and nearly the same speed as the fastest algorithm compared to other three step search algorithms, although theoretically it does not have the best search area region in terms of the highest computation complexity compared to others. While the leap three step search performs poorly. However, diamond and grid three step search is slightly better than the other in terms of quality. On average usage of the algorithm if trading-off between quality and complexity does not really matter, we recommend using the ordinal three step search.

Lastly, leap three step search may not have the best results and as a standalone algorithm is still insufficient. However, leap three step search can be further utilized in some case scenarios as a secondary algorithm, if the main algorithm did not find the matching block in the near search area region, then for a wider search area, leap three step search will be efficient finding that block.

References

1. Badawy W, Bayoumi M (2002) Algorithm-based low-power VLSI architecture for 2D mesh video-object motion tracking. IEEE Trans Circ Syst Video Technol 12(4):227–237
2. Kordasiewicz RC, Gallant MD, Shirani S (2007) Affine motion prediction based on translational motion vectors. IEEE Trans Circuits Syst Video Technol 17(10):1388–1394
3. Wei Y, Badawy W (2003) A new moving object contour detection approach. IEEE Int Workshop Comput Archit Mach Percept 2003:6–236
4. Badawy W, Bayoumi M (2000) A VLSI architecture for hybrid object-based video motion estimation. In: 2000 Canadian conference on electrical and computer engineering, vol 2. pp 1109–1113
5. Badaway W (2001) A structured versus unstructured 2D hierarchical mesh for video object motion tracking. In: Canadian conference on electrical and computer engineering 2001, vol 2. pp 953–956
6. Badawy W, Bayoumi M (2000) Low power video object motion-tracking architecture for very low bit rate online video applications. In: 2000 International conference on computer design 2000 Proceedings. pp 533–536
7. Goswami K, Hong G-S, Kim B-G (2011) Adaptive node dimension (AND) calculation algorithm for mesh based motion estimation. In: 2011 IEEE international conference on consumer electronics—Berlin (ICCE-Berlin), pp 91–95
8. Badawy W, Bayoumi M (2000) A VLSI architecture for 2D mesh-based video object motion estimation. In: International conference on consumer electronics 2000. ICCE. 2000 Digest of technical papers. pp 124–125
9. Badawy W, Bayoumi M (2001) A mesh based motion tracking architecture. In: The 2001 IEEE international symposium on circuits and systems 2001. ISCAS 2001, vol 4. pp 262–265
10. Liu X, Badawy W (2000) A novel adaptive error control scheme for real time wireless video streaming. In: International conference on information technology: research and education 2003 proceedings. ITRE2003. pp 460–463
11. Badawy W, Bayoumi M (2000) Low power VLSI architecture for 2D-mesh video object motion tracking. In: IEEE computer society workshop on VLSI 2000 proceedings. pp 67–72

12. Sayed M, Badawy W (2002) MPEG-4 synthetic video object prediction using embedded memory. In: The 2002 45th Midwest symposium on circuits and systems 2002. MWSCAS-2002, vol 1. pp I-591
13. Darwish T, Viyas A, Badawy W, Bayoumi M (2000) A low power VLSI prototype for low bit rate video applications. In: 2000 IEEE workshop on signal processing systems 2000. SiPS 2000. pp 159–167
14. Utgikar A, Badawy W, Seetharaman G, Bayoumi M (2003) Affine schemes in mesh-based video motion compensation. In: IEEE workshop on signal processing systems 2003. SIPS 2003. pp 159–164
15. Goswami K, Chakrabarti I, Sural S (2009) A probabilistic adaptive algorithm for constructing hierarchical meshes. IEEE Trans Consum Electron 55(3):1690–1698
16. Badawy W, Bayoumi MA (2002) A low power VLSI architecture for mesh-based video motion tracking. IEEE Trans Circ Syst II: Analog Digit Sign Process 49(7):488–504
17. Badawy W, Bayoumi M (2000) Low power VLSI prototype for motion tracking architecture. In: 13th Annual IEEE international ASIC/SOC conference 2000. Proceedings. pp 243–247
18. Munoz-Jimenez V, Zergainoh A, Astruc J-P (2006) Motion estimation method through adaptive deformable variable-sized rectangular meshes using warping function. 2006 IEEE international symposium on signal processing and information technology. pp 884–889
19. Liu X, Badawy W (2003) A novel error control scheme for video streaming over IEEE802.11 network. In: Canadian conference on electrical and computer engineering 2003. IEEE CCECE 2003, vol. 2. pp 981-984
20. Mohamed AT, Ezzat K, El Shal I, Badawy W (2020) On the application of hierarchical adaptive structured mesh "HASM®" Codec for Ultra Large Video Format. In: Proceedings of the 2020 9th International Conference on Software and Information Engineering (ICSIE) November 2020 pp 135–139. https://doi.org/10.1145/3436829.3436870
21. Badawy W (2020) On scalable video Codec for 4K and high definition video streaming – the hierarchical adaptive structure mesh approach "HASM-4k". In: 2020 IEEE 10th International Conference on Consumer Electronics (ICCE-Berlin), Berlin, Germany. pp 1–5. https://doi.org/10.1109/ICCE-Berlin50680.2020.9352175

Chapter 16
Rough Sets Crow Search Algorithm for Inverse Kinematics

Mohamed Slim, Nizar Rokbani, and Mohamed Ali Terres

1 Introduction

The forward kinematics of a robotic manipulator gives the position of its end-effector for a specific joint configuration; inverse kinematics is just opposite to forward kinematics, which is the process of obtaining the joint angles of the manipulator from known coordinates of the end-effector [3, 8]. Inverse kinematics is generally very difficult to solve, due to the difficulty of inverting the nonlinear forward kinematics equations. For this reason, ample heuristics and meta-heuristics methods for solving the inverse kinematics are welcomed for their simple iterative operations that gradually lead to a favorable approximation of the solution. They are also known for their low computational cost [6] which makes them able to return the configuration solution very quickly with respect of the joints constraints (limits). In the recent years, bio-inspired algorithms have been applied for inverse kinematics problem of different types of robotic arms; their application offered more accurate results than the traditional methods. From the most used bio-inspired meta-heuristic algorithm in inverse kinematics, we mention for example and not exclusively the genetic algorithm (GA) [13], particle swarm optimization (PSO) [10, 11], quantum-behaved particle swarm optimization (QPSO), [4] and firefly algorithm (FA) [12]; they have been used to solve the inverse kinematics of many robot systems.

M. Slim (✉) · N. Rokbani · M. A. Terres
Higher Institute of Applied Sciences and Technology of Sousse, University of Sousse, 4003 Sousse, TN, Tunisia

N. Rokbani
e-mail: nizar.rokbani@ieee.org

N. Rokbani
Research Group On Intelligent Machines, ReGim-Lab, National Engineering School of Sfax, University of Sfax, 3038 Sfax, TN, Tunisia

A. E. Hassanien et al. (eds.), *Enabling Machine Learning Applications in Data Science*, Algorithms for Intelligent Systems, https://doi.org/10.1007/978-981-33-6129-4_16

This paper presents an investigation on the rough sets crow search algorithm as an inverse kinematics solver; Sect. 2 briefly introduces crow search algorithm, CSA and the rough sets variant of this heuristic RCSA, and how to implement them for inverse kinematics; then, the detailed forward kinematics model of the industrial robot Kuka-Kr16 is developed in Sect. 3. Section 4 presents a comparative investigation of the performances of the proposed algorithm to the state-of-the-art contributions; finally, conclusions and perspectives are presented in Sect. 5.

2 Rough Crow Search Algorithm for Inverse Kinematics

2.1 The Crow Search Algorithm

The crow search algorithm (CSA) is an efficient meta-heuristic algorithm that was developed by Askarzadeh [1] for solving optimization problems. It is inspired from the intelligent behavior of crows in searching for food sources [9]. This manifests in the capability of a crow individual to tap into the food resources of other crows and chase them in order to know their food hiding spot. Each crow looks for a suitable hideout for its excessive food to use it as a pantry. Thus, each crow plays the role of the hider and the role of the seeker in the same time. However, a crow might feel that it has been followed by another. In that case, it tries to mislead the marauder crow by taking a fake path that leads to a deceptive hiding spot [2]. The CSA aims to simulate this unique behavior of crows, by considering the individual searches for hideouts with the best food resources as the global optima in terms of computational optimization. Therefore, the update of the individual positions in the search space is done according to two main features of crows: the chasing for other crows to find their hideout spots; and protection for their own. Initially, the crows spread randomly throughout the search space looking for the perfect hideout spot. It is important to mention that each crow can memorize the best one that it has encountered during its search phase, so far.

The algorithm processing starts by initializing a swarm of crows within the n-dimensional search space by assigning a random vector $X_i = \left(x_{i,1}, x_{i,2} \ldots, x_{i,n}\right)$ for the ith crow $i = 1, 2, \ldots, N$. Furthermore, each crow of the swarm is characterized by its memory $m_i = \left(m_{i,1}, m_{i,2}, \ldots, m_{i,n}\right)$ (initially $m_i = X_i$ because the crows have no experience about the food sources). Then, crow positions are evaluated according to the objective function $f(X)$.

The position update of the crow i could be mathematically described by Eq. (1):

$$x_{i,\text{Iter}+1} = \begin{cases} x_{i,\text{Iter}} + r_i \times fl_{i,\text{Iter}}\left(m_{j,\text{Iter}} - x_{i,\text{Iter}}\right) & a_j \geq AP_j \\ \text{random position} & \text{otherwise} \end{cases} \tag{1}$$

where r_i, a_i are random numbers with uniform distribution between 0 and 1, AP_j stands for the awareness probability of crow j and $fl_{i,\text{Iter}}$ is the flight length of crow

i at iteration Iter. $m_{j,\text{Iter}}$ is the memory of crow j at iteration Iter, where crow j is randomly selected crow followed by crow i.

The memory of each crow is updated only if the new position is better than the earlier one, see Eq. (2)

$$m_{i,\text{Iter}+1} = \begin{cases} x_{i,\text{Iter}+1} & f\left(x_{i,\text{Iter}}\right) > f\left(m_{i,\text{Iter}}\right) \\ m_{i,\text{Iter}} & \text{otherwise} \end{cases} \tag{2}$$

2.2 The Rough Searching Scheme

Rough searching scheme is a method suggested by Hassanien [5] to manage a rough information system of a certain problem in order to approach a global optimal solution. It uses the rough set theory RST methodology of approximations as it is introduced by [7]. It starts by defining a specific information system which is a non-empty finite set U of objects and a non-empty final set of attributes A; thus, the information system is defined as the pair (U, A). In the case of an optimization problem, the information system consists of a pair of sets, which are the problem's possible solutions (U) and their attributed values in each dimension (A). The next step is the approximation, which aims to encompass the target set through two types of approximations, the upper approximation and the lower approximation. The upper approximation is the set of objects that they possibly belong to a target set X, where the lower approximation is the set of objects that positively belongs to the target set X. In RSS, the rough approximation is done by defining a pair of U the non-empty finite set of object and C which is a reflexive relation on U that partitions U into N classes; i.e., let $\{x_i\}$ is the ith class $U/C = \{\{x_1\}, \{x_2\}, \ldots, \{x_i\}, \ldots, \{x_N\}\}$ in this case the lower approximation of x_i is detailed in Eq. (3):

$$\underline{\text{Apr}}(x_i) = \cup\{y \in U/C(y) \leq x_i\} \tag{3}$$

The upper approximation is denoted in Eq. (4):

$$\overline{\text{Apr}}(x_i) = \cup\{y \in U/C(y) \geq x_i\} \tag{4}$$

After determining the lower and upper approximation, it is possible now to define the rough interval of each dimension by defining its lower bound $x_{i,c}^{\text{LB}}$ and the upper bound $x_{i,c}^{\text{UB}}$ of the ith dimension, using Eqs. (5) and (6):

$$x_i^{\text{LB}} = \frac{1}{N_{\text{LB}}} \sum y | y \in \underline{\text{Apr}}(x_i) \tag{5}$$

$$x_i^{\text{UB}} = \frac{1}{N_{\text{UB}}} \sum y | y \in \overline{Apr}(x_i) \tag{6}$$

where N_{LB} is the size of $\underline{\mathrm{Apr}}(x_i)$; in other word, it is the number of elements in $\underline{\mathrm{Apr}}(x_i)$. And, so for N_{UB}, it is the size of $\overline{\mathrm{Apr}}(x_i)$. Therefore, these upper and lower bounds represent the boundaries of the new intervals called the rough boundary intervals; its range is simply calculated using Eq. (7):

$$\mathrm{RBI}(x_i) = x_i^{\mathrm{UB}} - x_i^{\mathrm{LB}} \tag{7}$$

The upper and lower bounds above are relative to one class or individual. Nonetheless, in order to generate new solution, they have to be generated in absolute intervals that depend on the upper and lower bounds of all the objects. These absolute intervals are called unified intervals for all the dimensions; they are assessed using Eqs. (8) and (9):

$$x_i^{\mathrm{LB}} = \frac{\left(x_{i1}^{\mathrm{LB}} + x_{i2}^{\mathrm{LB}} + \cdots + x_{iN}^{\mathrm{LB}}\right)}{N} \tag{8}$$

$$x_i^{\mathrm{UB}} = \frac{\left(x_{i1}^{\mathrm{UB}} + x_{i2}^{\mathrm{UB}} + \cdots + x_{iN}^{\mathrm{UB}}\right)}{N} \tag{9}$$

Thus, the unified rough intervals of all the dimensions for a class X are obtained and represented by Eq. (10):

$$\mathrm{RI} = \left\{\left[x_1^{\mathrm{LB}}, x_1^{\mathrm{UB}}\right], \left[x_2^{\mathrm{LB}}, x_2^{\mathrm{UB}}\right], \ldots, \left[x_n^{\mathrm{LB}}, x_n^{\mathrm{UB}}\right]\right\} \tag{10}$$

These reduced intervals provided by the RSS might be useful to guide the CSA for approaching the global optimal solution. This will be explained in the next paragraph.

2.3 The Rough Crow Search Algorithm

The rough crow search algorithm (RCSA) proposed by Hassanien [5] is a hybridization between rough searching scheme (RSS) and the CSA. The proposed RCSA operates in two phases in the first one; CSA is implemented as global optimization system to find an approximate solution of the global optimization problem. In the second phase, RSS is introduced to improve the solution quality through the roughness of the obtained optimal solution so far. By this way, the roughness of the obtained optimal solution can be represented as a pair of precise concepts based on the lower and upper approximations which are used to constitute the interval of boundary region. After that, new solutions are randomly generated inside this region to enhance the diversity of solutions and achieve an effective exploration to avoid premature convergence of the swarm.

Besides the hybridization between the RSS and the CSA, Hassanien [5] suggested another adjustment to the original CSA; the first is a dynamic flight length fl instead of a fixed value that cannot be changed during iterations to adjust the tendency of approaching the optimal solution. The dynamic change with iteration number of the flight length is expressed by Eq. (11):

$$fl_{\text{Iter}} = fl_{\max} \cdot \exp\left(\log\left(\frac{fl_{\min}}{fl_{\max}}\right) \cdot \left(\frac{\text{Iter}}{\text{Iter}_{\max}}\right)\right) \tag{11}$$

where fl_{Iter} is the flight length in each iteration, $fl_{\min}$ is the minimum flight length, $fl_{\max}$ is the maximum flight length, Iter is the iteration number, and $\text{Iter}_{\max}$ is the maximum iteration number. In addition to that Hassanien [5] suggested an opposition-based learning for updating the solutions to improve the diversity of solutions so Eq. (1) is modified to become as Eq. (12):

$$x_{i,\text{Iter}+1} = \begin{cases} x_{i,\text{Iter}} + r_i \times fl_{i,\text{Iter}}\left(m_{j,\text{Iter}} - x_{i,\text{Iter}}\right) \\ x_{i,\text{Iter}} - r_i \times fl_{i,\text{Iter}}\left(m_{j,\text{Iter}} - x_{i,\text{Iter}}\right) & \text{if } a_j \geq \text{AP}_{j,\text{Iter}} \\ LB + rand \times (UB - LB) & \text{otherwise} \end{cases} \tag{12}$$

After each update of the positions, the new positions are evaluated at each iteration until reaching a maximum number of iterations, or obtaining a satisfying solution in terms of objective function. The CSA phase yields an optimal solution $X^* = \left(x_1^*, x_2^*, \ldots, x_n^*\right)$ to the RSS phase, to create rough intervals arround it. These intervals will be the next search space for the CSA phase, to guide it for approaching the global optimal solution. In the RSS phase, an information system is formulated from the solutions of the CSA phase. Figure 1 represents the pseudocode of RCSA algorithm for inverse kinematics IK problem, IK-RCSA.

To solve the inverse kinematics using RSCSA, the following model is adopted:

- Crow stands for $Q = [q_1, q_2, \ldots, q_n]$.
- The search space stands for the number of dimensions n which is the number of degree of freedom of the model.
- The fitness function which is the objective functions of the inverse kinematics. It is the distance between target location (x_t, y_t, z_t) and the actual position reached by the end-effector (x, y) which is determined by applying the forward kinematics of a certain configuration of parameters $Q = [q_1, q_2, \ldots, q_n]$. It is measured by Eq. (13):

$$\varepsilon = \sqrt{(x_t - x)^2 + (y_t - y)^2 + (z_t - z)^2} \tag{13}$$

3 The Forward Kinematics Model of Kuka-Kr16

The Kuka-Kr16 is an industrial robot manipulator used essentially in welding and assembling operations in manufacturing. It is a 6-DOF revolute joints and its three-dimensional structure diagrams of 6-DOF and its mechanical size are shown in Fig. 2.

```
Initialize CSA Parameters: N, Iter_max, fl_max, fl_min
Generate N crows randomly in the search space X_i=(xi_1,...,xi_n)
Evaluate the intial positions
Initialize the memory of crows with the initial position m_i=X_i
Loop
    while Iter<Iter_max
        calculate fl_Iter using equation (3)
        for i=1 to N
         Update the position Xi using equation (4)
       end for
       Evaluate the obtained positions of the crows
       Update the memory of crows using (2)
    end while
    X_best=min(m)
    Formulate the information system
    Calculate the rough interval using (8) and (9)
    while Iter<Iter_max
       for i=1 to N
     if rand>=0.5
            X'=X^LB + rand.(X_best - X^LB)
         else
            X'=X_best + rand.(X^UB - X_best)
         End if
       End for
      Evaluate the obtained positions and update the best position
    end while
end looping
```

Fig. 1 Simplified pseudocode of IK-RCSA

The forward kinematic model, of the Kuka-Kr16 robotic manipulator, is obtained using Denavit–Hartenberg, D–H, method since it is considered as the most common method for describing the robot kinematics. The coordinate's frames from the kinematic scheme Fig. 3 were assigned to the D–H parameters: a_i, α_i, d_i and θ_i, to fill Table 1.

where:

$a_{i:}$ is link length which is the distance between O_i and the intersection of x_i and z_{i-1} axes, toward x_i, see Eq. (14).

$$a_i = \text{dist}(O_i, (x_i \cap z_{i-1})).\overrightarrow{x_i} \quad (14)$$

where $\cap$ means geometrical intersection.

$d_{i:}$ is the joint offset; it is the distance between O_{i-1} and the intersection of x_i and z_{i-1} axes, toward z_{i-1}, see Eq. (15).

$$d_i = \text{dist}(O_{i-1}, (x_i \cap z_{i-1})).\overrightarrow{z_{i-1}} \quad (15)$$

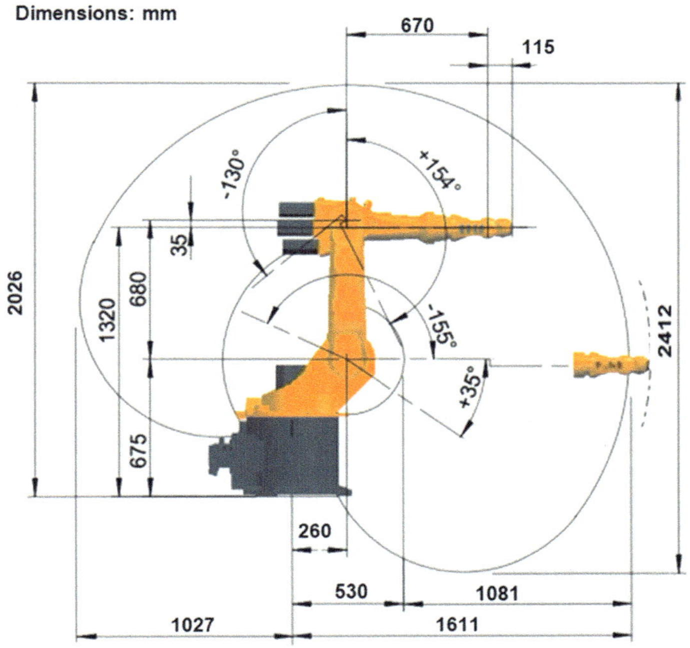

Fig. 2 Dimensions and joints limits of the Kuka-Kr16-2 manipulator

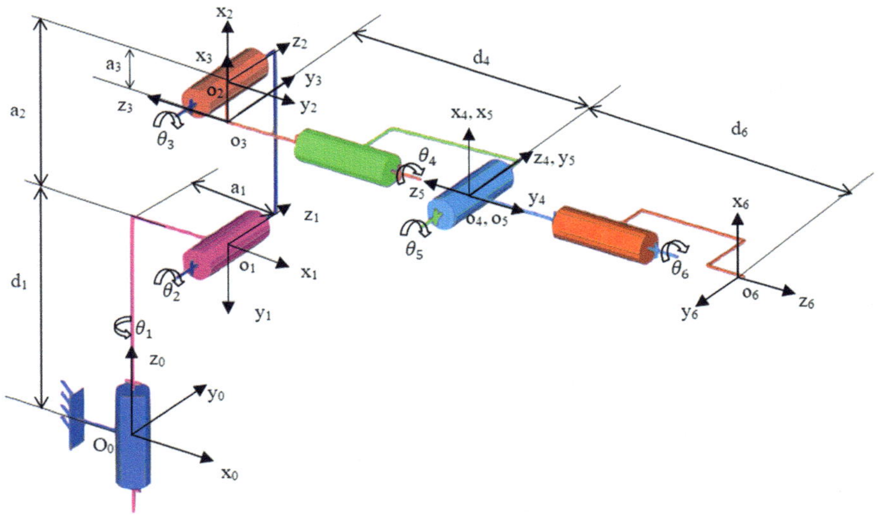

Fig. 3 Kinematic scheme of Kuka-Kr16-2 manipulator

Table 1 D-H parameters for the Kuka-Kr16-2 robot

n	a_i/mm	$\alpha_i/(°)$	d_i/mm	$\theta_i/(°)$	$\theta_n/(°)$
1	260	−90	675	0	±185°
2	680	0	0	−90	−155–35°
3	−35	90	0	0	−130–154°
4	0	−90	670	0	±350°
5	0	90	0	0	±130°
6	0	180	115	0	±350°

α_i is the angle between z_{i-1} and z_i. Its direction is determined according to the right-hand rule, and thus, its sign is according to its direction along x_i.

θ_i is the joint angle measured between x_{i-1} and x_i. Its direction is determined according to the right-hand rule, and thus, its sign is according to its direction along z_{i-1}. In our case, (Kuka-kr 16–2) θ_i is variable because all the joints are revolute joints.

θ_n is the range interval limited by mechanical joints constraints. The transformation between each two successive joints can be written by simply substituting the parameters from the parameters table into a standard homogeneous coordinate transformation matrix T developed in Eqs. (16)–(22);

$$T_{i+1}^{i} = \begin{bmatrix} \cos\theta_i & -\sin\theta_i\cos\alpha_i & \sin\theta_i\sin\alpha_i & a_i\cos\theta_i \\ \sin\theta_i & \cos\theta_i\cos\alpha_i & -\cos\theta_i\sin\alpha_i & a_i\sin\theta_i \\ 0 & \sin\alpha_i & \cos\alpha_i & d_i \\ 0 & 0 & 0 & 1 \end{bmatrix} \tag{16}$$

$$T_1^0 = \begin{bmatrix} \cos\theta_1 & 0 & -\sin\theta_1 & a_1.\cos\theta_1 \\ \sin\theta_1 & 0 & \cos\theta_1 & a_1.\sin\theta_1 \\ 0 & -1 & 0 & d_i \\ 0 & 0 & 0 & 1 \end{bmatrix} \tag{17}$$

$$T_2^1 = \begin{bmatrix} \cos\theta_2 & -\sin\theta_2 & 0 & a_2.\cos\theta_2 \\ \sin\theta_2 & \cos\theta_2 & 0 & a_2.\sin\theta_2 \\ 0 & 0 & 1 & d_1 \\ 0 & 0 & 0 & 1 \end{bmatrix} \tag{18}$$

$$T_3^2 = \begin{bmatrix} \cos\theta_3 & 0 & -\sin\theta_3 & a_3.\cos\theta_3 \\ \sin\theta_3 & 0 & \cos\theta_3 & a_3.\sin\theta_3 \\ 0 & 1 & 0 & 0 \\ 0 & 0 & 0 & 1 \end{bmatrix} \tag{19}$$

$$T_4^3 = \begin{bmatrix} \cos\theta_4 & 0 & -\sin\theta_4 & 0 \\ \sin\theta_4 & 0 & \cos\theta_4 & 0 \\ 0 & -1 & 0 & d_4 \\ 0 & 0 & 0 & 1 \end{bmatrix} \tag{20}$$

$$T_5^4 = \begin{bmatrix} \cos\theta_5 & 0 & \sin\theta_5 & 0 \\ \sin\theta_5 & 0 & -\cos\theta_5 & 0 \\ 0 & 1 & 0 & 0 \\ 0 & 0 & 0 & 1 \end{bmatrix} \tag{21}$$

$$T_6^5 = \begin{bmatrix} \cos\theta_6 & \sin\theta_6 & 0 & 0 \\ \sin\theta_6 & -\cos\theta_6 & 0 & 0 \\ 0 & 0 & -1 & d_6 \\ 0 & 0 & 0 & 1 \end{bmatrix} \tag{22}$$

Equation (23) is a multiplication of the transformations of all neighboring frames. It is used to calculate the forward kinematics of Kuka-Kr16

$$T_6^0 = T_1^0 T_2^1 T_3^2 T_4^3 T_5^4 T_6^5 = \begin{bmatrix} r_{11} & r_{12} & r_{13} & p_x \\ r_{21} & r_{22} & r_{23} & p_y \\ r_{31} & r_{32} & r_{33} & p_z \\ 0 & 0 & 0 & 1 \end{bmatrix} \tag{23}$$

The 3×3 matrix r_{ij} defines the orientation of the end-effector, while (p_x, p_y, p_z) defines the position. T_6^0 matrix produces a Cartesian coordinate for any six joint angles. Since the fitness of the proposed approach is the Euclidian distance in Cartesian space between the obtained and the target points, so T_6^0 is also useful in measuring the fitness. Considering (p_x, p_y, p_z) are the coordinates of the end-effector, therefore the fitness could be expressed by Eq. (24):

$$\varepsilon = \sqrt{(x_t - p_x)^2 + \left(y_t - p_y\right)^2 + (z_t - p_z)^2} \tag{24}$$

where (x_t, y_t, z_t) defines the target position. The coordinates (p_x, p_y, p_z) are evaluated by Eqs. (25), (26) and (27):

$$\begin{aligned} p_x = {} & \cos\theta_1 (a_1 + a_3 \cos(\theta_2 + \theta_3) + a_2 \cos\theta_2) + \sin(\theta_2 + \theta_3) \cos\theta_1 (d_4 + d_6 \cos\theta_5) \\ & - d_6 \sin\theta_1 \sin\theta_4 \sin\theta_5 + d_6 \cos(\theta_2 + \theta_3) \cos\theta_1 \cos\theta_4 \sin\theta_5 \end{aligned} \tag{25}$$

$$\begin{aligned} p_y = {} & \sin\theta_1 (a_1 + a_3 \cos(\theta_2 + \theta_3) + a_2 \cos\theta_2) + \sin(\theta_2 + \theta_3) \sin\theta_1 (d_4 + d_6 \cos\theta_5) \\ & + d_6 \cos\theta_1 \sin\theta_4 \sin\theta_5 + d_6 \cos(\theta_2 + \theta_3) \sin\theta_1 \cos\theta_4 \sin\theta_5 \end{aligned} \tag{26}$$

$$p_z = d_1 + \cos(\theta_2 + \theta_3)(d_4 + d_6 \cos\theta_5) - a_3 \sin(\theta_2 + \theta_3) - a_2 \sin\theta_2$$

$$- d_6 \sin(\theta_2 + \theta_3) \cos \theta_4 \sin \theta_5 \tag{27}$$

4 Experimental Analysis

Experimental investigations are based on a simulated industrial robot model, the Kuka-Kr16-2 manipulator. The experimentation is done using MATLAB simulations with and i5 machine of 8 GB of RAM.

4.1 Comparative Results

In this study, the RCSA is compared with the classical CSA, particle swarm optimization PSO and quantum particle swarm optimization QPSO as well the rough sets particle swarm optimization RPSO, which follows the same methodology as the RCSA; however, it uses classical PSO algorithm as a first phase instead of the CSA algorithm. All the mentioned techniques are compared in terms of precision and computing time in solving the inverse kinematic problem of the Kuka-Kr16. Two types of tests were used to obtain the error and computing time of each solver. Both tests are done using MATLAB simulations with an i5 machine of 8 GB of RAM.

4.1.1 One Target Position Test

This test consists of setting one target point in the workspace of the Kuka-Kr16-2. And each algorithm was tested to solve the inverse kinematic problem, in order to get the end-effector of the kuka-kr16 in the specific target point. Figure 4 shows different configurations of the solutions of each algorithm. Table 2 presents the angle values and the solution error of the used algorithms. The target points given in this test are (−0.4567, 1.1062, 0.9874 m) which are generated randomly from the search space. The maximum number of iterations in all the tests is 500. Knowing that for all the tests the number of population of each algorithm is 50.

For time comparison, a maximal error $e = 10.e^{-5}$ is given to all the algorithms in order to compare the amount of time needed to solve the inverse kinematics with a fitness value fitness $\leq$ e. Table 3 shows the calculation time as well as the number of iterations spent to get a satisfying solution in a range of the maximal accepted error $e = 10.e^{-5}$ Knowing that the max iteration count used in this test is 1000. Figure 5 shows the convergence rate of the compared algorithms during 500 iterations.

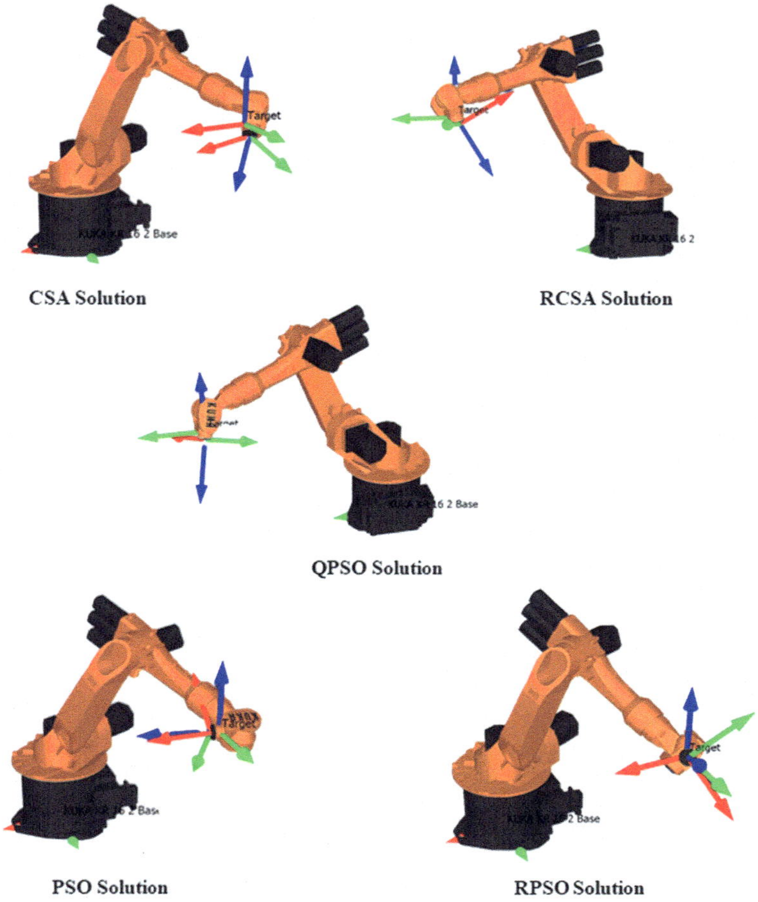

Fig. 4 Configurations of the inverse kinematic solution

Table 2 Angle values and errors of the solutions

Angles	CSA	RCSA	PSO	QPSO	RPSO
θ_1	118.0185	109.9702	117.2411	110.4539	116.1794
θ_2	−63.5829	−60.9629	−56.4755	−67.2262	−64.6946
θ_3	−6.8583	−15.8562	−13.8123	−7.1800	3.0254
θ_4	122.0571	−333.1483	−80.8563	−154.6191	−313.5283
θ_5	−64.4441	98.0057	120.6552	−57.0023	−69.6719
θ_6	141.2816	−325.0057	10.0000	−69.9536	−350
Error (m)	0.0452	1.1102e−16	0.0030	2.4458e−14	5.5511e−17
Convergence time (s)	2.266	1.676	2.093	1.996	2.303

Table 3 Time comparison results

Techniques	CSA	RCSA	PSO	QPSO	RPSO
Error (m)	0.0591	8.5827e−06	0.0066	8.8584e−06	4.9221e−06
Calculation time (s)	4.6781	0.3010	4.9243	0.2383	0.3471
Iterations	>1000	19	>1000	67	19

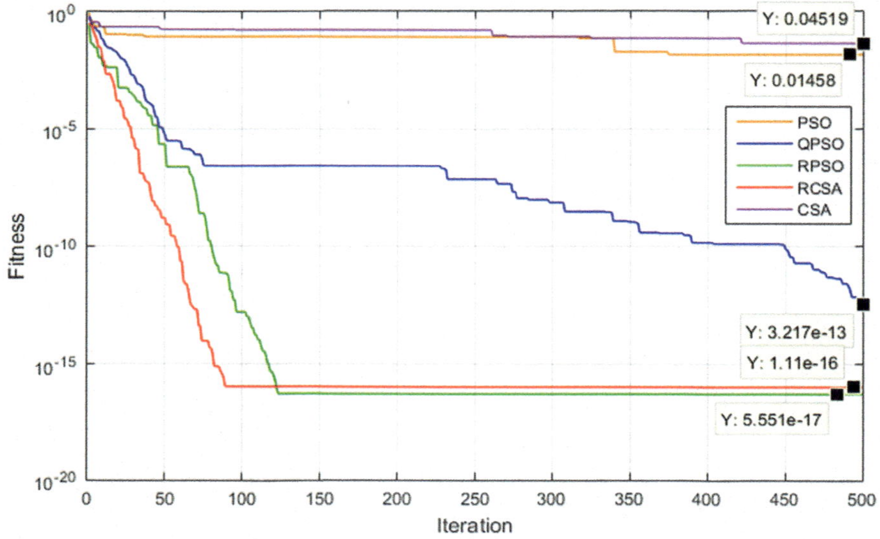

Fig. 5 Comparative evolution of the fitness of RSCSA to PSO, QPSO, CSA, and RSPSO

4.1.2 Multiple Targets Evaluation

This test is considered as a statistical analysis test, and a test of stability of the results of the different solvers in which 50 random points were generated in the working space of the robot as in [12]. Thus, each solver had run 50 times for each point. Therefore, the results are presented in terms of their mean and standard deviation with mentioning the minimal error and the maximal error among the 50 runs. Table 4 shows the results of the statistical analysis. Figure 6 shows the position error of the solutions for the 50 random points.

4.2 Non-parametric Statistical Analyses

The performance's evaluation of the proposed IK solver in comparison to the other techniques is done using the Wilcoxon signed ranks test proposed by Wilcoxon

Table 4 Statistical results

Technique	Mean	Std	Worst solution	Best solution
CSA	0.0510	0.0273	0.1084	0.0107
RCSA	0.0025	0.0175	0.1239	6.9389e−18
PSO	0.0178	0.0151	6.9095e−04	0.0630
QPSO	0.0017	0.0118	0.0838	5.3511e−15
RPSO	0.0012	0.0047	0.0242	5.2042e−18

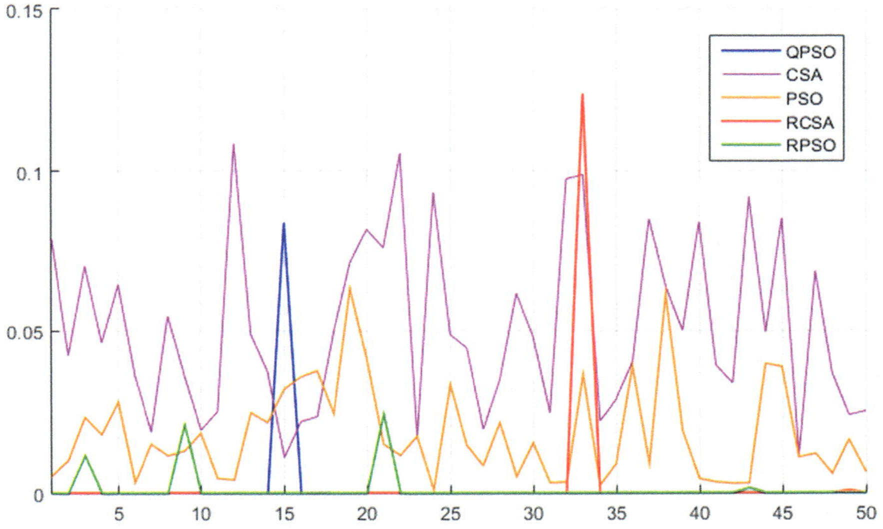

Fig. 6 Solution's error of random points generated in workspace of the Kuka-kr16-02

(Wilcoxon 1945). It is used to compare sample results of tested algorithms by assessing the difference in their mean ranks. In other word, this test is used to detect the difference in the median of the sample results and decide whether accept or reject the null hypothesis, according to the probability of getting a null hypothesis ρ assessed from the test. Therefore, if $\rho < 0.05$, then the null hypothesis is rejected, which means that there is a significant difference between the performance of the two tested methods, and that one method outperforms the other. However if $\rho > 0.05$, then the null hypothesis is accepted, meaning that there is no significant difference between both compared methods, since the medians of the sample results are almost equal. Another important indexes evaluated from the Wilcoxon signed rank test is the sum of positive ranks and negative ranks, respectively, R^{+} and R^{-}. They indicate which algorithm between the two compared algorithms gives better solutions. The number of search agents in all the compared algorithms is 50, and the size of result's sample is 30. Table 5 shows the comparative results using Wilcoxon signed rank test.

Table 5 Wilcoxon rank sum test results

Compared algorithms		Solution evaluation			
Algorithm 1	Algorithm 2	R^-	R^+	ρ	Best technique
RCSA	PSO	440	25	7.4086e−10	RCSA
RCSA	CSA	465	0	2.3408e−11	RCSA
RCSA	FA	270	195	0.0030	RCSA
RCSA	QPSO	290	175	0.0033	RCSA
RCSA	RPSO	34.5	31.5	0.5434	N/A

The results show a rejection of the null hypothesis with the outperformance of RCSA in comparison to the applied techniques in solving IK except for the comparison with RPSO where the null hypothesis is accepted according to the Wilcoxon test. This might prove the importance of the RSS phase in solving the IK problem, as its main role is to reduce the search space and thus the angle's intervals at each iteration.

5 Discussion and Perspectives

This paper investigated the new heuristic rough sets crow search, RCSA, to solve inverse kinematics with an heuristic search. Only the forward model is needed, which allows avoiding computed inverse models singularities with classical computing methods. In this paper, the crow search algorithm as well as rough sets crow search algorithm is applied to the inverse kinematics problem of the industrial mechanical manipulator, the Kuka-Kr16 robot. Obtained results showed that RSCSA is a potentially a fair solver for inverse kinematics, and the method showed time effectiveness with an error of e-16, which very precise millimeter range mechanical operations. RCSA showed also a faster convergence rate when compared to particle swarm optimization IK solvers.

More investigations are needed to better understand the impact of RCSA parameters on the quality of the inverse kinematics solutions; a statistical analysis of the convergence is also under-investigations as well as the extension of investigations to mechanical design problems.

References

1. Askarzadeh A (2016) A novel metaheuristic method for solving constrained engineering optimization problems: Crow search algorithm. Comput Struct 169:1–12
2. Clayton N, Emery N (2005) Corvid cognition. Curr Biol 15(3):R80–R81
3. Craig J (2005) Introduction to robotics: mechanics and control. Pearson/Prentice Hall, Upper Saddle River, NJ, USA

4. Dereli S, Köker R (2019) A meta-heuristic proposal for inverse kinematics solution of 7-DOF serial robotic manipulator: quantum behaved particle swarm algorithm. Artif Intell Rev 53(2):949–964
5. Hassanien AE, Rizk-Allah RM, Elhosney M (2018) A hybrid crow search algorithm based on rough searching scheme for solving engineering optimization problems. J Ambient Intell Hum Comput
6. Lee KY, El-Sharkawi MAE (2008) Modern heuristic optimization techniques: theory and applications to power systems. IEEE, 455 Hoes Lane Piscataway, NJ 08854
7. Pawlak Z (1982) Rough sets. Int J Comput Inform Sci 11(5):341–356
8. Popovic M (2019) Biomechatronics. Academic, s.l.
9. Prior H, Schwarz A, Güntürkün O (2008) Mirror-induced behavior in the magpie (Pica pica): evidence of self-recognition. PLoS Biol 6(8):e202
10. Rokbani N, Alimi A (2013) Inverse kinematics using particle swarm optimization, a statistical analysis. Procedia Eng 64:1602–1611
11. Rokbani N, Benbousaada E, Ammar B, Alimi AM (2010) Biped robot control using particle swarm optimization. IEEE, Istanbul, Turkey, pp 506–512
12. Rokbani N, Casals A, Alimi AM (2015) IK-FA, a new heuristic inverse kinematics solver using firefly algorithm. In: Computational intelligence applications in modeling and control. pp 369–395
13. Yang Y, Peng G, Wang Y, Zhang H (2007) A new solution for inverse kinematics of 7-DOF manipulator based on genetic algorithm. In: 2007 IEEE international conference on automation and logistics

Chapter 17
Machine Learning for Predicting Cancer Disease: Comparative Analysis

Bador Alqahtani, Batool Alnajrani, and Fahd Alhaidari

1 Introduction

Nowadays, different kinds of data have been widely available on the Internet [9]. In order to have a beneficial use of this data, data mining technology has been produced [9]. One data mining advantage is to get rid of the manual summarization and analysis for the data [9]. The use of data mining technology will assist in many different fields including health, industry, education, and others. Especially in the field of health care to prevent disease or detect it at least early and support decision-making as it provides image analysis for diagnosis. Numerous research has helped to predict various diseases such as heart disease, diabetes, and cancer.

According to the UK's Cancer Research Web site, between 2015 and 2017, the number of daily new cases of cancer diagnoses reached nearly 1000. Four types of cancer were 53%, representing more than half of the cancer. The four types of cancer are breast, prostate, lung, and intestine. In 2017, more than a quarter of the causes of death in the UK are cancer [4]. This paper compares the performance of three of the machine learning algorithms used to predict cancer: the support vector machine (SVM), artificial neural network (ANN), and K-nearest neighbor (KNN).

Cancer is the most common disease spread around the world which can cause death. Although it is a common disease, there is still no definite treatment pattern that doctors can follow for every cancer patient. This makes doctors spend greater

B. Alqahtani (✉) · B. Alnajrani · F. Alhaidari
Department of Computer Science, Imam Abdulrahman Bin Faisal University, Dammam, Saudi Arabia
e-mail: 2200500064@iau.edu.sa

B. Alnajrani
e-mail: 2200500057@iau.edu.sa

F. Alhaidari
e-mail: faalhaidari@iau.edu.sa

A. E. Hassanien et al. (eds.), *Enabling Machine Learning Applications in Data Science*, Algorithms for Intelligent Systems, https://doi.org/10.1007/978-981-33-6129-4_17

amount of time, energy, and money on diagnosing each patient and deciding the most suitable treatment procedure for them.

Therefore, this might complicate the condition of the patient since it will trigger emotional reactions, such as sadness and stress and it might extend to the patient's family. So, cancer not only causes death and physical illness, it also causes psychological illness. Early detection of the cancer can give an opportunity for patients to get cured. Since laboratory tests take a long time to determine the diagnosis, it is time to use the technology in order to get the diagnoses in early stage.

The remaining part of this work is organized as follow. Section 2 contains background. Section 3 contains the review of related literatures. Section 4 contains our discussion and analysis. Section 5 contains the conclusion and recommendation emanating from this work.

2 Background

Cancer is not only of one kind. There are more than 100 different kinds of cancers. It can be any place in the body. Several kinds can be in each part of the body such as the lungs, the breast, the colon, the blood, the skin, or the prostate cancer [13]. When cells die worn out or damaged, the body creates new cells needed, and it is a normal process. Cancer starts when a genetic disorder or change. Cells increase and grow out in the body uncontrollably than normal cells. They can get out of control if they continue to grow which is called a tumor. A tumor divided into cancerous or benign. A cancerous tumor increases and expands to each part in the body, but benign tumor only grows. Cancer patients can receive treatment and live after treatment [13].

There are many ways to treat the cancer in hospital like surgery, radiation therapy, chemotherapy, immunotherapy, targeted therapy, etc. Also, the mental health is important; so, you can go to support groups [6].

In 2019, an estimated 1.8 million new cases of cancer will be diagnosed in the United States and around one-third will die. The most popular cancers diagnosed are breast cancer, lung and bronchus cancer, and prostate cancer. However, deaths due to lung and bronchus cancer are the most common, then the second colorectal cancer and pancreatic cancer is the third.

The pie charts illustrate the new cancer cases and cancer death. This show for both female and male. From Fig. 1, we note that the most common type that has been cheaper in women is breast cancer, and on the other hand, prostate cancer in men, according to Fig. 2 [5].

In addition, Figs. 3 and 4 show deaths due to cancer, and the main cause of death among women is lung and bronchus cancer, which represents approximately one-third of deaths. Likewise, for men, lung and bronchial cancer also accounts for more than a third of deaths.

We can use machine learning (ML) to detect or predict cancer. This allows to classify the type and stage of cancer, and once finished treating cancer, we will know

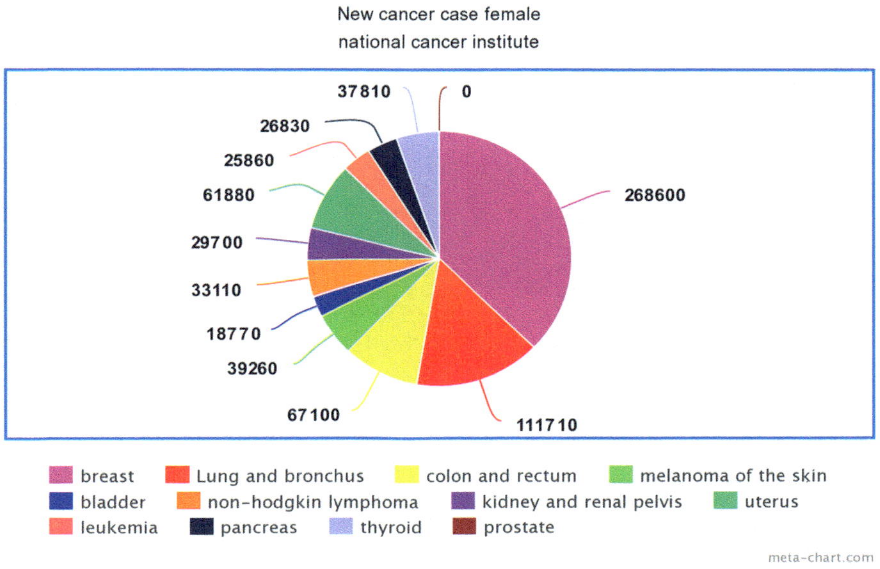

Fig. 1 Female new cancer cases

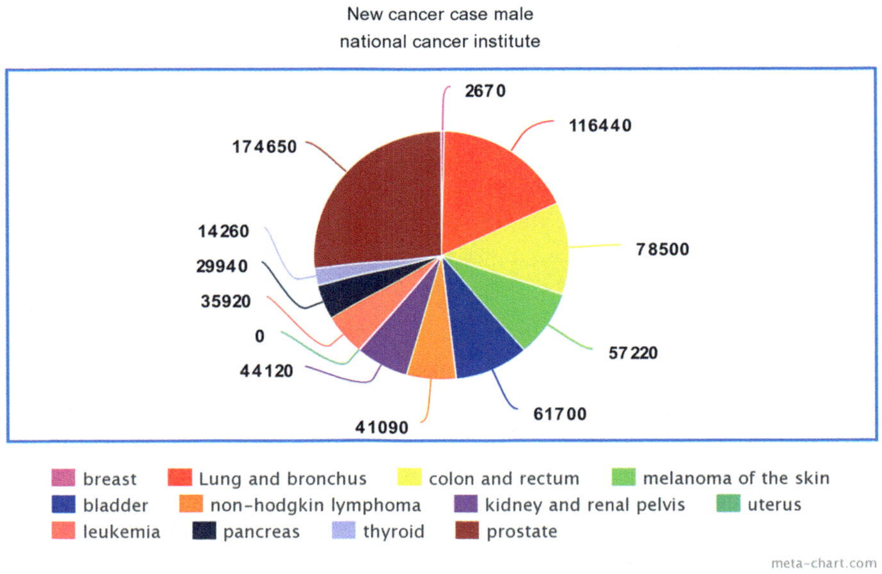

Fig. 2 Male new cancer cases

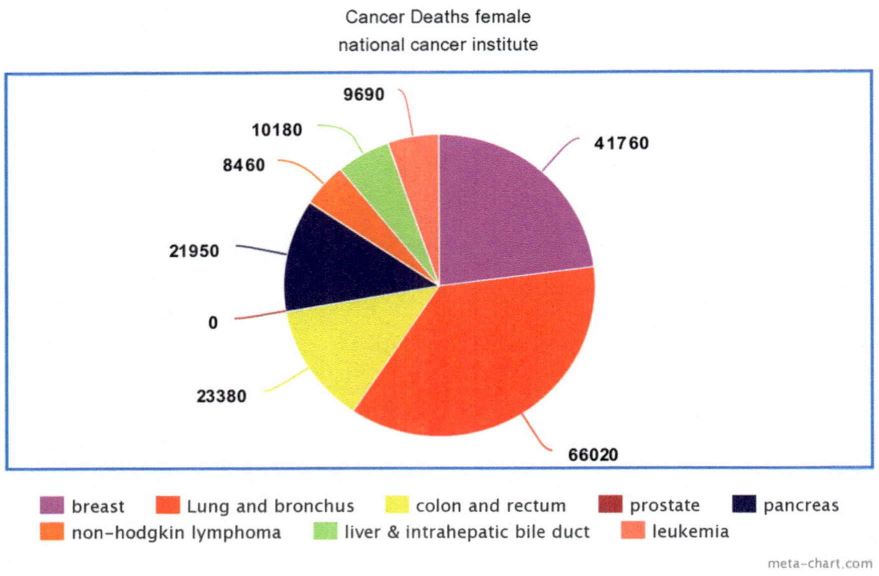

Fig. 3 Female deceased

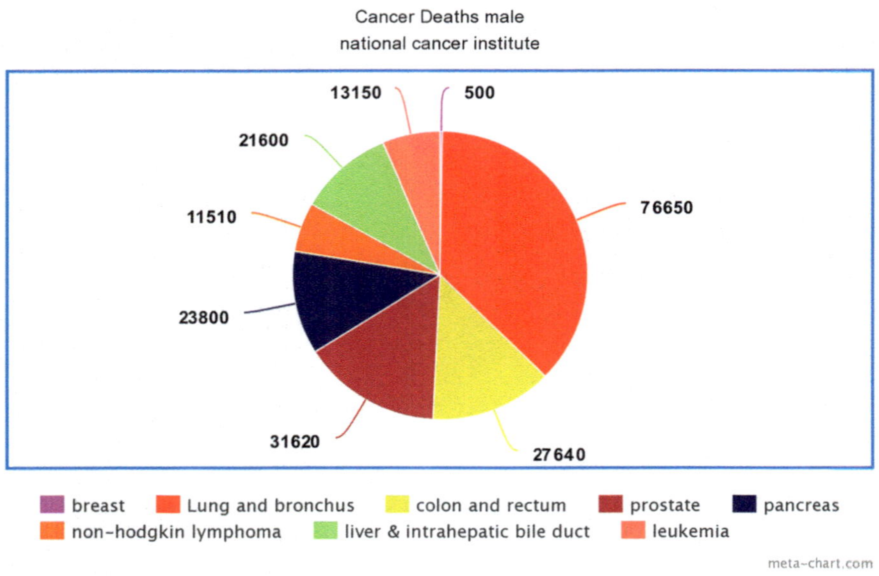

Fig. 4 Male deceased

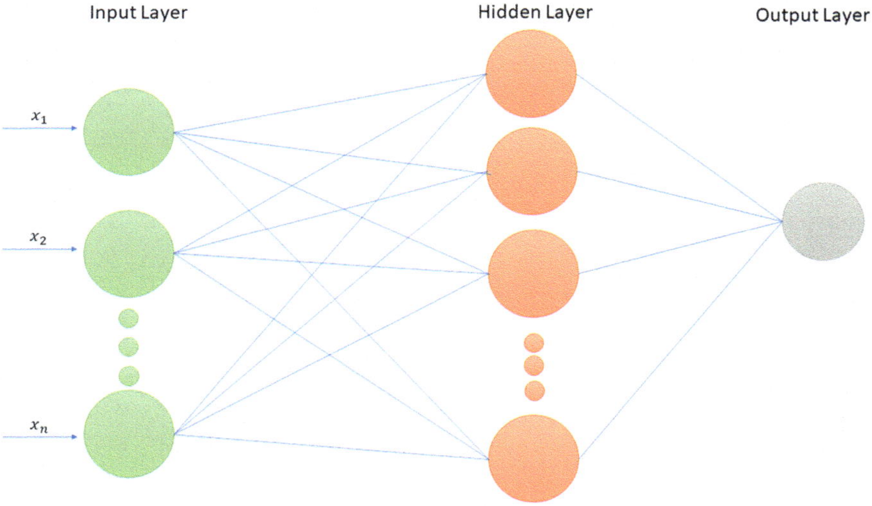

Fig. 5 Artificial neural networks

if it will re-occur or not. Consequently, there are different ways to use ML like ANN, SVM, and KNN to detect or predict cancer [18].

Artificial neural networks (ANNs) is one of the ML methods inspired by nervous systems of the brain which can be used when you have a complex, nonlinear pattern between input and output due to their adaptive nature. It can also be used to predict future recurrences depend on the previous dataset, take a difficult decision, monitoring data integrity, etc.

As presented in Fig. 5, the number of input layers represents a number of variables, and each node has a specific weight [11, 20].

Support vector machine (SVM) is to find a separating line or a hyperplane between data of two classes, it is one of the most powerful supervised classification method machine learning algorithms derived from statistical learning theory.

It gives good results for complex and noisy data. It can be use into groups of linear and nonlinear classification. Figure 6 illustrates the linear classification for the dataset via decision boundary [27].

K-nearest neighbor (KNN) is one of the simplest classification algorithms possible for supervised learning. To find similar data near feature space classification, it depends only on the nearest neighbor. Moreover, it can be used for regression (Fig. 7).

On the other hand, we need to exclude duplicated patterns that do not have added data to further enhance. Thus, the output is the average of values for all nearest neighbors [8, 10, 20].

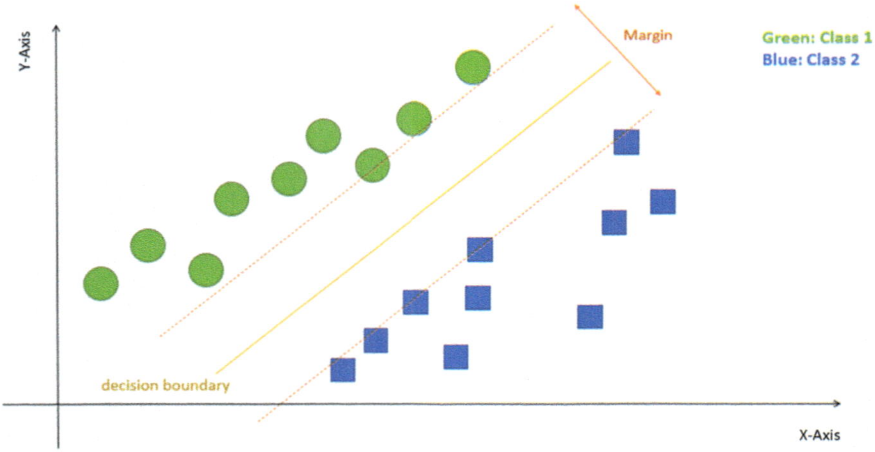

Fig. 6 Support vector machine linearly separable data into two classes

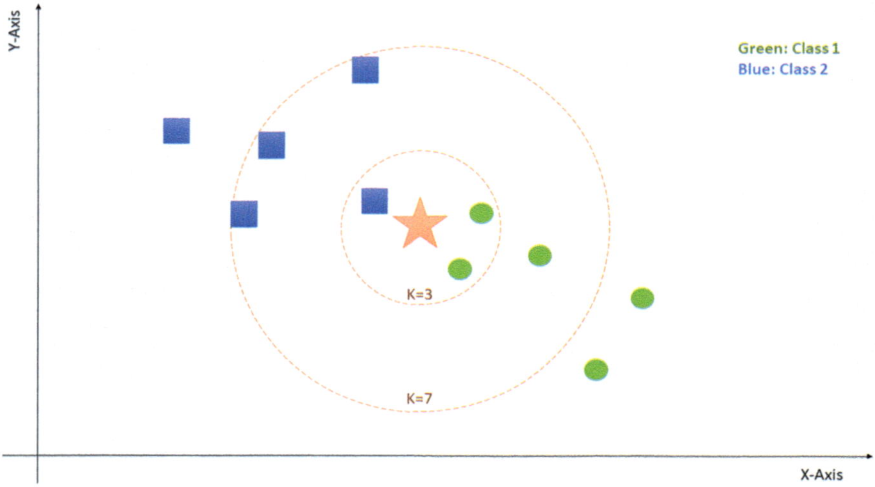

Fig. 7 K-nearest neighbor

3 Machine Learning: Review of Literature

Predicting early diseases such as breast cancer helps in healing and treating the patient with the lowest costs during the treatment stage [2]. Although there exist numerous data mining techniques, SVM, KNN, and ANN are considered the most powerful and superior techniques that are able to provide accurate results [1].

Below, summarization of the recent papers classified based on the classifier algorithms.

3.1 Artificial Neural Network (ANN)

In [24], researcher conducted an experiment to predict the breast cancer by using ANN and Naïve Bayes with accuracy of 86.95% and 83.54%, respectively. The parameters used to achieve the highest accuracy in ANN were 10 neurons in the input and hidden layers while 1 neuron in the output layer along with only one hidden layer. Also, learning rate (α) and coefficient of momentum (β) were changed to 0.2 and 0.3, respectively. Finally, they used Levenberg–Marquardt (trainlm) for the learning algorithm and logarithmic (logsig) sigmoid for the transfer function. While in [19], researcher was trying to find out the causes of lung cancer and also to detect lung cancer by applying ANN, and the accuracy he achieved was 96.67%. To reach that accuracy he used 15 neurons in the input, total of 7 neurons in the hidden layers and 1 neuron in the output layer along with three hidden layers.

Moreover, in [17], researchers conducted an experiment to predict pancreatic cancer by applying ANN. In their paper, the sample number was 800,114, and 18 characters were used for prediction. Researchers focused on sensitivity, specificity, and the area under the curve. It achieved 80.7% for sensitivity and specificity and 0.85 for the area below the curve. They were able to categorize patients into three levels of low, medium, and high cancer risk. Authors, in [15], had implemented affinity network model and a deep learning genre, which achieved satisfactory results were better than traditional neural network models, and the concentration in a number of the closest neighbors is $K = 2$ in kidney cancer and $K = 3$ with uterine cancer.

3.2 Support Vector Machine (SVM)

In [22], the researchers have conducted an experiment to detect the breast cancer disease. The data was obtained from the Gaza Strip hospitals of the Ministry of Health. The technique used is SVM where it achieved an accuracy of 94%. In Fernandez et al. [7], the author is going to comparative between seventh classification algorithms the highest accuracy when used SVM. These different methods to find the greatest algorithms are used to predict the cervical cancer result of the patient biopsy. The study was conducted on 858 randomly selected patients from 2012 to 2013.

Authors in [16] used different data mining techniques for prediction before it started or in the early stage of breast cancer. Data mining, classification algorithms such as KNN, ANN, SVM, decision tree, and logistic regression. Furthermore, for KNN test number of neighbors, SVM algorithm change kernel and C parameters, and logistic regression algorithm, we used solver parameter to change. Also, for decision tree (DT) algorithm, we tried to change min_samples_split and min_samples_leaf. In addition, for ANN algorithm, we tried to change both of the parameters activation and learning_rate. Based on this paper, KNN had better performance and predicted more accuracy SVM then KNN.

3.3 K-Nearest Neighbor Algorithm (KNN)

In [23], researcher conducted an experiment to predict lung cancer by using a lung dataset and applied three different algorithms the KNN, SVM, and DT along with classical featured selection, and the accuracies were 94.5%, 79.4%, and 95.3%, respectively. While in [28], researchers did a review study for six algorithms SVM, Naïve Bayes classifier (NBC), ANN, random forest (RF), KNN, and DT on breast cancer dataset and they found out that SVM and RF gave the highest accuracies of 97.2%, then ANN with 95.8% NBC 95.7%, then KNN with 93.7%, finally, DT 93.0%. Also, Naive Bayes classifier has the highest precision and recall of 97.2 & 97.1%. While in [21], researchers did an experiment on breast cancer dataset. The number of patient records was 198, and the number of features was 38, to help predict. The algorithm applied is KNN by Python languages with the K value of 19 and that gave an accuracy near to 73%.

Moreover, in [3], researchers have done the experiment on 20 different kinds of cancers in order to have clear perception. They used three algorithms KNN, J48, and logistic regression with ten-folds cross-validation. They concluded the paper with their findings of the highest accuracies for each algorithm. The accuracy for KNN was 97.1%, for J48 was 97.8%, and for logistic regression was 98.2%. Shailaja, Seetharamulu, and Jabbar in [25]. Used datasets of the UCI machine learning repository. After implementation of algorithms like KNN, SVM, decision tree, Naive Bayes, sequential minimal optimization (SMO), and expectation maximization algorithm, the (EM) classification algorithm was implemented. The outcome of these was greatest in KNN with future selection achieving 98.14% accuracy.

Additionally, researchers in [26], looking for classified the kind of cancer and the severity of the disease by working with microarray data. They applied KNN with feature particle swarm optimization (PSO). Their approach obtained an accuracy of 69.8%. In addition, the semi-supervisory algorithms are 10% better. In Liu et al. [14], it was also discussed in the study using a hybrid procedure and chooses a small group of informative genes. The results show that the KNN using PCC came out with the best predictor 96.83% accuracy. It used different types of data mining techniques in order to predict and detect breast cancer [12]. In this paper, they the researcher applied several algorithms, so that the random forest (RF) algorithm comes with better performance. As well as in the IBK algorithm, number of nearest neighbors $K = 3$.

Tables 1, 2 and 3 show an overview of predicting cancers by three algorithms.

4 Discussion and Analysis

In this paper, we reviewed the most recent works that have been using three specific algorithms for predicting cancer diseases. Table 4 summarizes the accuracy for the prediction classified based on the classifier.

Table 1 Summary of papers in predicting cancers by using artificial neural network

Reference	Year	Classifier and accuracy
Saritas and Yasar [24]	2019	ANN: 86.95% Naïve Bayes: 83.54%
Nasser and Abu-Naser [19]	2019	ANN: 96.67%
Muhammad et al. [17]	2019	ANN: 85%
Ma and Zhang [15]	2019	Affinity network model: Achieved satisfactory outcomes
Kaya Keleş [12]	2019	Bagging: 90.9% Ibk: 90% Random Committee: 90.9% RF: 92.2% Simplecart: 90.1%

Table 2 Summary of papers in predicting cancers by using support vector machine

Reference	Year	Classifier and accuracy
Priya et al. [22]	2019	SVM: 94%
Fernandes et al. [7]	2018	SVM: Greatest than other algorithms
Meraliyev et al. [16]	2017	KNN: 99.0% SVM: 100% Logistic Regression: 98.0% DT: 98.0% ANN: 98.0%

Table 3 Summary of papers in predicting cancers by using K-nearest neighbor algorithm

Reference	Year	Classifier and accuracy
Rashid et al. [23]	2019	KNN: 94.5% SVM: 79.4% DT: 95.3%
Yadav et al. [28]	2019	SVM: 97.2% NBC: 95.7% ANN: 95.8% RF: 97.2% KNN: 93.7% DT: 93%
Bhuiyan et al. [3]	2019	KNN: 97.1% J48: 97.8% Logistic Regression: 98.2%
Sneka and Palanivel [26]	2019	KNN with PSO: 69.8%
Shailaja et al. [25]	2018	KNN: 98.14%
Liu et al. [14]	2015	KNN: 96.83% SVM: 93.65% RF: 91.27%
Pawlovsky and Nagahashi [21]	2014	KNN: 73%

Table 4 Summarize the accuracy of the recent work for the three algorithms

Algorithms	Accuracy (%)	Cancer dataset
SVM	100	Breast cancer
	97.2	
	94.0	
	79.4	
	93.65	Generic data
KNN	99.0	Breast cancer
	98.14	
	94.5	
	93.7	
	90.9	
	73.0	
	97.1	Lung cancer
	99.7	Liver cancer
	97.0	Kidney cancer
	97.8	Prostate cancer
	69.8	Generic data
	96.83	
ANN	98.0	Breast cancer
	95.8	
	86.95	
	96.67	Lung cancer

Our focus was on reviewing the accuracy of each classifier. The accuracy is testing the performance of the classifier in predicting cases correctly [28]. From that point of view, we notice that the SVM gives the highest accuracy.

After reviewing papers, it has been drawn to our attention that researchers have been focusing on breast cancer as summarized in Sect. 2. Researchers have achieved 100% accuracy for predicting breast cancer disease by using SVM, 99.0% by using KNN and 98.0% by using ANN in [16].

Moreover, in the first section, we mentioned in the UK that the four most recent cancers are breast, lung, intestine, and prostate. According to the National Cancer Institute [4].

In the United States of America, as shown in Figs. 1 and 2, the most common types of cancer are breast cancer for women and prostate for men. Moreover, in Figs. 3 and 4, lung cancer was the most common cause of death for both sexes [5].

From this point, applying and comparing the performance of the three KNN, ANN, and SVM algorithms in an updated dataset to predict early lung cancer diseases. This may help detect the disease at an early stage and reduce the risk of the advanced stage, which helps reduce the death rate for this type of cancer.

5 Conclusion and Future Works

In this paper, we reviewed the most recent work for predicting cancer diseases. After reviewing the papers, we found out that the researchers have been focusing heavily on breast cancer. Until now, our paper is focusing on the past studies. Currently, the finding of this stage is a comparison between three algorithms. This comparison will help researchers expand the database of founded information as well as increase the added value to the research field.

The challenges that we faced during our research were that, first, most of the recent papers that have been found were focused on breast cancer, and when looking for the breast cancer dataset in order to do enhancements, the dataset was old. Likewise, for the first cause of death, lung cancer, we also did not find a recent dataset to apply the three algorithms on it and compare performance among them ANN, SVM, and KNN.

Also, the comparison that we did in this paper did not focus on one type of cancer which makes it more difficult in charge of the scope.

In the following points, the most important recommendations:

- More research is needed on other cancers.
- Find and use an updated dataset for breast cancer.
- We recommend obtaining lung patient data.
- Reducing the parameter for any dataset by applying feature identification may help to obtain better accuracy when forecasting.
- Focus in the confusion matrix as well as accuracy.

References

1. Ravanshad A (2018) How to choose machine learning algorithms?—Abolfazl Ravanshad—Medium. Available at: https://medium.com/@aravanshad/how-to-choose-machine-learning-algorithms-9a92a448e0df. Accessed 9 Dec 2019
2. Alyami R et al (2017) Investigating the effect of correlation based feature selection on breast cancer diagnosis using artificial neural network and support vector machines. In: 2017 international conference on informatics, health and technology, ICIHT 2017. https://doi.org/10.1109/ICIHT.2017.7899011
3. Bhuiyan AI et al (2019) EasyChair preprint predicting cancer disease using KNN, J48 and logistic regression algorithm
4. Cancer Research UK (2019) Cancer Statistics for the UK, Cancer Research UK. Available at: https://www.cancerresearchuk.org/health-professional/cancer-statistics-for-the-uk#heading-Zero. Accessed 12 Nov 2019
5. Common Cancer Sites—Cancer Stat Facts (2019) National Cancer Institute. Available at: https://seer.cancer.gov/statfacts/html/common.html. Accessed 10 Nov 2019
6. Complementary and Alternative Medicine (CAM)—National Cancer Institute (2019) National Cancer Institute. Available at: https://www.cancer.gov/about-cancer/treatment/cam. Accessed 10 Nov 2019
7. Fernandes K et al (2018) (2018) Supervised deep learning embeddings for the prediction of cervical cancer diagnosis. PeerJ Comput Sci 5:1–20. https://doi.org/10.7717/peerj-cs.154

8. Goel A, Mahajan S (2017) Comparison: KNN & SVM algorithm. Int J Res Appl Sci Eng Technol (IJRASET). Available at: www.ijraset.com. Accessed 9 Nov 2019
9. Goyal A, Mehta R (2012) Performance comparison of Naïve Bayes and J48 classification algorithms. Int J Appl Eng Res 7(11 Suppl):1389–1393
10. Hemphill GM (2016) A review of current machine learning methods used for cancer recurrence modeling and prediction intended for: report for pilot 3 of NCI Project. Available at: https://permalink.lanl.gov/object/tr?what=info:lanl-repo/lareport/LA-UR-16-27376
11. Karayiannis NB, Venetsanopoulos A (1993) Neural network architectures and learning schemes. In: Artificial neural networks. The Springer international series in engineering and computer science
12. Kaya Keleş M (2019) Breast cancer prediction and detection using data mining classification algorithms: a comparative study. Tehnicki Vjesnik 26(1):149–155. https://doi.org/10.17559/TV-20180417102943
13. Listing A, Diseases OF (2009) Handbook of autopsy practice. Humana Press
14. Liu YX et al (2015) Prediction of core cancer genes using a hybrid of feature selection and machine learning methods. Genet Mol Res 14(3):8871–8882. https://doi.org/10.4238/2015.August.3.10
15. Ma T, Zhang A (2018) AffinityNet: semi-supervised few-shot learning for disease type prediction. Available at: https://arxiv.org/abs/1805.08905
16. Meraliyev M, Zhaparov M, Artykbayev K (2017) Choosing best machine learning algorithm for breast cancer prediction. Int J Adv Sci Eng Technol (5):2321-9009. Available at: https://iraj.in
17. Muhammad W et al (2019) Pancreatic cancer prediction through an artificial neural network. Front Artif Intell 2(May):1–10. https://doi.org/10.3389/frai.2019.00002
18. Mulatu D (2017) Survey of data mining techniques for prediction of breast cancer recurrence. Int J Comput Sci Inf Technol 8(6):599–601
19. Nasser IM, Abu-Naser SS (2019) Lung cancer detection using artificial neural network
20. Nikam SS (2015) A comparative study of classification techniques in data mining algorithms. Oriental J Comput Sci Technol 8(1):13–19. Available at: https://www.computerscijournal.org/vol8no1/a-comparative-study-of-classification-techniques-in-data-mining-algorithms/. Accessed 10 Nov 2019
21. Pawlovsky AP, Nagahashi M (2014) A study on the use of the KNN algorithm for prognosis of breast cancer. Available at: https://www.mhlw.go.jp/toukei/saikin/hw/jinkou/kakutei12/. Accessed 10 Nov 2019
22. Priya AM et al (2019) Identify breast cancer using machine learning algorithm. Int J Eng Sci Math 8
23. Rashid S et al. (2019) Lung cancer classification using data mining and supervised learning algorithms on multi-dimensional data set. 7(2):438–447. Available at: https://pen.ius.edu.ba. Accessed 10 Nov 2019
24. Saritas M, Yasar A (2019) Performance analysis of ANN and Naive Bayes classification algorithm for data classification. Int J Intell Syst Appl Eng
25. Shailaja K, Seetharamulu B, Jabbar MA (2018) Prediction of breast cancer using big data analytics. Int J Eng Technol 7(46):223. https://doi.org/10.14419/ijet.v7i4.6.20480
26. Sneka T, Palanivel K (2019) Pattern similarity based classification using K-nearest neighbor and PSO model for cancer prediction with genetic data. (8):27–31
27. Wen T, Edelman A (2003) Support vector machines: algorithms and applications. pp 249–257
28. Yadav A et al (2019) Comparative study of machine learning algorithms for breast cancer prediction-a review. https://doi.org/10.32628/CSEIT1952278

Chapter 18
Modeling and Performance Evaluation of LoRa Network Based on Capture Effect

Abdessamad Bellouch, Ahmed Boujnoui, Abdellah Zaaloul, and Abdelkrim Haqiq

1 Introduction

The Internet of things (IoT) is one of the technologies of the fourth Industrial Revolution that is speedily gaining ground in the scenario of modern wireless communication systems. The IoT devices are growing remarkably and expected to reach 41.6 billion by 2025 [1]. These smart devices are characterized by their low complexity, their low power consumption, and also by their ability to interact without human involvement. IoT technology offers the interconnection between physical objects (devices) that surround us, to send and/or receive information to or from the Internet. Accordingly, many technologies can make these devices connect to the network backbone, such as cellular communication (e.g., 3G, 4G, 5G) for long-reach situations [2, 3], short-range IoT network solutions (e.g., Wi-Fi, Bluetooth, ZigBee, NFC/RFID) for short-range wireless communication, and long power wide area networks (LPWAN) solutions for addressing the various requirements of long-range and low-power IoT applications [4].

Long-range low-power communication technologies such as LoRa [5], Sigfox [6], NB-IoT [7], weightless [8] represent a new and fast fashion in the evolution of wire-

A. Bellouch (✉) · A. Boujnoui · A. Haqiq
Computer, Networks, Mobility and Modeling laboratory: IR2M, Faculty of Sciences and Techniques, Hassan First University of Settat, Settat 26000, Morocco
e-mail: a.bellouch@uhp.ac.ma

A. Boujnoui
e-mail: a.boujnoui@uhp.ac.ma

A. Haqiq
e-mail: abdelkrim.haqiq@uhp.ac.ma

A. Zaaloul
Complex System and Interactions research team, Ibn Zohr University, F.S.J.E.S. Ait Melloul, Agadir, Morocco
e-mail: a.zaaloul@uiz.ac.ma

A. E. Hassanien et al. (eds.), *Enabling Machine Learning Applications in Data Science*, Algorithms for Intelligent Systems, https://doi.org/10.1007/978-981-33-6129-4_18

less communication technologies. These technologies offer long-range connectivity (up to several tens of kilometers) for low-power (years of battery operation) and low-rate devices (tens of kbps), not provided by legacy technologies and are thus promising and significant technologies in various IoT application scenarios, such as smart grids, smart cities, especially in monitoring field. Among these technologies, LoRa has gained great interest in recent years due to its reliable physical layer (PHY) design which provides low power consumption for long-range communications, and also, it enables resilience to noise [9].

LoRa is an acronym for long range, and it typically uses Chirp Spread Spectrum (CSS) modulation. Due to this modulation, signals with different spreading factors (SF) can be received simultaneously, even if they were transmitted at the same time over the same channel [10]. CSS scheme maintains the same low power characteristics as frequency shift keying (FSK) scheme, but this latter significantly increases the communication range. LoRa is usually used with LoRaWAN where this latter is essentially designed to define MAC protocols used to wirelessly connect LoRa devices to the gateways. LoRa network is characterized by a large number of end-devices that are connected to the gateway, and this has profound consequences for the efficiency of such networks. As such, one of the main technical challenges with LoRa technology is the collision that occurs at the gateway when two or more end-devices (LoRa network nodes) transmit their data simultaneously using the same SF and over the same channel. The collision's result is that losing all colliding packets. However, thanks to the capture effect (CE) technique [11], the receiver can sometimes successfully decode one packet even in the case of a collision. In other words, the capture effect is defined as the receiver's ability to receive one out of two (or more) colliding frames correctly.

Our main contribution in this paper is three-fold. Firstly, we evaluate the impact of the collision caused by simultaneous transmission using the same SF and over the same channel in LoRa networks, showing that such collision drastically reduces the performance of a LoRa network. Secondly, we measure network performance gains by introducing the CE technique. Thirdly, the numerical results show that our proposed technique largely outperforms the conventional LoRa jointly in terms of network throughput and delay of transmitted packets. Most importantly, our results prove that our proposed technique outperforms the zigzag decoding (ZD) approach [12, 13] due to the large amount of end-devices connected in the LoRa network.

In this work, we model a LoRa network consisting of a gateway and end-devices operating on different power levels. Each end-device can transmit packets (new arrival or collide packets) using one of the available power levels. When a collision happens, the gateway may be able to decode the packet transmitted at the highest power among all the concurrent transmissions so that the LoRa network performance is improved.

The rest of this paper is structured as follows. Section 2 describes briefly the related work. In Sect. 3, we present an essential background on the LoRa network. Section 4 introduces the problem addressed by our study. In Sect. 5, we develop the Markov model and the team formulation of the problem. We evaluate the performance of the model numerically in Sect. 6, and Sect. 7 concludes the paper.

2 Related Work

Although LoRa technology is relatively new and still under development, it has gained the attention of many researchers. Several efforts are focusing on evaluating LoRa performance in terms of collision rate, capacity, energy consumption, and scalability. They are based either on simulations results [14–16], or mathematical modeling in [17, 18] or even on real measurements [19–23].

In [14], the authors presented a simulation analysis of a LoRa network. They showed that when the fraction of nodes that requires confirmed traffic grows excessively, the performance degrades in a LoRa network. Numerical results showed that carefully choose the maximum number of transmission attempts for confirmed packets, based on the node density and traffic load, can enhance the system performance. Then, they discussed some realistic issues such as duty cycle, data rate, and multiple receive paths at the gateways were addressed to exalt the advantages of LoRaWAN technology.

In [15], the authors assessed the performance of a LoRa-based IoT network consisting in tens of thousands of end-devices by implementing a new system-level simulator in ns–3. Simulation results demonstrated that partial orthogonality between the spreading factors can improve LoRa network performance. Moreover, they proved that when a gateway is serving a number of end-devices in the arrangement of 104, a packet success rate above 95% is attained.

In [16], the authors analyzed the LoRa performance only by simulation. They showed that the capture effect technique and serial interference cancelation method [17] can drastically improve the LoRa performance.

To improve the LoRa network performance, the authors in [18] proposed two methods. The first is message replication to exploit time diversity, and the other is multiple receive antennas to exploit spatial diversity. They evaluated these two methods through theoretical analysis and computer simulations. Results show that the proposed methods can enhance LoRa network performance and also the combination of both methods can be more beneficial.

In [19], the authors presented a theoretical study and an experimental evaluation on the performance of LoRa technology. They assessed via a numerical model the LoRa performance in additive White Gaussian noise (AWGN) channels, to clarify its communication abilities in negative signal-to-noise ratio conditions. They also proposed an analytic bit error-rate expression to examine the impact of the code rate parameter in the LoRa performance.

Based on real measurements, the authors in [20] evaluated the indoor propagation performance of LoRa technology. Then, they presented all the models that best identify the indoor propagation of LoRa technology. They proved that multi-wall has the best overall performance among all the models.

In [21], the authors evaluated the impact of inter-network interference on LoRa networks, and they have shown that such interference can considerably reduce the performance of a LoRa network. They have proposed to use multiple base stations and directional antennae to improve LoRa network performance. The numerical results show that using multiple base stations functions well than the use of directional antennae.

Authors in [22] evaluated the impact of LoRa transmission parameter selection on communication performance, especially on energy consumption and communication reliability where a bad choice of transmission parameter setting for LoRa nodes can reduce energy consumption. They have presented an algorithm that is capable of quickly finding the good transmission parameter choice among 6720 transmission parameter combinations available for each node.

In another related paper [23], the authors evaluated how data rate affects the packet loss. They considered the packet loss, RSSI, SNR, delay, power consumption, and packet error to verify how the average current consumption of one end-device impacts in the performance of the LoRa network with only one end-device and one gateway.

In [24], the authors evaluated LoRa network performance in a tree farm. They showed that there is a big impact of PHY factors configurations, spreading factor, coding rate, and different distances on LoRa performance.

3 LoRa Overview

This section gives an overview of LoRa physical layer and LoRaWAN technology, and the LoRa system architecture.

3.1 LoRa Physical Layer

LoRa is a physical layer identifications based on Chirp Spread Spectrum (CSS) modulation [25]. CSS modulation offers low sensitivity required for long communication ranges while maintaining low power. CSS was suggested for the first time for communication systems [26] and for application to digital communication [27]. Carrier frequency, transmitted power, spreading factor, bandwidth, and coding rate characterize the LoRa transmission. The selection of these five configuration parameters determines energy consumption, transmission range, and resilience to noise.

- Bandwidth (BW): It is the range of transmission frequencies. It can be set from 7.8 up to 500 kHz. Larger BW increases the data rate (the chips are sent out at a rate equivalent to the bandwidth), but it gives a lower sensitivity (due to integration of additional noise).
- Spreading factor (SF) defines the number of raw bits that can be encoded in each symbol, and it represents also the ratio ($SF = \log_2(R_c/R_s)$) between the symbol rate R_s and chip rate R_c. SF can be selected from 7 to 12. A larger SF such as SF12 increases the transmission duration (time on-air), which increases energy consumption, reduces the transmission rate, and improves communication range.

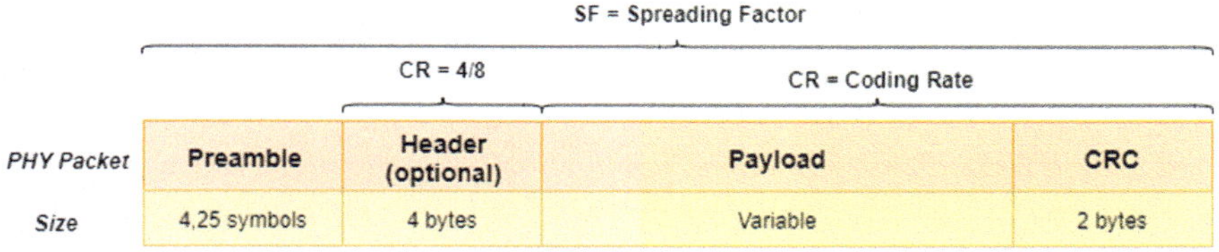

Fig. 1 LoRa packet structure: PHY packet combining a preamble, a header, a payload, and an error-detecting code (CRC) in one structured frame [16]

- Coding Rate (CR): Coding rate (or information rate) corresponds to the rate of forward error correction (FEC) used to improve LoRa performance in terms of packet error rate. A higher coding rate results in more protection against burst interference, but it increases the time on-air. The available values of CR are $4/5$, $4/6$, $4/7$, and $4/8$.
- Transmitted Power (TP): It is the power needed to send a specific data packet. TP has a direct influence on the throughput performance of LoRa network. Since LoRa employs bidirectional communication, for uplink (sending data from an end-device to the gateway), the maximum TP is limited to 25 mW (14 dBm). For downlink (sending responses from the gateway to an end-device), the maximum TP is limited to 0.5 W (27 dBm).
- Carrier Frequency (CF): CF represents the medium transmission frequency.

Figure 1 shows the structure of a LoRa packet. The LoRa packet comprises three elements, and it starts with a preamble followed by an optional header. After the header, there is a payload and at the end of the packet is the cyclic redundancy check (CRC). When a packet is transmitted from an end-node, it takes a certain amount of time before a gateway receives this packet. This time is called time on-air, and it can be defined as

$$T_{\text{air}} = T_s(n_{\text{preamble}} + n_{\text{payload}} + 4.25), \tag{1}$$

where n_{preamble} and n_{payload} are the preamble and payload lengths. T_s denotes symbol period, and $T_s = \frac{1}{R_s}$ with $R_s = \frac{\text{BW}}{2^{\text{SF}}}$ is symbol rate. The constant value 4.25 shows the minimum length of preamble.

3.2 LoRaWAN

LoRaWAN is an open standard administered by the LoRa Alliance. LoRaWAN defines the communication protocol on top of the LoRa physical layer. It also determines the system framework. LoRaWAN has three different classes of end-devices, which are [28]:

- Class A: At any time, an end-device of this class can broadcast a packet (ALOHA-type of protocol). After an uplink transmission (tx), the end-device will listen for a response from the gateway where it opens two receive windows at t1 and t2 seconds. The gateway can respond within the first receive slot or the second receive slot, but not both. This class is considered more appropriate for energy-constrained end-devices.
- Class B: In addition to Class A receive windows, class B end-devices open additional receive windows at defined durations. The end-device receives a response from the gateway for the time synchronization, permitting the gateway to know when the end-device is listening.
- Class C: In addition to Class A receive windows, a class C end-device will listen continuously for responses from the gateway (closed when transmitting). Class C end-devices are more energy consuming as compared with Class A and Class B counterparts.

3.3 LoRa Network Architecture

The architecture of LoRa network relies on a star-of-stars topology as shown in Fig. 2. As can be seen, gateways transfer packets between end-devices and the network server where the end-devices connect to one or several gateways through LoRa wireless link, and the gateways connect to the network server through an IP-based network and act as a bridge to the IP world.

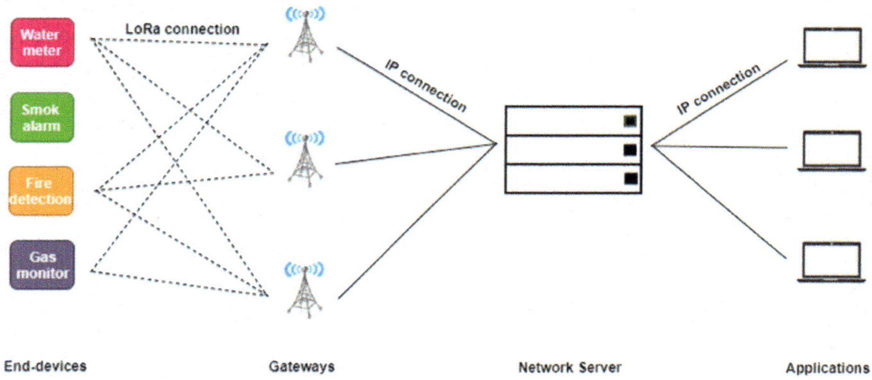

Fig. 2 LoRa network architecture implemented as a star-of-stars topology, in which a central network server and end-devices are connected through the gateways [29]

4 Model Scenario and Problem Statement

We consider a LoRa network consisting of m end-devices connected to a single gateway. Since most collisions occur at the gateways due the large number of deployed end-devices, we are interested only in the interference during the uplink transmissions (transmission path from a node to the gateway). For simplicity, we consider all these end-devices have no queue, i.e., they do not generate a new packet till the current one is successfully transmitted.

Each end-device generates a new packet with a Bernoulli process with parameter q_a, and transmits it over a common channel to the gateway with probability one. Moreover, a collision occurs when two or more end-devices transmit packets simultaneously using the same spreading factor. All collided packets are assumed to be corrupted, and they are re-transmitted in the following slots with probability q_r. We assume that all end-devices use slotted ALOHA as a random access mechanism. Hence, the time axis is divided into slots, all end-devices are synchronized, and they are allowed to transmit only at the beginning on a slot time.

We consider a scenario, when two or more packets transmitted simultaneously with the same spreading factor, using the capture effect technique, the receiver might be able to perform the successful reception of the packet transmitted with high power. In fact, all concurrent packets are treated as interfering noise, and they are lost, but the packet with the highest power is successfully received due to that its SINR is higher than the target SINR. We assume that each end-device can transmit its packet (new arrival packets or backlogged packets) using one of N distinct available power levels. Let $T = \{T_1, T_2, \ldots, T_N\}$ be the power levels available to the end-devices to transmit their packets.

5 Markov Model

In this section, we develop a Markov model of the LoRa network senario described in Sect. 4. We consider n_t the number of backlogged end-devices at the beginning of the slot time t. n_t can be modeled using a discrete-time Markov chain where its transition diagram is depicted in Fig. 3. Our Markov chain is irreducible and includes a finite set of states $\{0, 1, \ldots, m\}$. Therefore, it has a unique stationary distribution for any value $q_r > 0$ (see Remark 1). The transition probabilities are given below.

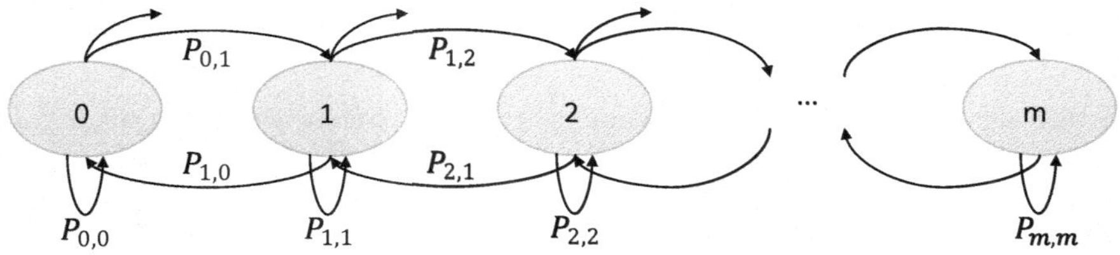

Fig. 3 Markov transition diagram

$$P_{n,n+i}(q_a, q_r) = \begin{cases} Q_a(i,n)[\sum_{j=0}^{n} Q_r(j,n)(1-B_{j+1})], \quad i = m-n\,,\, i \geq 2 \\ Q_a(i,n)[\sum_{j=0}^{n} Q_r(j,n)(1-B_{j+1})] \\ \quad +Q_a(i+1,n)[\sum_{j=0}^{n} Q_r(j,n)B_{j+i+1}], \quad 2 \leq i \leq m-n \\ Q_a(1,n)[\sum_{j=0}^{n} Q_r(j,n)(1-B_{j+1})] \\ \quad +Q_a(2,n)[\sum_{j=0}^{n} Q_r(j,n)B_{j+2}], \quad i = 1 \\ Q_a(0,n)[Q_r(0,n) + \sum_{j=2}^{n} Q_r(j,n)(1-B_j)] \\ \quad +Q_a(1,n)[Q_r(0,n) + \sum_{j=1}^{n} Q_r(j,n)B_{j+1}], \quad i = 0 \\ Q_a(0,n)[Q_r(1,n) + \sum_{j=2}^{n} Q_r(j,n)B_j], \quad i = -1 \\ 0 \quad \text{Otherwise.} \end{cases} \tag{2}$$

where $Q_r(j,n) = \binom{n}{j}(1-q_r)^{n-j}q_r^n$ represents the probability that j out of the n ($n = 0, 1, \ldots, m$) backlogged packets are re-transmitted in the current slot, and $Q_a(j,n) = \binom{m-n}{j}(1-q_a)^{m-n-j}q_a^j$ represents the probability that j unbacklogged end-devices transmit packets in a given slot. And B_s represents the probability of a successful packet among $s \geq 2$ simultaneous packet transmissions, and it is given by

$$B_s = s\sum_{l=2}^{N} X_l \left(1 - \sum_{i=l}^{N} X_i\right)^{s-1}, \tag{3}$$

with $B_0 = 0$ and $B_1 = 1$. The X_k is the probability that an end-device (with new arrival or backlogged packet) (re)transmits selecting power level T_k among N available power levels.

5.1 Performance Parameters for Unbacklogged Packets

In this subsection, the performance parameters of the proposed technique have been derived. Since the considered Markov chain is irreducible and the state space is finite, then it admits a stationary distribution. Additionally, let $\pi(q_a, q_r)$ denotes the vector of steady-state probabilities, where its n-th entry is $\pi_n(q_a, q_r)$. It denotes the probability of n backlogged packets at the beginning of a slot. Subsequently, the steady state of our Markovian process is given by the following set of linear equations:

$$\begin{cases} \pi(q_a, q_r) = \pi(q_a, q_r)P(q_a, q_r) \\ \pi_n(q_a, q_r) \geq 0 \,, \qquad n = 0, 1, \ldots, m \\ \sum_{n=0}^{m} \pi_n(q_a, q_r) = 1. \end{cases} \tag{4}$$

π can be calculated by solving the system (4) using a simple iterative method. Therefore, the average number of backlogged packets is given by:

$$S(q_a, q_r) = \sum_{n=0}^{m} \pi_n(q_a, q_r)n. \tag{5}$$

The average throughput is defined as the average number of packets that are successfully transmitted (new packets and backlogged packets) which is given by:

$$Th(q_a, q_r) = \sum_{n=1}^{m} P_{\text{succ}}^{n} \pi_n(q_a, q_r) + Q_a(1, 0)\pi_0(q_a, q_r), \tag{6}$$

where $P_{\text{succ}}^{n} = \sum_{i=0}^{m-n} Q_a(i, n) \sum_{j=0}^{n} Q_r(j, n)B_{i+j}$, $n = 1, \ldots, m$, represents the probability of a successful transmission which is the expected number of successful transmissions. Therefore, using the normalization formula $\sum_{n=0}^{m} \pi_n(q_a, q_r) = 1$, the average throughput can be rewritten as follow:

$$Th(q_a, q_r) = q_a \left(m - S(q_a, q_r)\right). \tag{7}$$

We define the expected delay as the number of slots used by a packet going from its source to its destination. It is obtained using the Little formula [30].

$$D(q_a, q_r) = \frac{Th(q_a, q_r) + S(q_a, q_r)}{Th(q_a, q_r)} = 1 + \frac{S(q_a, q_r)}{q_a(m - S(q_a, q_r))}. \tag{8}$$

5.2 Performance Parameters for Backlogged Packets

Another way to evaluate the performance of LoRa network is to analyze the ability to serve the backlogged packets. Thus, the average throughput for backlogged packets is given by

$$Th_{\text{backlog}}(q_a, q_r) = Th(q_a, q_r) - Th_{\text{succ}}(q_a, q_r), \tag{9}$$

where $Th_{\text{succ}}(q_a, q_r)$ is the throughput of packets which are successfully transmitted at the first attempt, is calculated by

$$Th_{\text{succ}}(q_a, q_r) = \sum_{n=0}^{m} \sum_{i=1}^{m-n} \sum_{j=0}^{n} \left(\frac{i}{i+j} Q_a(i, n) Q_r(i, n) Q_r(j, n) B_{i+j} \right) \pi_n(q_a, q_r). \tag{10}$$

Thereafter, we can compute the expected delay of backlogged packets, which is defined as the average time (in slots) that a backlogged packet takes to go from the source to the receiver. It is obtained using the Little's formula.

$$D_{\text{backlog}}(q_a, q_r) = 1 + \frac{S(q_a, q_r)}{Th_{\text{backlog}}(q_a, q_r)}. \tag{11}$$

5.3 Finding the Social Optimum

The analysis of the Eqs. (7) and (8) shows that maximizing the throughput is equivalent to minimizing the expected delay of transmitted packets. Thus, the optimal solution of the team problem is obtained by resolving the following optimization problem:

$$\max \text{objective}(q_a, q_r) \quad \text{s.t.}$$
$$\begin{cases} \pi(q_a, q_r) = \pi(q_a, q_r) P(q_a, q_r) \\ \pi_n(q_a, q_r) \geq 0\,, \qquad n = 0, 1, \ldots, m \\ \sum_{n=0}^{m} \pi_n(q_a, q_r) = 1, \end{cases} \tag{12}$$

where the objective function objective(q_a, q_r) could be replaced by any performance metric function (e.g., the average throughput or minus the expected delay).

Remark 1 We emphasize that for $q_r = 0$ and $q_a \neq 0$ the Markov chain introduced in 5 may end up in one of the two absorbing states ($n = m$ and $n = m - 1$) where the throughput is, respectively, 0 and q_a. These absorbing states are reached with a non-null probability from the non-absorbing states and represent a deadlock state for the overall system. Therefore, we shall exclude the case of $q_r = 0$ and consider only the range $[\epsilon, 1]$, where $\epsilon = 10^{-4}$.

Remark 2 The steady-state probabilities $\pi(q_r)$ are continuous over $0 < q_r \leq 1$. This is not a closed interval, and therefore, a solution does not necessarily exist. However, as we restrict to the closed interval $q_r \in [\epsilon, 1]$ where $\epsilon > 0$, an optimal solution indeed exists. Therefore, for any $\gamma > 0$, there exists some $q_r > 0$ which is γ-optimal. $q_r^* > 0$ is said to be γ-optimal for the average throughput maximization if it satisfies $Th(q_r^*) \geq Th(q_r) - \gamma$ for all $q_r \in [\epsilon, 1]$. A similar definition holds for any objective function (e.g., average delay minimization).

Stability: A qualitative approach to deal with the performance of our protocol is to study its stability. The expected number of successful transmissions is just the probability of a successful transmission P_{succ}. Let us define the $drift$ in state n (D_n) as the expected change in backlog over one slot time, starting in state n. Thus, D_n is the expected number of new arrivals accepted into the system [i.e., $q_a(m-n)$] minus the expected number of successful transmissions in the slot P_{succ}. So, $D_n = q_a(m-n) - P_{\text{succ}}$.

If $D_n < 0$, then the system tends to become less backlogged, and we are in a good situation. Otherwise, if $D_n > 0$, then the system tends to become more backlogged, and the situation is not good.

6 Numerical Results

In this section, we compare the performance of the LoRa network with and without the proposed CE technique, and we also compare our approach with the ZD approach reported in [31]. To accurately capture the massive connectivity of LoRa network, we vary the number of devices from 100 to 1000. We assume that all nodes operate in the same wireless channel and use the same SF.

In LoRa network, the end-devices are usually battery powered devices (e.g., IoT wireless sensors, …). This kind of devices is known for low power consumption and low data transmission rate [32]. To this end, we set $q_a = \alpha$ and $0 \leq q_r \leq \alpha$, where α can be adjusted according to the amount of data generated by the end-device. Throughout this work, we set $\alpha = 0.1$. All numerical results are conducted with $N = 10$ (the number of power levels) and with an uniform distribution for selecting a power level among tens available power levels.

We show in Fig. 4 the optimal retransmission probability q_r^*, which is obtained by solving the team problem (12). The results show that q_r^* decreases as the number of devices increases. Indeed, it is required to lower the transmission rate when a large number of devices attempt to access the wireless medium. To have a better insight on how a large number of devices affect the performance of the system, we consider the example of 1000 nodes transmitting with a probability of 0.1 $(packet/slot)$. Even though this probability is low, the amount of traffic generated by all devices combined

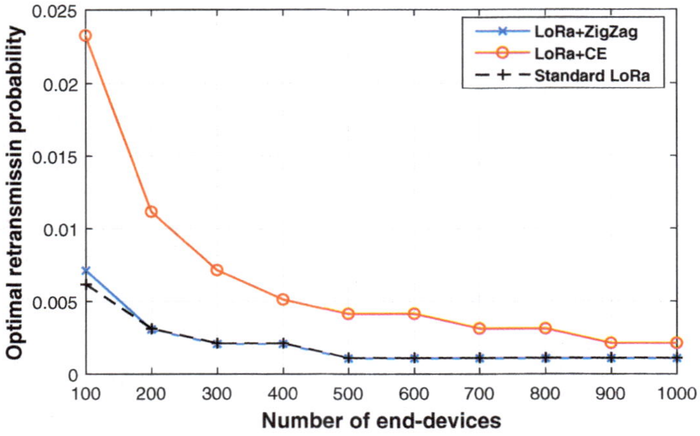

Fig. 4 Optimal retransmission probability as a function of number of connected devices

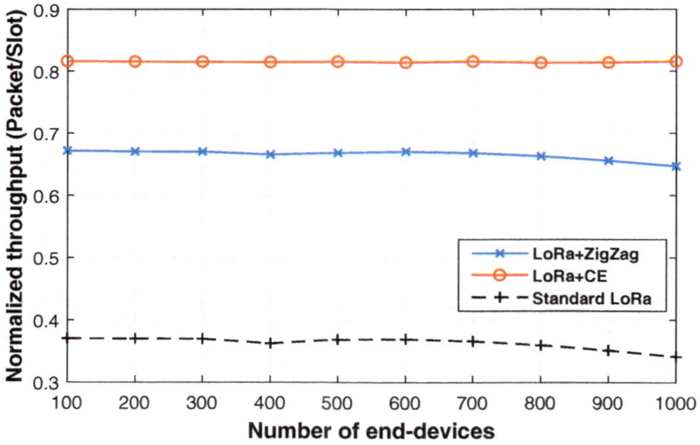

Fig. 5 Average throughput Th versus the number of devices

yields a traffic congestion and a dramatic decrease in the performance of the overall system. This is why the retransmission probability should be lower when we deal with a large number of devices. Our results also show that the optimal retransmission probability in our case of the CE is slightly higher than the one of ZD. This is due to the fact that a receiver implementing the CE technique can successfully decode the packet from N simultaneous transmission (as soon as it is the only one transmitted at the higher power). However, the ZD technique is able to successfully decode the packet only from two simultaneous transmissions.

Figure 5 shows the normalized throughput as a function of the number of nodes. The results show that the CE maintains the throughput above 0.8 for any number of

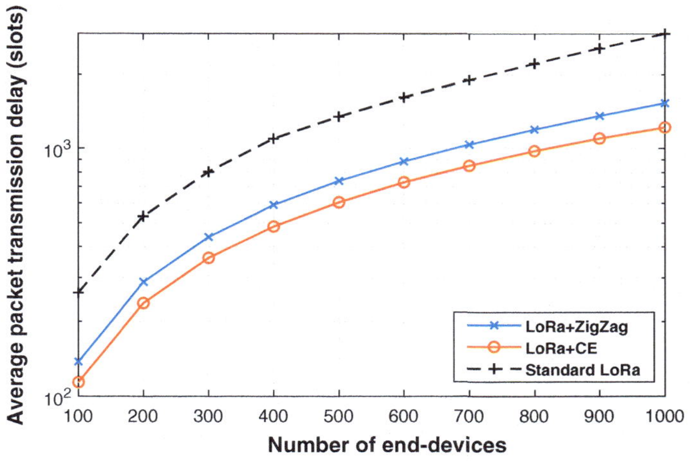

Fig. 6 Delay D versus the number of devices

nodes. However, for ZD approach, a slight decrease in the throughput is observed when the number of devices gets close to 1000. The reason behind is quite simple: The ZD approach is beneficial when a collision occurs between exactly two simultaneous transmissions. Therefore, when three or more packets collide, the LoRa network improved with ZD behaves like the standard LoRa, and all collided packets get lost. This latter case is more likely to happen when a large number of devices compete for channel access. However, unlike ZD, the CE is beneficial for any number of collided packets, even for 1000 simultaneous transmissions. This is why the throughput remains higher. The results also show a significant improvement of 44% in the throughput compared to the standard LoRa.

We present in Fig. 6 the packet delay for different number of connected nodes. The delay is depicted in terms of slots and represents the time elapsed from the transmission of the packet to the moment that it is successfully received. As the number of nodes increases, the delay tends to increase due to the collisions arising from simultaneous transmissions. Furthermore, the CE proves very effective in term of delay compared to ZD. We emphasize that Figs. 5 and 6 show, respectively, the optimal throughput and delay when all nodes use the optimal transmission probability depicted in Fig. 4.

7 Conclusion and Perspectives

In this paper, we investigated the impact of packet collisions in LoRa network where a large number of devices attempt to access channel with the same spreading factor. To address this issue, we proposed to introduce the capture effect (CE): a technique capable of decoding one packet among multiples simultaneous transmissions. We modeled our system using a discrete-time Markov chain, then we derived the performance metrics at the steady state. To show the effectiveness of our proposal, a comparative study is carried out with the zigzag decoding (ZD) approach. Our results show that LoRa with CE performs way better than standard LoRa in terms of all performance parameters of interest. Moreover, it outperforms the ZD approach introduced in the literature. Our future work includes the implementation of LoRa technology in the healthcare field and for wireless body area network (WBAN).

References

1. Framingham M (2019) The growth in connected IoT devices is expected to generate 79.4 ZB of data in 2025, according to a new IDC forecast
2. Akpakwu GA, Silva BJ, Hancke GP, Abu-Mahfouz AM (2017) A survey on 5G networks for the internet of things: communication technologies and challenges. IEEE Access 6:3619–3647
3. Zayas AD, Pérez CAG, Pérez ÁMR, Merino P (2018) 3GPP evolution on LTE connectivity for IoT. In: Integration, interconnection, and interoperability of IoT systems. Springer, Cham, pp 1–20
4. Asif YM, Rochester EM, Ghaderi M (2018) A low-cost LoRaWAN testbed for IoT: Implementation and measurements. In; 2018 IEEE 4th world forum on internet of things (WF-IoT). IEEE, pp 361–366
5. Lora alliance (2020). https://lora-alliance.org/
6. Sigfox (2020). https://www.sigfox.com/
7. Landström S, Bergström J, Westerberg E, Hammarwall D (2016) NB-IoT: a sustainable technology for connecting billions of devices. Ericsson Technol. Rev. 4:2–11
8. Weightless. Available http://www.weightless.org/
9. Semtech (2015). LoRa™Modulation Basics, AN1200. 22, Revision 2
10. Georgiou O, Raza U (2017) Low power wide area network analysis: can LoRa scale? IEEE Wirel Commun Lett 6(2):162–165
11. Leentvaar K, Flint J (1976) The capture effect in FM receivers. IEEE Trans Commun 24(5):531–539
12. Zaaloul A, Haqiq A (2015) Enhanced slotted aloha mechanism by introducing ZigZag decoding. arXiv:1501.00976
13. Boujnoui A, Zaaloul A, Haqiq A (2018) Mathematical model based on game theory and markov chains for analysing the transmission cost in SA-ZD mechanism, IJCISIM
14. Capuzzo M, Magrin D, Zanella A (2018) Confirmed traffic in LoRaWAN: pitfalls and countermeasures. In: 2018 17th annual mediterranean ad hoc networking workshop (Med-Hoc-Net). IEEE, pp 1–7
15. Magrin D, Centenaro M, Vangelista L (2017) Performance evaluation of LoRa networks in a smart city scenario. In: 2017 IEEE international conference on communications (ICC). IEEE, pp 1–7
16. Noreen U, Clavier L, Bounceur A (2018)Lora-like css-based phy layer, capture effect and serial interference cancellation. In: European wireless 2018; 24th European wireless conference. VDE, pp 1–6
17. Yu Y, Giannakis GB (2007) High-throughput random access using successive interference cancellation in a tree algorithm. IEEE Trans Inf Theory 53(12):4628–4639
18. Hoeller A, Souza RD, López OLA, Alves H, de Noronha Neto M, Brante G (2018) Analysis and performance optimization of LoRa networks with time and antenna diversity. IEEE Access 6:32820–32829

19. Faber MJ, vd Zwaag KM, Dos Santos WG, Rocha HR, Segatto ME, Silva JA (2020) A theoretical and experimental evaluation on the performance of LoRa technology. IEEE Sens J
20. Hosseinzadeh S, Larijani H, Curtis K, Wixted A, Amini A (2017) Empirical propagation performance evaluation of LoRa for indoor environment. In: 2017 IEEE 15th international conference on industrial informatics (INDIN). IEEE, pp 26–31
21. Voigt T, Bor M, Roedig U, Alonso J (2016) Mitigating inter-network interference in LoRa networks. arXiv:1611.00688
22. Bor M, Roedig U (2017) LoRa transmission parameter selection. In: 2017 13th international conference on distributed computing in sensor systems (DCOSS). IEEE, pp 27–34
23. Neumann P, Montavont J, Noél T (2016) Indoor deployment of low-power wide area networks (LPWAN): a LoRaWAN case study. In: 2016 IEEE 12th international conference on wireless and mobile computing, networking and communications (WiMob). IEEE, pp 1–8
24. Yim D, Chung J, Cho Y, Song H, Jin D, Kim S, Ko S, Smith A, Riegsecker A (2018) An experimental LoRa performance evaluation in tree farm. In: 2018 IEEE sensors applications symposium (SAS). IEEE, pp 1–6
25. Springer A, Gugler W, Huemer M, Reindl L, Ruppel CCW, Weigel R (2000) Spread spectrum communications using chirp signals. In: IEEE/AFCEA EUROCOMM 2000. Information systems for enhanced public safety and security (Cat. No. 00EX405). IEEE, pp 166–170
26. Winkler MR (1962) Chirp signals for communication. IEEE WESCON Convention Record, p 7
27. Berni A, Gregg WO (1973) On the utility of chirp modulation for digital signaling. IEEE Trans Commun 21(6):748–751
28. Polonelli T, Brunelli D, Marzocchi A, Benini L (2019) Slotted aloha on lorawan-design, analysis, and deployment. Sensors 19(4):838
29. de Carvalho Silva J, Rodrigues JJ, Alberti AM, Solic P, Aquino AL (2017) LoRaWAN—a low power WAN protocol for internet of things: a review and opportunities. In: 2017 2nd international multidisciplinary conference on computer and energy science (SpliTech). IEEE, pp 1–6
30. Nelson R (2013) Probability, stochastic processes, and queueing theory: the mathematics of computer performance modeling. Springer Science & Business Media
31. Boujnoui A, Zaaloul A, Haqiq A (2017) A stochastic game analysis of the slotted ALOHA mechanism combined with zigzag decoding and transmission cost. In: International conference on innovations in bio-inspired computing and applications. Springer, Cham, pp 102–112
32. Alliance L (2015) A technical overview of LoRa and LoRaWAN. White Paper

Part II
Deep Learning Applications

Chapter 19
E_Swish Beta: Modifying Swish Activation Function for Deep Learning Improvement

Abdulwahed Salam and Abdelaaziz El Hibaoui

1 Introduction

Currently, neural networks and deep learning affords different solutions to many problems in image recognition and many other domains [1]. Simply, a neural network contains a chain of stacked layers. The first layer receives the data and then it proceeds the values to the last layer which presents the outputs [2]. Each layer has many neurons [3]. The number of neurons on the input layer depends on the amount of input data. And, the number of categories is represented by neurons of the output layer [4]. The activation function has an essential role in the neural network generally and deep learning specially. A correct selection improves accuracy and speed of neural network. Therefore, many activation functions were produced to achieve better improvements [5].

Here in this work, we present an activation function. It improves the result of deep learning models and outperforms other activation functions. We call it E_Swish Beta, and it is defined as:

$$f(x) = \frac{\beta x}{\left(1 + e^{-\beta x}\right)} \tag{1}$$

We concentrate on the different activation functions not on the models. We can apply them to many models that are used in the deep network. Due to limitations of our resources, we focused on some models, which did not require high resources to figure out the best performance of activation functions.

A. Salam (✉) · A. El Hibaoui
Faculty of Science, Abdelmalek Essaadi University, Tetouan, Morocco
e-mail: a.sallam@uae.ac.ma

A. El Hibaoui
e-mail: aelhibaoui@uae.ac.ma

A. E. Hassanien et al. (eds.), *Enabling Machine Learning Applications in Data Science*, Algorithms for Intelligent Systems, https://doi.org/10.1007/978-981-33-6129-4_19

We organized this paper as follows: In Sect. 2, we present the most works related to ours. In Sect. 3, we present E_Swish Beta properties and compare them to closed activation functions. We cover experiments and their results in Sect. 4. Section 5 concludes the paper.

2 Related Works

Sigmoid [6] was a widely used activation function and is still used until now. The output values are between zero and one. Its gradient is very high between −3 and 3 but gets much flatter in other regions. The gradient approaches to zero, and the network is not really learning. Its equation is:

$$f(x) = \sigma(x) = \frac{1}{(1 + \mathrm{e}^{-x})} \tag{2}$$

Hyperbolic tangent activation function [7] is just a scaled version of the sigmoid. However, it is zero centered and its output between −1 and 1

$$f(x) = \tanh(x) = 2{*}\sigma(2x) - 1 \tag{3}$$

SoftSign [7] is similar to hyperbolic tangent activation function.

$$f(x) = \frac{x}{1 + |x|} \tag{4}$$

With the deep neural network models, it becomes difficult to train with Sigmoid, hyperbolic tangent, and SoftSign.

In [8], authors defined a rectified linear unit (ReLU) as:

$$\mathrm{ReLU} = x^{+} = \begin{cases} x \text{ if} x \geq 0 \\ 0 \text{ if} x < 0 \end{cases} \tag{5}$$

This function is sample and fast to learn due to its ability to flow in a positive way. For those reasons, ReLU is the most used function in deep learning. Since, other works present adjusted functions to replace the horizontal line with a zero value to a non-zero, namely leaky ReLU [9].

$$\mathrm{LReLU} = \begin{cases} x & \text{if} x \geq 0 \\ ax & \text{if} x < 0 \end{cases} \tag{6}$$

Parametric ReLU [10] is similar to LReLU. The only difference is α, a learnable parameter. SoftPlus [11] is a smooth function with positive value. It is defined as:

$$\text{SoftPlus} = \log(1 + e^x) \tag{7}$$

Exponential linear unit (ELU) [12] modifies the negative part of ReLU:

$$\text{ELU} = \begin{cases} x & \text{if}\, x \geq 0 \\ a(e^x - 1) & \text{if}\, x < 0 \end{cases} \tag{8}$$

Scaled exponential linear unit (SELU) [13] is a modified type of ELU, and it is an internal normalization. Its parameters are fixed.

$$\text{SELU} = \gamma \begin{cases} x & if\ x \geq 0 \\ \alpha(e^x - 1) & if\ x < 0 \end{cases} \tag{9}$$

where $\alpha \approx 1.67326$ and $\gamma \approx 1.0507$.

For a long time, a lot of activation functions have been suggested but no one substitutes ReLU as a default activation function of the neural network. We cite SoftExponential [14], Gaussian error linear unit (GELU) [15], S-shaped rectified linear activation units [16], randomized leaky rectified linear unit (RReLU) [9], and many more activation functions.

Google brain proposes automated search techniques to search for activation functions and observes new activation functions, and the best one founded according to others is Swish [17]. The experimental results of this function are greatly amazing and gain the raise.

$$f(x) = x * \sigma(\beta x) \tag{10}$$

where

$$\sigma(\beta x) = \frac{1}{\left(1 + e^{-\beta x}\right)}.$$

The difference between Swish and ReLU is the non-monotonic of Swish, when x is less than zero. The authors of [18] introduce E_Swish which is an adjusting Swish. The unique difference between Swish and E_Swish is the constant or parameter Beta at the beginning of the equation:

$$f(x) = \beta x * \sigma(x) \tag{11}$$

The properties of Swish and E_Swish are almost the same and different from ReLU in the negative direction, where E_Swish and E_Swish have none zero values if x is different to zero.

3 E_Swish Beta

By studying and examining activation functions, where we modify Swish and E_Swish to discover a new activation function, which enhances their results. Here, we propose an adjusted Swish Beta function (E_Swish Beta) we modify Swish by multiply it by its parameter as long as its parameter is not equal to one.

$$f(x) = \beta x * \sigma(\beta x)$$

where

$$\sigma(\beta x) = \frac{1}{\left(1 + e^{-\beta x}\right)}$$

In the same way, the first derivative of E_Swish Beta is therefore:

$$\begin{aligned} f'(x) &= \beta\sigma(\beta x) + \beta x{*}\sigma(\beta x)(\beta - \beta\sigma(\beta x)) \\ &= \beta\sigma(\beta x) + f(x)(\beta - \beta\sigma(\beta x)) \\ &= \beta\sigma(\beta x) + \beta f(x) - \beta f(x)\sigma(\beta x) \\ &= \beta f(x) + \beta\sigma(\beta x)(1 - f(x)) \end{aligned}$$

This function is smooth and its properties, like Swish and E_Swish, are unbounded up and bounded down. The parameter β can be any real positive number. Therefore, when $\beta = 1$, the three functions are similar and equal to Swish function:

$$f(x) = x{*}\sigma(x)$$

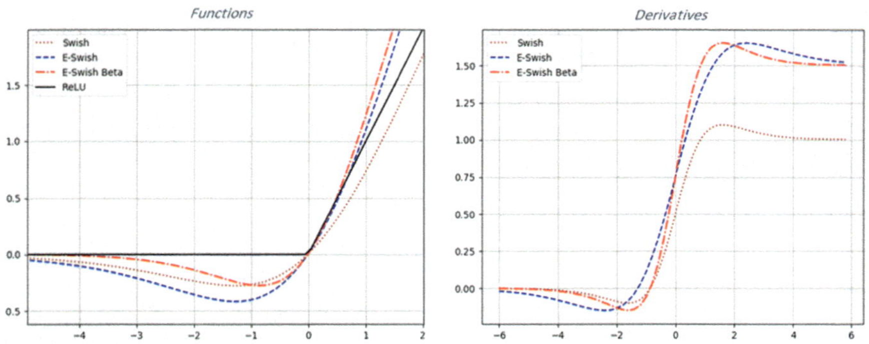

Fig. 1 Functions and derivatives of Swish, E_Swish and E_Swish Beta where $\beta = 1.5$

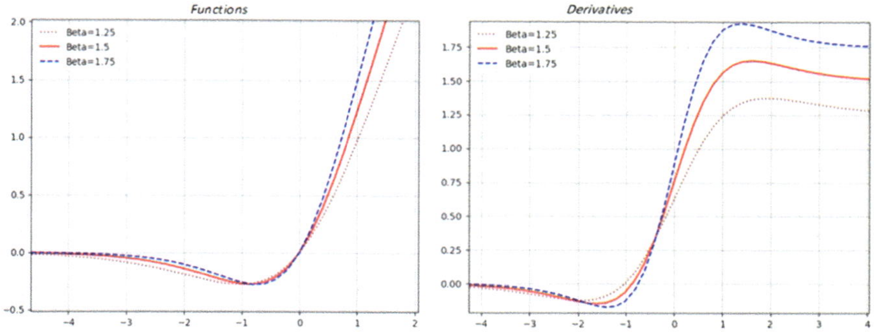

Fig. 2 E_Swish Beta function with first derivatives where beta equals to 1.25, 1.5, and 1.75

According to the experimentation results, we remark that the best choice of β is being greater than one and less or equal to two. Figure 1 shows the difference between the three functions and their derivative functions when $\beta = 1.5$.

The non-monotonic of Swish, E_Swish, and E_Swish Beta is an important aspect, and β determines the shape of the function. Figure 2 shows the E_Swish Beta function shape for three values of β. The non-monotonic and smooth aspect make the result get better compared to ReLU.

4 Experiments and Results

In the work presented here, we compare only E_Swish Beta to Swish, adjusting Swish and ReLU ignoring other activation functions. For the other comparisons, the reader can refer to the original paper of Swish [17] where authors compare his work to other activation functions in more detail. We use three different datasets: MNIST, CIFAR10, and CIFAR100. The same environment and the same parameters are used for the experiments.

4.1 MNIST Dataset

MNIST is a dataset of handwritten digits which has 60K examples for a training set, and 10K examples for a testing set [19]. The different activation functions are compared by the same model and parameters that are used in E_Swish experimentation [18]. They are trained by using a fully connected feed forward neural networks of five layers with a dropout of 0.2. The algorithm is stochastic gradient descent (SGD) and we set learning rate equals to 0.1. The number of neurons in the five layers is 200, 100, 60, 30, and 10. The initialization of training is uniform [7]. We run five experiments with twenty epochs and compare their medians. ReLU gets 97.94%, and Swish gets 97.92% when $\beta = 1$. It is obviously shown in Fig. 3 that E_Swish Beta achieves the best result where β equals to 1.25, 1.35, 1.5, 1.75 and 1.85.

Moreover, the E_Swish Beta learns faster than others do as it is shown in Fig. 4.

Now, we examine the ability of deep network on a simple convolutional neural network (CNN) with two convolutional layers and two fully connected layers. The convolution layers use dropout of 0.25 and 0.5, respectively. It is trained on 60K and validated on 10K. Also, we use SGD optimizer with 0.1 learning rate, 128 batch size, Glorot uniform initialization [7], and 20 epochs of training. The result is getting better and proves the ability of deep learning models.

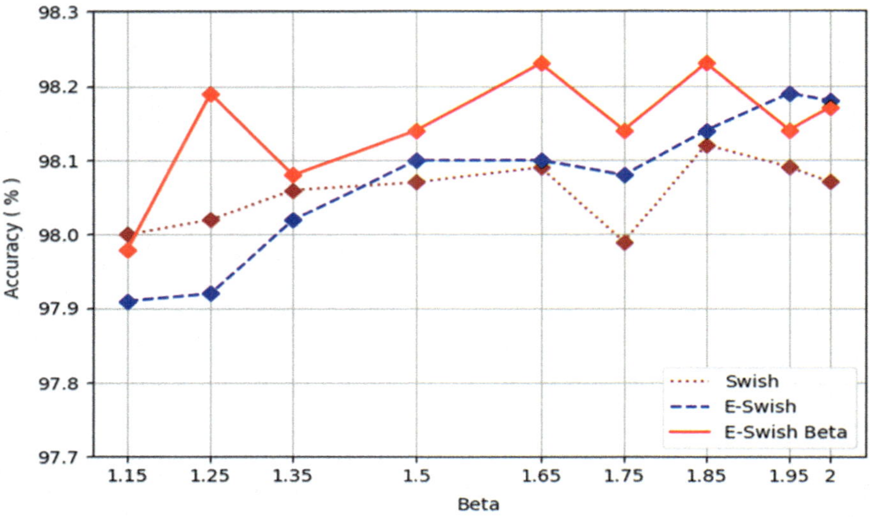

Fig. 3 Test accuracy for Swish, E_Swish, and E_Swish Beta with 20 epochs

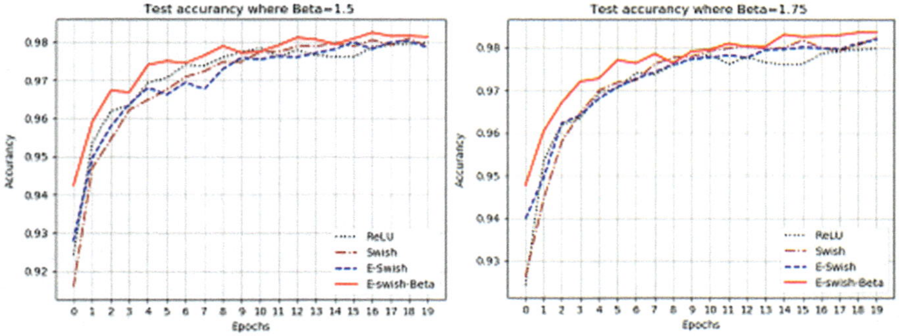

Fig. 4 Test accuracy for the four activation functions using MNIST dataset

Table 1 Accuracy for MNIST dataset on simple CNN

Activation function	Accuracy (%)	Activation function	Accuracy (%)
ReLU	99.09	Swish ($\beta = 1$)	99.1
E_Swish ($\beta = 1.15$)	99.16	E_Swish-Beta ($\beta = 1.15$)	99.12
E_Swish ($\beta = 1.25$)	99.15	E_Swish-Beta ($\beta = 1.25$)	99.15
E_Swish ($\beta = 1.35$)	99.06	E_Swish-Beta ($\beta = 1.35$)	99.17
E_Swish ($\beta = 1.5$)	99.17	E_Swish-Beta ($\beta = 1.5$)	99.27
E_Swish ($\beta = 1.65$)	99.21	E_Swish-Beta ($\beta = 1.65$)	**99.28**
E_Swish ($\beta = 1.75$)	99.01	E_Swish-Beta ($\beta = 1.75$)	99.2
E_Swish ($\beta = 1.85$)	99.17	E_Swish-Beta ($\beta = 1.85$)	99.22
E_Swish ($\beta = 1.95$)	99.12	E_Swish-Beta ($\beta = 1.95$)	99.2
E_Swish ($\beta = 2$)	99.2	E_Swish-Beta ($\beta = 2$)	99.23

Bold indicates the best result

As shown in Table 1, E_Swish Beta when β equals to 1.65 performs the best result. It provides 0.19% relative to ReLU, 0.18% relative to Swish when β equals to 1 and 0.07 relative to the highest value of E_Swish when β equals to 1.65. With all values of β, the E_Swish Beta gets the best result except the case when $\beta = 1.25$. In this case, E_Swish Beta and E_Swish are the same. Moreover, when $\beta = 1.15$ E_Swish Beta gets less than E_Swish.

4.2 *CIFAR 10 Dataset*

The CIFAR 10 dataset [20] contains 60K small images with 32 × 32 pixels and 3 RGB colors. It consists of 50K images for training and 10K images for testing with ten different classes.

4.2.1 Wide Residual Network Model with 10–2 Structure

At the beginning, we examine the different activation functions on a simple convolutional neural network (CNN). We use wide residual networks [21], which has many

Table 2 Test accuracy for Cifar10 with 125 epochs on WRN 10–2 architecture

Activation function	Accuracy (%)	Activation function	Accuracy (%)
ReLU	89.19	Swish ($\beta = 1$)	91.25
E_Swish ($\beta = 1.375$)	91.05	E_Swish-Beta ($\beta = 1.25$)	91.32
E_Swish ($\beta = 1.25$)	91.26	E_Swish-Beta ($\beta = 1.375$)	90.88
E_Swish ($\beta = 1.5$)	90.88	E_Swish-Beta ($\beta = 1.5$)	**91.4**
E_Swish ($\beta = 1.75$)	91.15	E_Swish-Beta ($\beta = 1.75$)	91.29
E_Swish ($\beta = 2$)	90.89	E_Swish-Beta ($\beta = 2$)	91.19

Bold indicates the best result

architectures. It raises the width and reduces the depth of residual [22]. In this experiment, we use WRN of ten depth and two width. To do so, we use the same code that is used by adjusted Swish paper. We implement the stochastic gradient descent (SGD) as the optimizer. The epochs that we use are 125 with different learning rates. We use 0.125, 0.025, 0.005, 0.001, and 0.0005 learning rate for 44, 84, 104, 124, and 125 epochs, respectively. According to the original work presented in [21], this structure of WRN 10–2 accomplishes 89% accuracy on Cifar10 dataset and this by using the ReLU function with 310K parameters. The result in Table 2 shows that E_Swish Beta with β equals to 1.5 outperforms others. The achievement of E_Swish Beta when β equals to 1.5 supplies 2.21% relative to ReLU, 0.15% relative to Swish when β equals to 1, and 0.14 relative to the highest value of E_Swish when β equals to 1.25.

4.2.2 Wide Residual Network Model with 16–4 Structure

To go deeper, we use WRN16-4, to test E_Swish Beta with the other activation functions on Cifar10 dataset. We use this model with the same parameters used both in [13] and [18]. We train the model with 125 epochs, no dropout, and SGD optimizer with 0.9 momentum. We use 0.125, 0.025, 0.005, 0.001, and 0.0005 learning rate for 44, 84, 104, 119, and 125 epochs, respectively.

As shown in Table 3, E_Swish Beta when $\beta = 1.75$ provides 0.78, 0.39, and 0.38% relative to ReLU, Swish ($\beta = 1$), and E_Swish ($\beta = 1.75$), respectively.

Figure 5 shows the performance of the comparative activation function in training and validation.

Now, we increase the number of epochs to 200. We only use two different values of Beta (1.5 and 1.75) which provide the highest values in the previous experiments. The learning rate is set to 0.1 and dropped by 0.2 at 60, 120, 160, and 180 epochs.

The result in Table 4 shows the high performance of E_Swish Beta. Comparing this result to that one of Table 3, E_Swish Beta is still better than others.

Table 3 Test accuracy for Cifar10 on WRN 16–4 with 125 epochs

Activation function	Accuracy (%)	Activation function	Accuracy (%)
ReLU	93.27	Swish ($\beta = 1$)	93.66
E_Swish ($\beta = 1.25$)	93	E_Swish-Beta ($\beta = 1.25$)	93.64
E_Swish ($\beta = 1.5$)	93.64	E_Swish-Beta ($\beta = 1.5$)	93.76
E_Swish ($\beta = 1.75$)	93.67	E_Swish-Beta ($\beta = 1.75$)	**94.05**
E_Swish ($\beta = 2$)	93.43	E_Swish-Beta ($\beta = 2$)	93.89

Bold indicates the best result

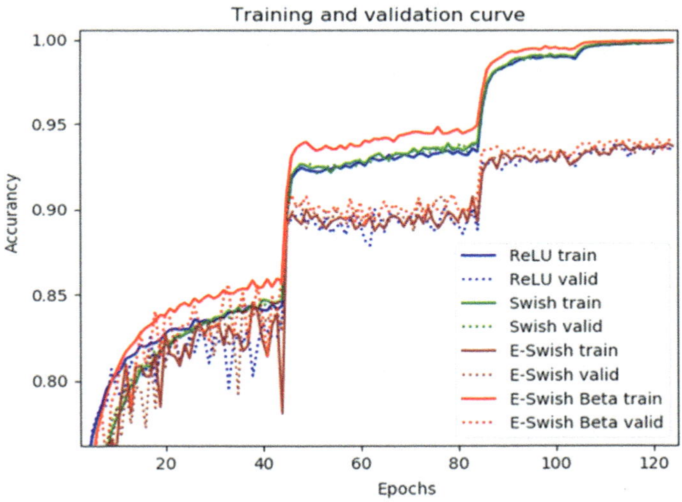

Fig. 5 Test accuracy for Cifar10 on WRN 16–4 with 125 epochs

Table 4 Test accuracy for Cifar10 on WRN 16–4 with 200 epochs

Activation function	Test accuracy (%)
ReLU	94.08
Swish ($\beta = 1$)	94
E_Swish ($\beta = 1.5$)	94.19
E_Swish ($\beta = 1.75$)	94.02
E_Swish-Beta ($\beta = 1.5$)	**94.61**
E_Swish-Beta ($\beta = 1.75$)	94.57

Bold indicates the best result

4.2.3 CNN: Simplenet

This model is a deep CNN model. It contains 13 convolutional layers. It is designed to utilize 3 × 3 kernels for convolutional and 2 × 2 for pooling operations [23]. It uses 5,497,226 parameters. We run the model in Keras with TensorFlow backend. Batch size of 128 and data augmentation similar to that used in [23]. The optimizer that we use is SGD in place of Adadelta, and the epochs are reduced to 100 instead of 200. We utilize 0.1, 0.02, 0.004, and 0.0008 learning rate for 40, 60, 80, and 100 epochs, respectively.

As Table 5 shows, E_Swish Beta where $\beta = 1.125$ outperforms all. Even though this model designs to use ReLU as default activation. It provides 0.46% relative to ReLU, 0.27 relative to Swish and 0.21 relative to E_Swish where $\beta = 1.125$. We do not go for Beta greater than 1.25 because of gradient exploding.

Table 5 Test accuracy for Cifar10 on deeper CNN model

Activation function	Test accuracy (%)
ReLU	92.87
Swish ($\beta = 1$)	93.06
E_Swish ($\beta = 1.125$)	93.12
E_Swish ($\beta = 1.25$)	93.02
E_Swish-Beta ($\beta = 1.125$)	**93.33**
E_Swish-Beta ($\beta = 1.25$)	92.95

Bold indicates the best result

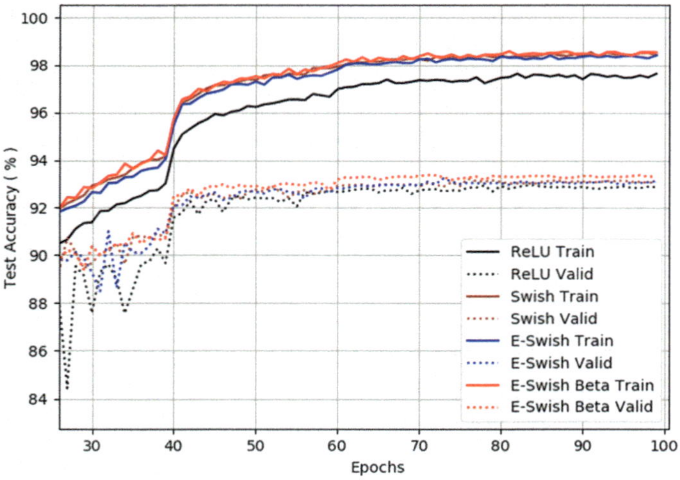

Fig. 6 Test accuracy for ReLU, Swish, E_Swish and E_Swish Beta where $\beta = 1.125$ on CNN Simplenet

Table 6 Test accuracy for Cifar100 dataset on WRN 10–2 model

Activation function	Accuracy (%)	Activation function	Accuracy (%)
ReLU	62.8	Swish ($\beta = 1$)	67.46
E_Swish ($\beta = 1.125$)	67.48	E_Swish-Beta ($\beta = 1.125$)	66.95
E_Swish ($\beta = 1.25$)	66.45	E_Swish-Beta ($\beta = 1.25$)	67.2
E_Swish ($\beta = 1.375$)	67.12	E_Swish-Beta ($\beta = 1.375$)	67.2
E_Swish ($\beta = 1.5$)	66.34	E_Swish-Beta ($\beta = 1.5$)	**67.5**
E_Swish ($\beta = 1.65$)	66.54	E_Swish-Beta ($\beta = 1.65$)	66.94
E_Swish ($\beta = 1.75$)	66.91	E_Swish-Beta ($\beta = 1.75$)	66.39
E_Swish ($\beta = 1.85$)	66.08	E_Swish-Beta ($\beta = 1.85$)	66.77
E_Swish ($\beta = 2$)	67.02	E_Swish-Beta ($\beta = 2$)	67.08

Bold indicates the best result

Figure 6 shows the performance of the comparative activation function in training and validation on CNN Simplenet model.

4.3 CIFAR 100 Dataset

This dataset [20] contains 60K small images with $32 \times 32 \times 3$. It has 100 classes containing 600 images each. We train this model on 50,000 images and validate it on 10,000.

4.3.1 Wide Residual Network 10–2

This test is the same as that one used for the Cifar 10 dataset. We apply the same parameters except the number of classes which is here 100. The result in Table 6 shows good performance for E_Swish Beta when $\beta = 1.5$. There is no big difference but it is still on top of the others.

4.3.2 Wide Residual Network 16–4

This experience is executed with WRN 16–4 model. We train the model with 130 epochs, no dropout, and SGD optimizer with 0.9 momentum. We use 0.125, 0.025, 0.005, 0.001, and 0.0005 learning rate for 44, 84, 104, 124, and 130 epochs.

Each activation function has an upper value and a lower value with a middle line representing the median of three times of execution. Brown, green, blue, and red colors are used for ReLU, Swish, E_Swish, and E_Swish Beta, respectively. E_Swish Beta provides 1.77, 0.69, and 0.27% relative to ReLU, Swish, and E_Swish (Fig. 7).

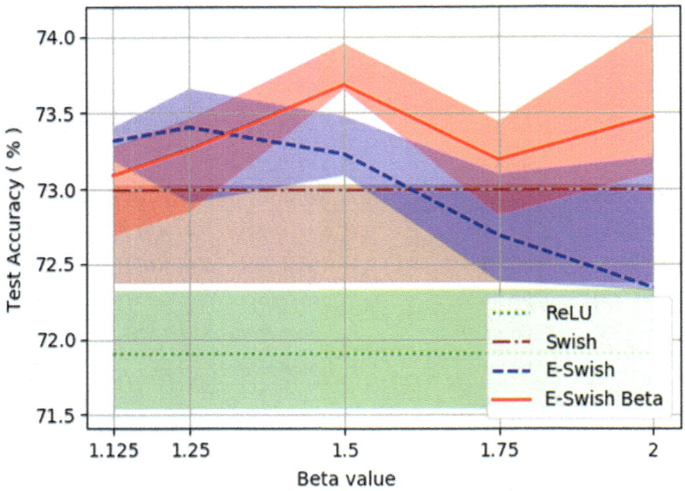

Fig. 7 Test accuracy for Cifar100 with 130 epochs of training on WRN 16–4

5 Conclusion

According to the important role of activation function in the deep network, we presented a new activation function which is defined as $f(x) = \beta x$*Sigmoid (βx) and we called it E_Swish Beta. We studied E_Swish Beta, E_Swish, Swish, and ReLU on different models, which were designed for ReLU activation. E_Swish Beta outperforms other activation on varied models and datasets. E_Swish Beta achieves the best performance compared to its counterparts. Moreover and in the term of experimentation, we found that the best choice of Beta is between one and two.

References

1. Schmidhuber J (2015) Deep learning in neural networks: an overview. Neural Netw 61:85–117
2. Salam A, El Hibaoui A (2018) Comparison of machine learning algorithms for the power consumption prediction—case study of Tetouan city. In: Proceedings of 2018 6th international renewable and sustainable energy conference, IRSEC 2018
3. LeCun Y, Bengio Y, Hinton G (2015) Deep learning. Nature 521(7553):436–444
4. Maarof M, Selamat A, Shamsuddin SM (2009) Text content analysis for illicit web pages by using neural networks. J Teknol Siri 50(D):73–91
5. Ding B, Qian H, Zhou J (2018) Activation functions and their characteristics in deep neural networks. In: Proceedings of the 30th Chinese control and decision conference, CCDC 2018. pp 1836–1841
6. Minai AA, Williams RD (1993) On the derivatives of the sigmoid. Neural Netw 6(6):845–853
7. Glorot X, Bengio Y (2010) Understanding the difficulty of training deep feedforward neural networks. In: Proceedings of the thirteenth international conference on artificial intelligence and statistics. pp 249–256

8. Hahnloser RHR, Sarpeshkar R, Mahowald MA, Douglas RJ, Seung HS (2000) Erratum: digital selection and analogue amplification coexist in a cortex-inspired silicon circuit. Nature 405(6789):947–951
9. Xu B et al (2015) Empirical evaluation of rectified activations in convolutional network. arXiv preprint arXiv:1505.00853
10. He K, Zhang X, Ren S, Sun J (2015) Delving deep into rectifiers: surpassing human-level performance on imagenet classification. In: Proceedings of the IEEE international conference on computer vision, pp 1026–1034
11. Nair V, Hinton GE (2010) Rectified linear units improve restricted Boltzmann machines. In: ICML
12. Clevert DA, Unterthiner T, Hochreiter S (2015) Fast and accurate deep network learning by exponential linear units (elus). arXiv preprint arXiv:1511.07289
13. Klambauer G, Unterthiner T, Mayr A, Hochreiter S (2017) Self-normalizing neural networks. In: Advances in neural information processing systems. pp 971–980
14. Godfrey LB, Gashler MS (2015) A continuum among logarithmic, linear, and exponential functions, and its potential to improve generalization in neural networks. In: 2015 7th international joint conference on knowledge discovery, knowledge engineering and knowledge management (IC3K), vol 1. IEEE, pp 481–486
15. Hendrycks D, Gimpel K (2016) Bridging nonlinearities and stochastic regularizers with gaussian error linear units
16. Jin X et al (2015) Deep learning with s-shaped rectified linear activation units. arXiv preprint arXiv:1512.07030
17. Ramachandran P, Zoph B, Le QV (2017) Searching for activation functions. arXiv preprint arXiv:1710.05941
18. Alcaide E (2018) E-Swish: adjusting activations to different network depths. arXiv preprint arXiv:1801.07145
19. Lecun Y, Bottou L, Bengio Y, Haffner P (1998) Gradient-based learning applied to document recognition. Proc IEEE 86(11):2278–2324
20. Krizhevsky A, Sutskever I, Hinton GE (2012) Imagenet classification with deep convolutional neural networks. In: Advances in neural information processing systems. pp 1097–1105
21. Zagoruyko S, Komodakis N (2016) Wide residual networks. arXiv preprint arXiv:1605.07146
22. He K, Zhang X, Ren S, Sun J (2016) Deep residual learning for image recognition. In: Proceedings of the IEEE conference on computer vision and pattern recognition. pp 770–778
23. Hasanpour SH, et al. (2016) Lets keep it simple, using simple architectures to outperform deeper and more complex architectures. arXiv preprint arXiv:1608.06037

Chapter 20
Online Arabic Handwriting Recognition Using Graphemes Segmentation and Deep Learning Recurrent Neural Networks

Yahia Hamdi, Houcine Boubaker, and Adel M. Alimi

1 Introduction

Nowadays, online handwriting recognition represents a hard problem, especially for cursive script such as Arabic language which serves as a challenging task due to some property. Indeed, the crucial challenges include the huge variability of human handwriting style, the great similarities between character shapes, etc.

More than 440 million people in the world use Arabic as a native language. Its alphabet is utilized in about 27 languages, including Kurdish, Persian, Jawi, and Urdu [1]. It is always cursive in both handwriting and print and still needs an accurate handwriting recognition system [2].

Generally, the recognizers of Arabic script can be divided into two main branches: holistic and analytic approaches. The words are treated as a whole without any segmentation into smaller components in holistic approaches, while in analytic approaches, the words are segmented into simple parts such as characters or strokes before passed to the recognizer [3]. The development of efficient algorithms for the segmentation of cursive scripts is a difficult task and requires an in-depth knowledge of expert language.

Lately, the performance of handwriting recognition system has been greatly improved with the progress of deep learning technologies such as deep belief network (DBN) [4], convolutional neural networks (CNNs) [5], and recurrent neural network

Y. Hamdi (✉) · H. Boubaker · A. M. Alimi
Research Groups in Intelligent Machines, REGIM-Laboratory, National School of Engineers, University of Sfax, BP 1173, 3038 Sfax, Tunisia
e-mail: yahia.hamdi.tn@ieee.org

H. Boubaker
e-mail: houcine-boubaker@ieee.org

A. M. Alimi
e-mail: adel.alimi@ieee.org

A. E. Hassanien et al. (eds.), *Enabling Machine Learning Applications in Data Science*, Algorithms for Intelligent Systems, https://doi.org/10.1007/978-981-33-6129-4_20

(RNN) specifically with BLSTM [6]. Despite the advancement of deep learning technology in the handwriting recognition field, we announce that the research work in online Arabic handwriting recognition area is very limited compared to other scripts such as English, Chinese, Indic.

In this paper, we propose an effective system for online Arabic handwriting recognition words. The presented system combines an efficient segmentation algorithm, feature extraction, and sequence labeling with recurrent neural networks to achieve the best results.

The main contributions are the following:

- A novel segmentation algorithm consists of dividing the cursive words into a continuous part called graphemes based on baseline detection and their ligature bottoms neighboring points. We also evaluate the employed segmentation algorithm using constructed grapheme database and present its influence on word recognition level.
- A selection of a pertinent set of features that combine dynamic and geometric parameters for stroke modeling using advanced beta-elliptic model [7], geometric features that contain Fourier descriptors for trajectory shape modeling and other normalized parameters representing the grapheme dimensions, positions with respect to the baseline and the assignment diacritics codes.
- A demonstration that our proposed-based segmentation approach, which is characterized by powerful feature extraction models and deep learning RNN, gives better word recognition rates that exceed the best result for other state-of-the-art systems.

The rest of this paper is organized as follows: Sect. 2 gives a brief description of Arabic handwriting specificity. An outline of the most prominent related works in Arabic handwriting recognition is presented in Sect. 3. Section 4 summarizes our proposed method that includes preprocessing techniques applied in our work, baseline detection and grapheme segmentation algorithm, feature extraction, and sequence recognition. Section 5 presents the experimental results. Finally, Sect. 6 provides conclusions and plans for future work.

2 Arabic Script Specificity

In this section, we briefly give an overview of Arabic script supported by our system, considering their writing diversity compared to other scripts. The Arabic script is employed in multiple languages such as Persian, Kurdish, Urdu, and Pashto. More than 420 million people of the word use the Arabic as main language [8]. It is generally written cursive, and most of these letters are connected to their neighboring. It generally contains 28 basic letters with other 10 marks like dot and delayed stroke that can change the meaning of the word. As a comparison, there are 10 digits for usual digit recognition tasks, and there are 26 alphabetic letters for English, while there are over 28 for Arabic.

Table 1 Some of Arabic characters in different forms

Characters	Beginning	Middle	Isolated	End
Alif (أ)	أ	ـا	أ	ـا
Baa (ب)	بـ	ـبـ	ب	ـب
Taa (ت)	تـ	ـتـ	ت	ـت
Thaa (ث)	ثـ	ـثـ	ث	ـث
Raa (ر)	ر		ر	ـر

The latter are handwritten from right to left, encompass small marks and dots. In addition, the letters have multiple shapes that depend on their position in the word in which they are found. As shown in Table 1, we can distinguish four different forms (beginning, middle, isolated, and end form) of the Arabic letters according to their position within a word. Different letters might share the same beginning and isolated shapes such as Alif 'أ', Raa 'ر'. Also, several letters having the same body (e.g., ب, ت, ث) but distinct only in position and the number of diacritics (dots and marks). Further detail of the Arabic characteristics and difficulties of writing is presented in [9].

3 Related Work

As mentioned earlier, prior studies on the topic of online Arabic handwriting word recognition have been still restricted. We discussed below some available works in both Arabic and non-Arabic scripts.

Related work on non-Arabic scripts: Graves et al. [10] developed an unconstrained online Latin handwriting recognition system based on BLSTM and connectionist temporal classification (CTC) labeling layer in order to avoid the need for positions respect pre-segmented data. Another work for Latin script has been introduced by [11]. In this article, the CNN network is used for recognizing online handwritten document after online words segmentation.

Recently, Ghosh et al. [12] presented a system for Bengali and Devanagari word recognition of cursive and non-cursive scripts based on LSTM and BLSTM recurrent networks. In this method, each word is divided into three horizontal zones (upper, middle, and lower). Then, the word portions of the middle zone are re-segmented into elementary strokes in each directional, and structural features are extracted before carrying out training LSTM and BLSTM. In another study for multi-lingual scripts, Keyseres et al. [13] investigated an online handwriting recognition system based on BLSTM and CTC loss function that supported 102 languages.

Related work on Arabic scripts: Many researches are available for online Arabic handwriting recognition script [14]. Most of these studies are focused on characters' recognition [15, 16]. Izadi et al. [17] introduced an analytic approach based on segmentation for online Arabic handwriting recognition. It decomposes the online trajectory into elementary segment known as convex/concave. Daifallah et al. [18] proposed an algorithm for stroke segmentation founded on four main steps: arbitrary and enhancement segmentation, connecting consecutive joints followed by locating segmentation points.

A similar approach that divides the online signal into grapheme depending on the local writing direction has been developed by [19]. After baseline and delayed strokes detection, the characters are constructed from the main graphemes according to a set of rules. Another approach that supported a large vocabulary for online Arabic handwriting recognition has been developed by [20]. A novel preprocessing method based on delayed strokes rearrangement algorithm is introduced dealing with handling delayed strokes problem. Furthermore, the combination of online/offline features to improve system recognition rate is also investigated by [21]. Directional and aspect features are extracted from the online signal while offline features are described by the number of foreground pixels and gradient features.

Recently, the combination of beta-elliptic and convolutional deep belief network (CDBN) is presented by [22] using SVM classifier. Despite the success of these approaches in online Arabic handwriting recognition, we notice that the research in this field still remains.

4 System Overview

In this section, we present an overview of our proposed approach which develops an online handwriting recognition system for cursive and non-cursive Arabic words based on grapheme segmentation and deep learned recurrent neural network (RNN).

As shown in Fig. 1, the architecture of our system proceeds as follows:

First, the input signal (x, y) is denoised and normalized using preprocessing technique. Next, the developed algorithm detects the script baseline by considering the

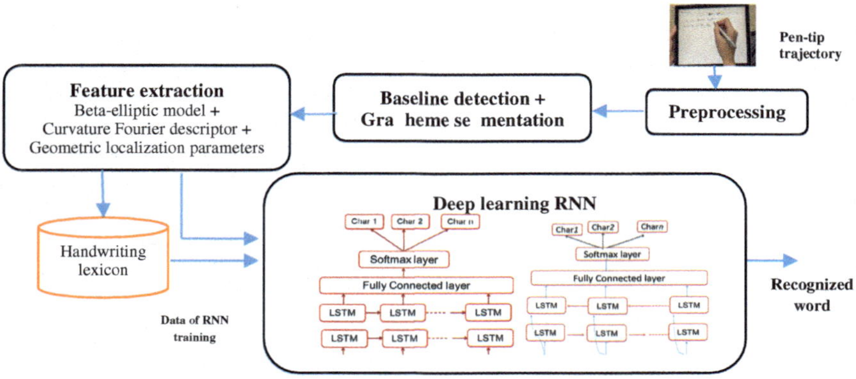

Fig. 1 System architecture

conformity between the alignment of its trajectory points and tangent directions. Then, the handwritten word or pseudo-words are segmented in continuous part called graphemes delimited by the points of ligature bottoms neighboring the baseline. Various types of features are extracted for each grapheme combines beta-elliptic model, Fourier descriptors, and other parameters of grapheme dimensions, positions with respect to the baseline and the assignment diacritics codes. These graphemes features are then studied using LSTM and BLSTM version of RNN. In the following subsections, we introduce each module in detail.

4.1 Preprocessing

The word data handwriting trajectories are collected online via a digitizing device. It is characterized by a high variation that requires some geometric processing steps to *minimize the variabilities of handwriting, reduce noise, and normalize handwriting size dimensions*. The total number of coordinate points within the same basic stroke may vary. This problem can be solved by using smoothing operations which replace every point by the average value of its neighboring. Also, the Chebyshev type II low-pass filtering with a cut-off frequency of fcut = 10 Hz is employed to eliminate the effect of noise and errors caused by spatial and temporal quantification produced by the acquisition system (see Fig. 2). The horizontal fulcrums level of handwriting trajectory divides their drawing area into three distinct zones called upper, core, and lower regions, respectively. Therefore, a normalizing procedure of the handwriting size is applied to adjust its height to a fixed value $h = 128$, while keeping the same ratio length/height [23].

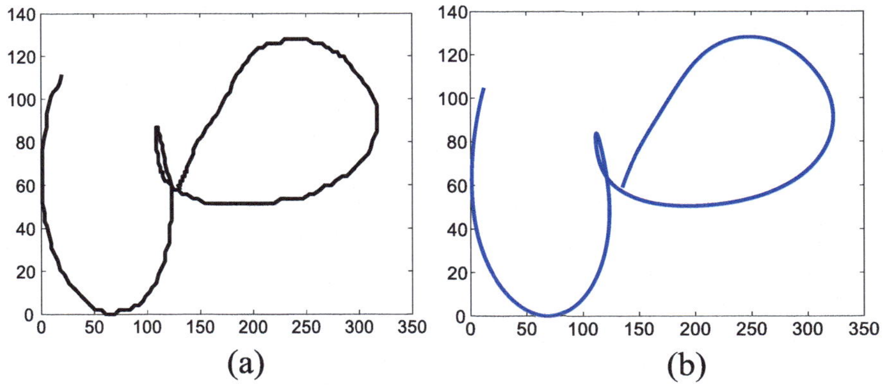

Fig. 2 **a**) Acquired grapheme "Sad" and **b**) after low-pass filtering and smoothing

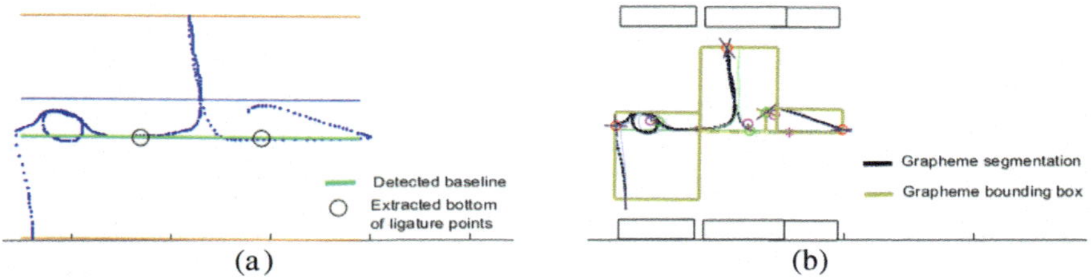

Fig. 3 Baseline detection and grapheme limits

4.2 Baseline Detection and Grapheme Segmentation

The handwriting segmentation tasks are one of the most important steps for recognition of cursive script. Our segmentation technique consists of dividing the word or pseudo-words into graphemes based on the detection of the baseline direction and grapheme limits.

4.2.1 Baseline Detection

The baseline detection of the online handwriting script is identified for several objectives such as the segmentation of trajectory and feature extraction. In our case, we employed geometric algorithm for straight or curved handwriting baseline detection (GLoBD) developed by [24]. Indeed, the baseline (see Fig. 3a) is detected in accordance with two steps: the first geometric step determines the nearly aligned sets of candidate points respecting the conformity between the fit of the path points and their tangent directions. To correct the detection results, a topologic evaluation of the candidates' sets is executed.

4.2.2 Grapheme Limits

In our work, the segmentation technique of the Arabic word into graphemes is based on identifying two points that delimiting the graphemes position. These points correspond to the bottoms of ligature valleys neighboring to the baseline and to the vertical angular points (see Fig. 3a).

- Bottom of the ligature valleys: represent the point of trajectory grapheme moving from right to left verifying a local nearest position to the baseline with trajectory tangents parallel to its direction.
- Angular point: top point representing the extremum of a vertical shaft trajectory turning back.

After segmentation step, each grapheme can be classified into five classes, namely beginning grapheme (BG); middle grapheme (MG); isolated grapheme (IG); end

Table 2 Different forms of graphemes according to their position in the pseudo-words

Beginning graphemes	
Isolated graphemes	
Middle graphemes	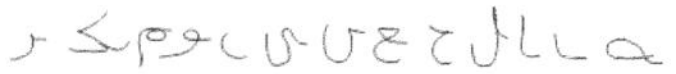
End graphemes	
Diacritic	

grapheme (EG) and diacritics according to their position in the word. Table 2 shows the different forms of graphemes obtained from the Arabic script. This step is important in our work; it generally focuses on the creating of the graphemes database collected from the word of ADAB database which allows us to determine the efficiency of the segmentation algorithm both at the grapheme classification level as well as word recognition level. Afterward, the segmented graphemes are presented by multiple features combining the beta-elliptic model, Fourier descriptors, and geometric localization parameters. In the next paragraph, we describe each model in detail.

4.3 *Feature Extraction*

In the proposed system, multiple types of features are extracted for each grapheme in the word such as beta *velocity* and *elliptic arc* for stroke modeling obtained by using beta-elliptic model, *curvature Fourier descriptor* parameters for grapheme trajectory shape modeling, *grapheme dimensions, positions respect to the baseline,* and the *assignment diacritics codes.*

4.3.1 Beta Velocity Profile and Elliptic Arc Modeling

The handwriting movement is expressed as the neuromuscular system response which is represented by the sum of impulse signals [25] as the beta function [14]. The beta-elliptic model derives from the kinematic beta model with a juxtaposed analysis of the spatial profile. It is represented by the combination of two online handwriting modeling aspects: the velocity profile (dynamic features) and an elliptic arc modeling the static profile (geometric features) for each segmented trajectory.

(a) **Velocity profile:**
In the dynamic profile, the curvilinear velocity $V_\sigma(t_c)$ of the handwriting trajectory shows a signal that alternates between three extremums type of points

(minima, maxima, and inflection points) which delimit and define the number of trajectory strokes. For beta-elliptic model, $V_\sigma(t_c)$ can be reconstructed by overlapping Beta signals where each stroke corresponds to the production of one beta impulse described by the following expression:

$$\text{pulseB}(t, q, p, t_0, t_1) = \begin{cases} k.\left(\frac{t-t_0}{(t_c-t_0)}\right)^p.\left(\frac{t_1-t}{(t_1-t_c)}\right)^q & \text{if } t \in [t_0, t_1] \\ 0 & \text{elsewhere} \end{cases} \text{ with } t_c = \frac{p*t_1+q*t_0}{p+q} \tag{1}$$

where t_0 and t_1 are the starting and the ending times of the produced impulse delimiting the continuous stroke (segment), respectively. t_c represents the instant when the beta function reaches its maximum value, K is the impulse amplitude, and p and q are intermediate shape parameters.
Indeed, four dynamic features are extracted for each stroke such as: K, $\Delta t = (t_1 - t_0)$ the beta impulse duration, $p/p + q$ the rapport of beta impulse asymmetry or culminating time, p: coefficient of impulse width.
The velocity profile of the handwritten trajectory can be reconstructed by the algebraic addition of the successive overlapped beta signals as shown in Eq. (2).

$$V_\sigma(t) = \sum_{i=1}^{n} V_i(t - t_{0i}) \approx \sum_{i=1}^{n} \text{pulse}\beta i(K_i, t, q_i, p_i, t_{0i}, t_{1i}) \tag{2}$$

As depicted in Fig. 4a, the number of stroke building the global handwriting trajectory of the Arabic word "حلم" is the number of beta impulses that they constitute.

(b) **Elliptic arc modeling**
Each elementary beta stroke located between two successive extrema speed times can be modeled by an elliptic arc in the space domain. This latter is characterized by four geometric parameters: *the half dimensions of the large and the small axes of the elliptic arc, the angle of the ellipse major axe inclination*, and *the trajectory tangent inclination at the minimum velocity endpoint*. These parameters reflect the geometric properties of the end effector (pen or finger) trace, dragged by the set of muscles and joints involved in handwriting. An example of geometric profile of the same Arabic word is shown in Fig. 4b.

4.3.2 Curvature Fourier Descriptor

Fourier descriptors are considered one of the most appropriate tools of closed paths modeling which can be represented by a 2π periodic signature function [26]. In order to exploit its strong capacity of periodic function approximation in handwriting grapheme modeling, the signatures equivalent to the open trajectories must be transformed into periodic functions. Each segmented trajectory is represented by an angular signature modeled by a periodic sinusoidal function. As shown in Fig. 5, the chosen function as a stroke trajectory signature$\theta_i = f(\ell_i)$ describes the variation

Fig. 4 Online handwriting modeling of Arabic word“ حلم”

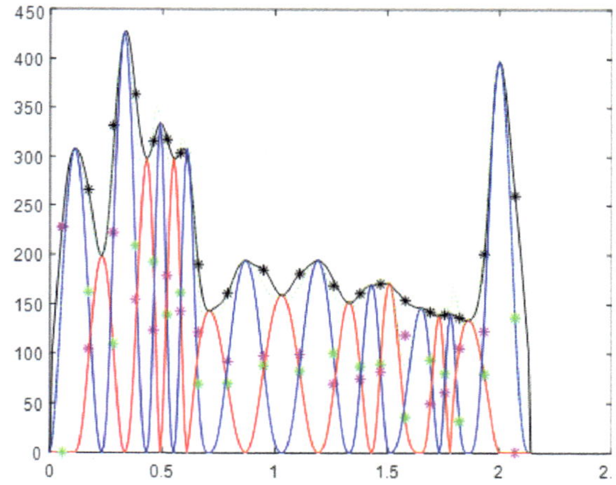

(a) Velocity profile modeling

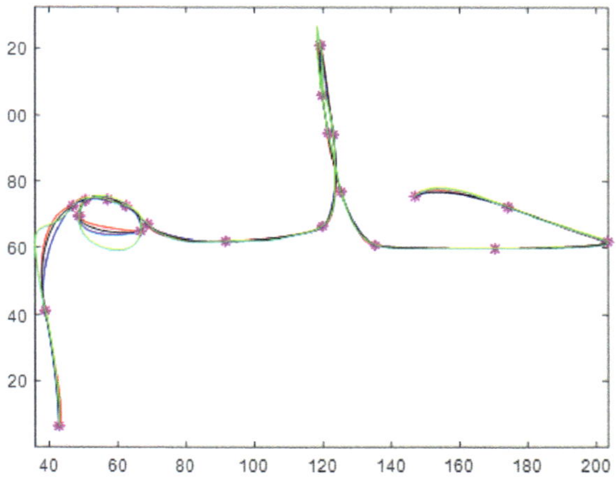

(b) Geometric profile

of the inclination angle θ_i of the trajectory tangent at a point M_i depending on its corresponding curvilinear abscissa ℓ_i:

$$\ell_i = \sum_{j=1}^{i} dL_j \quad for\ i = 1, \ldots, 2n \tag{3}$$

where dL_i represents the distance between the current point M_i and its previous.

Further, the Fourier descriptors coefficients a_0, a_j, and b_j of the Fourier series that approximates the signature function $\theta_i = f(\ell_i)$ at the kth harmonic are computed as follows:

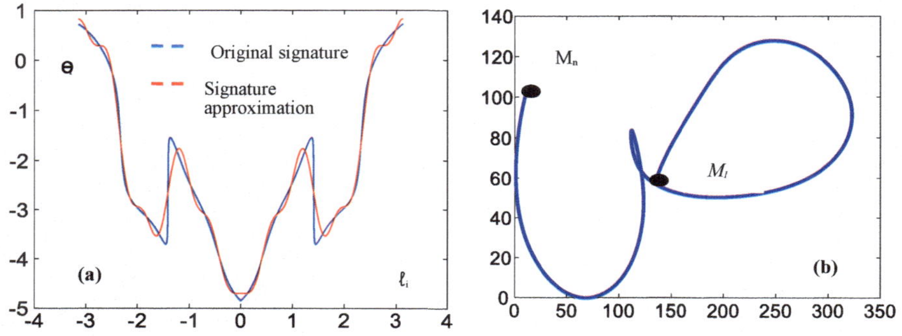

Fig. 5 Signature functions approximated at the eighth harmonic (**a**) and correspondent trajectory of Arabic grapheme "ص" (**b**)

$$a_0 = \frac{1}{2\pi} \cdot \sum_{i=1}^{2n} \theta_i \cdot dL_i$$

$$\begin{cases} a_{j=\frac{1}{\pi} \cdot \sum_{i=1}^{2n} \theta_i \cdot cos\left(j \cdot \frac{2 \cdot \pi \cdot \ell_i}{\ell_{2n}}\right) \cdot dL_i} \\ b_{j=\frac{1}{\pi} \cdot \sum_{i=1}^{2n} \theta_i \cdot \sin\left(j \cdot \frac{2 \cdot \pi \cdot \ell_i}{\ell_{2n}}\right) \cdot dL_i} \end{cases} \quad j = 1, \cdots k, \tag{4}$$

To reconstruct the trajectory segment and the correspondent signature, we use the approximation function described by the following equation:

$$\theta_i = f(\ell_i) \approx a_0 + \sum_{j=1}^{k} \left[a_j \cdot cos\left(j \cdot \frac{2 \cdot \pi \cdot \ell_i}{\ell_{2n}}\right) + b_j \cdot \sin\left(j \cdot \frac{2 \cdot \pi \cdot \ell_i}{\ell_{2n}}\right) \right] \tag{5}$$

In our case, we calculate the Fourier descriptor coefficient for each segmented grapheme at the eighth harmonic yielding a total 17 parameters for grapheme shape modeling.

4.3.3 Geometric Localization Parameters

In addition, we extended our feature vector by other 21 geometric parameters (*bounding box measurements, position to the bounding box, positions of the reference points, grapheme curvature, and assignment of diacritics*) describing the graphemes rapports dimensions, the local position of their trajectories respect to the baseline and the interior bounding box.

- *Bounding box measurements:* As shown in Fig. 6, each grapheme can be described by their measurements that represent the horizontal and vertical dimensions L_V, $L_{H,}$ respectively relative to the bounding box (quadrilateral rectangles).

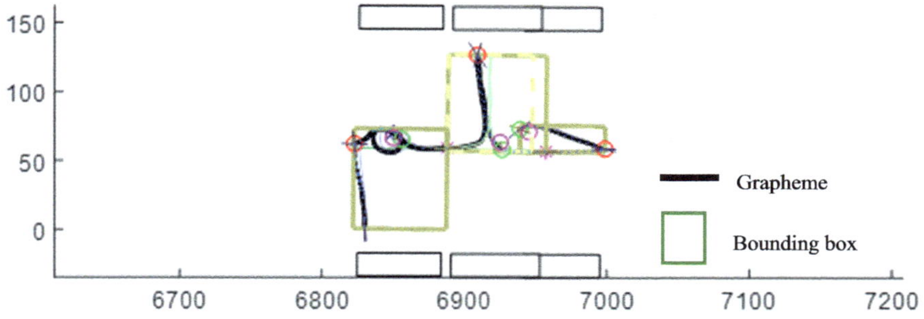

Fig. 6 Graphemes bounding box delimitation

- *Position to the bounding box:* According to the vertical level of its bounding box, a grapheme can be written in over of the baseline (e.g., the second grapheme" ـلـ" of Arabic word " حلم"), descend underneath the baseline (e.g., last grapheme " ـم"), or drawn outwards the baseline (e.g., diacritics).
- *Position of the reference points*: Three marking points are extracted for each graphemes or character trajectory: starting point, point of arrival, and point of minimum curvature (see Fig. 7). Moreover, we accumulate the grapheme trajectory shape feature according to the position of these points in the grapheme bounding box.
- *Assignment of diacritics:* The diacritics stroke (single dot, two or three merged dots, 'shadda: " ّ") are initially filtered from the main handwriting shape based on their position relative to the baseline. Then, the identified diacritics are presented by Fourier descriptor and classified using LSTM network.

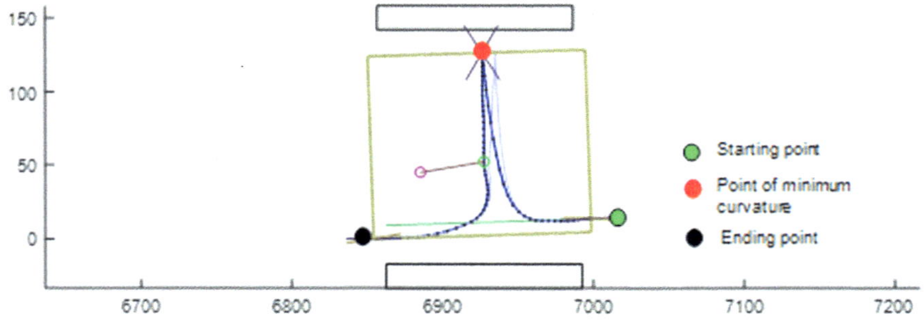

Fig. 7 Position of the graphemes reference points

4.4 Sequence Recognition

For the sequence recognition process, we have used LSTM and BLSTM version of RNN. The network settings are fixed after several tests. The topology of our network architecture is as follows: The input layer accepts the same dimensional feature vectors of 46 parameters modeling grapheme trajectory for both networks. LSTM is formed by one forward hidden layer while BLSTM constituted by two separate hidden layers to treat the input sequence in both forward and backward direction. 400 LSTM memory blocks are utilized in the hidden layer for each network with sigmoid as a gate activation function. Hyperbolic tangent has been employed as output activation function. Dropout is used in fully layer with probability 0.3. We train the networks using its parameters: a stochastic gradient descent "sgdm" with momentum and mini-batch of size 200. We also start the training with an initial learning rate of 0.001 and 1000 maximum number of epochs. A categorical cross-entropy loss function is utilized to optimize the networks. After each epoch, we shuffle the training data to make different mini-batches. The number of neurons in the output layer is the number of labeling classes and has been activated with the soft-max activation function.

5 Experimental Results

In this section, we describe our experimentation which is carried out on online handwriting recognition. The utilized datasets are initially presented followed by the conducted experiments and discussed results. Next, we compare our method with those of the state-of-the-art approaches using the same database.

5.1 Dataset

One of the most difficult attitudes of online handwriting recognition is the requirement for a standard database that serves a variety of writing styles and includes various classes. In order to test the robustness of our proposed system, we have used ADAB benchmark available datasets [27]. It consists of more than 33,000 Arabic words handwritten by more than 170 different writers. The text written is from 937 Tunisian town/village names. This database is divided into six distinct sets from the ICDAR 2011 for online Arabic handwriting recognition competition [28]. We use set 1, 2, 3, and 4 for training process. We tested our system using set 5, 6.

5.2 Results and Discussion

To understand the effectiveness of our approach, we have design two groups of experiments: The first is carried out using the constructed grapheme database to evaluate the performance of the segmentation technique and the second on global word recognition level using the two versions of RNN, LSTM, and BLSTM.

5.2.1 Evaluation of Segmentation Algorithm

In order to evaluate the segmentation algorithm, we have used a constructed graphemes database which contains more than 100,000 grapheme samples.

Figure 8 shows an example of segmentation algorithm of the Arabic word“ سيدي بوزيد'” from the online ADAB database. Most of the letters are segmented into one grapheme such as end “ـد,ـو” isolated “ز,ي”; some other characters are segmented into two or three graphemes like character “ سـ” which is formed by the combination of a single beginning grapheme called “nabra” (ـ) and two middle graphemes “nabra” (ـ ـ).

Table 3 presents the evaluation of graphemes database samples using LSTM and BLSTM. The average of the grapheme recognition accuracy was about 97.34% using LSTM and 98.79% with BLSTM, respectively, which exceed our previous work [29] using MLP network. Test results show the precision of the segmentation algorithm on the one hand, and the used complementary feature vectors are very relevant for graphemes classification on the other hand.

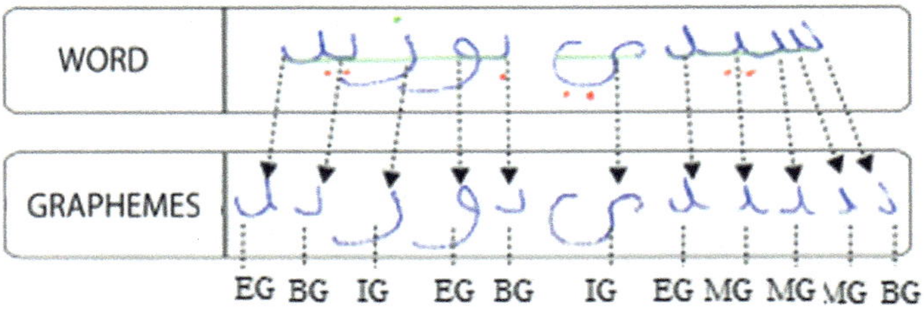

Fig. 8 Examples of the segmentation algorithm result of Arabic word “ سيدي بوزيد”

Table 3 Graphemes classes' recognition rate

	LSTM (%)	BLSTM (%)
Beginning grapheme	97.45	99.30
Middle grapheme	96.55	98.12
Isolated grapheme	98.40	99.76
End grapheme	96.99	97.87
Diacritic grapheme	97.35	98.93

5.2.2 Word Recognition Results

The training and testing data process was performed by varying the number of epoch and batch size for each model (LSTM and BLSTM). "Epoch" indicates the total number of iterations necessary during training dataset process, while "batch size" denotes the total number of input–output operations performed at each epoch, before updating the weights. Table 4 summarizes the word recognition performances obtained by our system employing LSTM and BLSTM.

It may be noted from this table that our proposed system provides a better recognition rate of 98.73% when we use BLSTM for low batch sizes of 40 and 50 epochs.

Further, Fig. 9 shows the word recognition results of different subsets of features using LSTM and BLSTM. We note that the eight features extracted using beta-elliptic model (BEM) which contain dynamic and geometric parameters are very useful for grapheme representation. Indeed, the addition of Fourier descriptor (FD) parameters as well as the geometric localization parameters (GLP) for grapheme shape modeling makes it possible to boost the word recognition accuracy. The best recognition rate 98.73% using BLSTM is achieved by combining the three subsets of features which

Table 4 Word recognition rate

Epoch	Batch size	Accuracy	
		LSTM (%)	BLSTM (%)
50	20	17	68.03
50	30	86.18	92.13
50	**40**	94.55	**98.73**
50	60	**96.65**	98.73
50	70	96.07	97.18

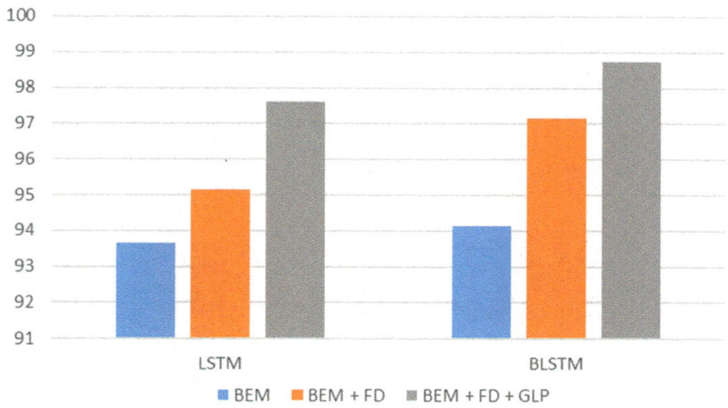

Fig. 9 Word recognition results of different subsets of features using LSTM and BLSTM

Table 5 Comparative word recognition results with existing works

System	Feature	Classifier	Result (%)
Ahmed et al. [21]	Directional features + offline features	HMM	97.78
Abdelaziz et al. [20]	Chain code + curliness + loop detection + curvature + aspect ratio + writing direction	HMM	97.13
Elleuch et al. [22]	Beta-elliptic model + CDBN	SVM	97.51
Our system	BEM + GLM + FDM	LSTM-BLSTM	98.73

explains the strong complementary between them which leads to quite together high recognition rates.

5.3 Comparison with the State of the Art

As already mentioned in related work, little research studies are available for recognition of online Arabic handwriting words. All of these existing works are not comparable directly because they have used different datasets. Indeed, we have compared only with studies that employed the same database. Table 5 depicted comparative analysis. It may be noted from this table that our system provides better results compared to other available works that used HMM model [20, 21]. Also, our system has outperformed the recently existing work-based SVM [22]. This is due to the use of the complementarity models for handwriting representation on one hand and the strength of a deep learning RNN model which represents the discriminating power to differentiate Arabic graphemes on the other hand.

6 Conclusion

This study presents an online Arabic handwriting recognition system using grapheme segmentation and deep learned recurrent neural networks. The proposed system divides the word into graphemes using baseline detection algorithm and identification of topological points delimiting the grapheme trajectory. Then, each grapheme is described by a set of pertinent features that are studied by LSTM and BLSTM version of RNN.

The efficiency of the proposed system was evaluated using ADAB database. Experimental results show that the proposed method for grapheme segmentation and the adoption of RNN outperform the best existing word recognition systems. It also suggests that our system may be appropriate for unlimited vocabulary recognition.

In addition, the obtained results show that the hybrid features extracted models are useful in graphemes classification as well as word labeling. This demonstrates that our method is promising. Further, we have observed that the used features extraction models are rather generic and their application in the other scripts as Persian, Kurdish, English, etc. is interesting, and we envisage as future work. Again, we schedule to accomplish our method of text recognition.

References

1. Lewis MP (ed.) (2009) Ethnologue: languages of the world. SIL International
2. Margner V, El Abed H (2011) ICDAR 2011—Arabic handwriting recognition competition. In: International conference on document analysis and recognition. pp 1444–1448
3. Lorigo L, Govindaraju V (2006) Offline Arabic handwriting recognition: a survey. IEEE Trans Pattern Anal Mach Intell 28(5):712–724
4. Tagougui N, Kherallah M (2017) Recognizing online Arabic handwritten characters using a deep architecture. In: Proceedings of SPIE 10341, ninth international conference on machine vision
5. Xie Z, Sun Z, Jin L, Ni H, Lyons T (2017) Learning spatial-semantic context with fully convolutional recurrent network for online handwritten text recognition. In: IEEE transactions on PAMI
6. Hamdi Y, Boubaker H, Dhieb T, Elbaati A, Alimi A (2019) Hybrid DBLSTM-SVM based Beta-elliptic-CNN Models for online Arabic characters recognition. In: International conference on document analysis and recognition (ICDAR). pp 803–808
7. Boubaker H, Kherallah M, Alimi A (2007) New strategy for the on-line handwriting modelling. In: Proceedings of the 9th international conference on document analysis and recognition. Curitiba, Brazil, pp 1233–1247
8. UNESCO. World Arabic language day. https://www.unesco.org/new/en/unesco/events/prizes-andcelbrations/celebrations/internationaldays/world-arabiclaguageday/
9. Al-Helali BM, Sabri AM (2016) A statistical framework for online Arabic character recognition. Cybern Syst 47(6):478–498
10. Graves A, Liwicki M, Fernandez S, Bertolami R, Bunke H, Schmidhuber J (2009) A novel connectionist system for unconstrained handwriting recognition. IEEE Trans Pattern Anal Mach Intell 31(5):855–868
11. Yuan A, Bai G, Yang P, Guo Y, Zhao X (2012) Handwritten English word recognition based on convolutional neural networks. In: Proceedings of the 13th international conference on frontiers in handwriting recognition2. Bari, Italy, pp 207
12. Ghosh R, Chirumavila V, Kumar P (2019) RNN based online handwritten word recognition in Devanagari and Bengali scripts using horizontal zoning. Pattern Recogn. https://doi.org/10.1016/j.patcog.2019.03.030
13. Keysers D, Deselaers T, Rowley H, Wang LL (2020) Fast multi-language LSTM-based online handwriting recognition. Int J Document Anal Recogn (IJDAR) 23:89–102
14. Alimi AM (1997) An evolutionary neuro-fuzzy approach to recognize online Arabic handwriting, In: Document analysis and recognition, 1997, Proceedings of the Fourth International Conference on. IEEE, pp 382–386
15. Omer MA, Ma SL (2010) Online Arabic handwriting character recognition using matching algorithm. In: Proceedings of the 2nd international conference on computer and automation engineering (ICCAE), vol. 2. IEEE/ICCAE, pp 259–262
16. Ismail SM, Abdullah SNHS (2012) Online Arabic handwritten character recognition based on a rule-based approach. J Comput Sci 8(11):1859–1868. https://doi.org/10.3844/jcssp.1859.1868
17. Izadi S, Haji M, Suen CY (2008) A new segmentation algorithm for online handwritten word recognition in Persian script. In: ICHFR 2008. pp 1140–1142

18. Daifallah K, Zarka N, Jamous H (2009) Recognition-based segmentation algorithm for on-line Arabic handwriting. In: Proceedings of the international conference on document analysis and recognition. Barcelona, Spain, pp 877–880
19. Abdelazeem S, Eraqi H (2011) On-line Arabic handwritten personal names recognition system based on HMM. In: Proceedings of ICDAR 2011. pp 1304–1308
20. Abdelaziz I, Abdou S (2014) AltecOnDB: a large-vocabulary Arabic online handwriting recognition database. arXiv preprint arXiv:1412.7626
21. Ahmed H, Azeem SA (2011) On-line arabic handwriting recognition system based on hmm, In: International conference on document analysis and recognition (ICDAR). IEEE, pp 1324–1328
22. Elleuch M, Maalej R, Kherallah M (2016) A new design based—SVM of the CNN classifier architecture with dropout for offline arabic handwritten recognition. Procedia Comput. Sci. 80:1712–1723
23. Bezine H, Alimi AM, Derbel N (2003) Handwriting trajectory movements controlled by a bêta-elliptic model. In: Proceedings of 7th international conference on document analysis and recognition, ICDAR 2003, vol 2003-January. Edinburgh, UK, pp 1228–1232, 3–6 August 2003. Article number 1227853
24. Boubaker H, Chaabouni A, El-Abed H, Alimi AM (2018) GLoBD: geometric and learned logic algorithm for straight or curved handwriting baseline detection. Int Arab J Inf Technol 15(1)
25. Plamondon R (1995) A kinematics theory of rapid human movements. Part I: Movement representation and generation. Biol Cybernet 72:295–307
26. Persoon E, Fu KS (1986) Shape discrimination using Fourier descriptors. IEEE Trans Pattern Anal Mach Intell 388–397
27. Boubaker H, Elbaati A, Tagougui N, ElAbed H, Kherallah M, Alimi AM (2012) Online Arabic databases and applications. In: Guide to OCR for Arabic Scripts. Springer, pp 541–557
28. Kherallah M, Elbaati A, Abed HE, Alimi AM (2008) The on/off (LMCA) dual Arabic handwriting database. In: 11th International conference on frontiers in handwriting recognition (ICFHR). Montréal, Québec, Canada
29. Hamdi Y, Chaabouni A, Boubaker H, Alimi AM (2017) OffLexicon online Arabic handwriting recognition using neural network. In: Proceedings of SPIE, vol. 10341. pp 103410G-1

Chapter 21
On the Application of Real-Time Deep Neural Network for Automatic License Plate Reading from Sequence of Images Targeting Edge Artificial Intelligence Architectures

Ibrahim H. El-Shal, Mustafa A. Elattar, and Wael Badawy

1 Introduction

A vehicle plate recognition system is a vehicle identification tracking system that monitors the vehicle via the database server. License plate recognition, as a challenging research subject, it has received huge interest in recent years. This is because the conditions and styles of license plates vary between locations (e.g., light, color, dirt, shadow, character sharpness, language, and so on). The unit of recognition is usually installed at the gate of the residential area, toll gates, or other highly protected facilities like defense institutes, nuclear plant facilities [1, 2].

Vehicle re-identification system can easily reach the destination of the vehicle in the city via the ubiquitous monitoring network. The target vehicle can be identified automatically, located, and monitored via multiple cameras with the assistance of the vehicle re-identification device, thereby saving labor and expense. There are three main steps of a traditional license plate recognition system. Primarily, locate license plates on the basis of handmade features in the complete picture; second, segmenting the plate identified into single blocks of character; finally, the segmented characters with a pre-designed classifier [3–5].

I. H. El-Shal (✉) · M. A. Elattar
School of Information Technology and Computer Science, Center for Informatics Sciences, Nile University, 26th of July Corridor, City of Sheikh Zayed, Giza, Egypt
e-mail: i.hamed@nu.edu.eg

M. A. Elattar
e-mail: melattar@nu.edu.eg

W. Badawy
School of Engineering and Technology, Badr University in Cairo, Cairo, Egypt
e-mail: Wael@waelbadawy.com

A. E. Hassanien et al. (eds.), *Enabling Machine Learning Applications in Data Science*, Algorithms for Intelligent Systems, https://doi.org/10.1007/978-981-33-6129-4_21

The traditional, handmade characteristics have their features [6–8], but their common disadvantage is the weak ability for generalization in:

1. This is only useful for specific tasks and cannot be adapted to various application cases, like a color histogram. It is efficient to identify images, but it does not help to segment images.
2. The features, focused on handmade works, only depend on certain aspects of the image, for example, the SIFT, the local image appearance, the HOG, the image, the edge detail, the LBP, the texture of the image, etc.

In order to overcome above-mentioned methods of machine learning drawbacks, deep convolutional neural networks (CNNs) introduce many hidden layers in order to learn high-level features to increase their capacity to generalize not only the target re-identification function but also other computer vision tasks such as picture recognition, object detection, semantic segmentation, video tracking can be well generalized. In recent years, vehicle re-identification methods based on deep learning have become a hub for research [9, 10].

The core aspect of smart transport access control and security systems is automatic plate recognition. It is important to develop an efficient real-time system for automated plate recognition with an increasing number of vehicles [11]. Computer vision methods for this role are typically used. Recently, computer vision in the field of object detection, comprehensive work has been performed for industrial automation, medical, military, and video surveillance. A combination of the highest accuracy and low processing time should be considered. Two phases are required: detection and recognition. In the object recognition phase, the raw input data is represented in the pixels matrix form, in which the pixels are abstracted and the edges are covered in the first-pixel representation layer; the edge layer is formed and encoded by the next, eyes and noses are encoded in the next layer up, and a face is recognized in the final layer. In the recognition stage, features are extracted using a histogram of the directed gradients method from the area of interest [12–15].

The specific objectives of object detection are to obtain high precision and efficiency through the development of robust algorithms for object detection. To achieve high detection accuracy and high efficiency, we have to overcome the problems associated with them. The high accuracy problems are represented in intra-class variations, visual noise, image conditions, unconstrained environments, and many structured and unstructured real-world object categories. On other hand, the challenges related to efficiency are failure to handle items that were previously unknown, large image, or video data size, low-end mobile devices where memory, speed, and low calculation capabilities are limited [16]. The performance of recognition system relies on the accurateness of the object recognition algorithm or image segmentation methods and the quality of the image acquisition [17, 18].

The purpose of this paper is to provide a systematic survey of the existing license plate recognition algorithms and categorize these algorithms according to the process, compare the advantages and disadvantages of detection algorithms and recognition algorithms, respectively.

This paper presented accordingly. In Sect. 2, data collection and annotation. Section 3 demonstrates license plate extraction methods that are classified with a detailed review from training models. Section 4 discusses optimization hyperparameters. Section 5 presents our experiments for tested algorithms. Then, we show our performance results and discussions in Sect. 6. In Sect. 7, we summarize the paper and discuss future research.

2 Data Collection and Annotation

In machine learning problems, data collection and labeling is the most important step. Therefore, images should be collected in-context of where the model will perform in production by ensuring the frame have similar angles, lighting, quality, and objects. The more situations of training sets take into account that the model may encounter in production, performance will be better. At labeling, create bounding boxes that contain the entirety of the objects, even if there is a small amount of space between the object and the bounding box in order to enhance the quality of the model which the result of the model depends on the labels we feed in during its training phase.

2.1 Annotation Types

Various types of image annotations are available such that you select the appropriate type for use:

a. **Bounding boxes**: Bounding boxes are the most frequently used form of annotation of rectangular cases to define the target object's location in computer vision. These boxes can be specified by the coordinates of the x-axis and y-axis in the upper-left corner and in the lower-right corner of the rectangle. Bounding boxes are typically used to identify and locate objects. Representation of bounding boxes by either two coordinates (x_1, y_1) and (x_2, y_2) or by one coordinate (x_1, y_1) and width (w) and height (h) of the bounding box.
b. **Polygonal Segmentation**: The advantages of polygonal segmentation over bounding boxes are that they minimize a significant amount of noise or unnecessary pixels that can potentially mislead the classification around the object.
c. **3D cuboids**: A powerful type, same as bounding boxes with additional object details. It helps the system to differentiate features such as volume and locations in 3D space with 3D cuboids. A case of 3D cuboids is used in self-driving vehicles to measure the distance between objects and the car with depth information.
d. **Keypoint and Landmark**: Used by generating dots around the image to identify tiny objects and shape changes. This form of annotation is helpful for the

identification of facial features, hair, emotions, and physical parts of the human body.

e. **Semantic Segmentation**: Is a pixel-wise annotation, where pixels in the image are assigned to a class and each pixel carries a semantic meaning, and classification being on every pixel in the desired region rather than an object. It is used for any task where large, discrete regions must be categorized or recognized. For example, field analyses to detect weeds and particular types of crops, or for self-driving cars which classes could be traffic lights, cars, and road lines.
f. **Lines and Splines**: involves the creation with lines and splines, designed for cases such as lane detection and recognition in autonomous vehicles, or for the training of robots in warehouses to understand variations in transport belt sections.

2.2 Annotation Formats

There is no single standardized format, and the following are commonly used annotation formats.

a. **COCO**: Using JSON files, the annotations are handled. It has five types for object detection, panoptic segmentation, keypoint detection, stuff segmentation, and image captioning.
b. **Pascal VOC**: Specifies annotations as an XML format description.

2.3 Annotation Tool

LabelImg is an annotation tool for graphical images. It is written in Python and uses Qt for its graphical interface. Annotations will be saved to the format used by ImageNet as PASCAL VOC XML files. We have prepared the dataset for our application. This dataset contains images of vehicle license plates with their bounding box labels distributed images and labels folders.

3 Traning Models

Deep neural networks (DNN) have recently been applied, like other computer vision applications in order to increase the accuracy compared to statistical and classical image processing techniques. However, the real-time response and effective accuracy yet to be assessed in addition to the use of a simple video stream localized at the traffic intersection where each camera is covering a lane of each direction is not explored. We illustrated different object detection algorithms, implementation details, and performance comparison for our custom dataset.

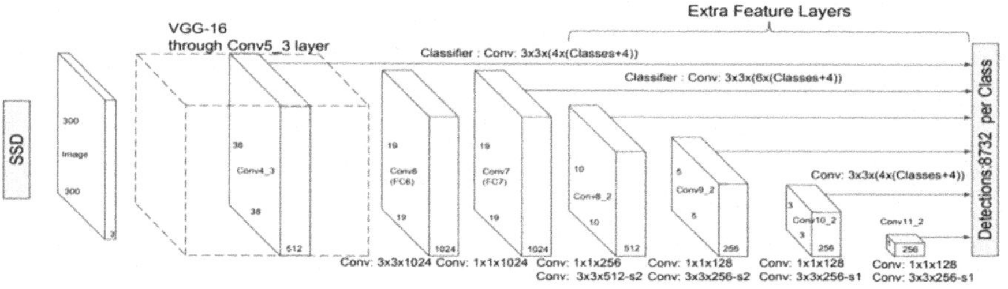

Fig. 1 SSD. *Source* (SSD: Single-Shot MultiBox Detector [19])

3.1 Single-Shot Detector (SSD)

SSD is designed for object detection in real time. The whole process runs at seven frames per second, and it is considered the start-of-the-art inaccuracy. SSD accelerates the process by removing the need for a network of region proposals. SSD enhanced to recover the drop inaccuracy, including multi-scale features and default boxes. These enhancements enable SSD to suit the precision of the faster R-CNN using images with low resolution that drive the speed further up. Input image fed to CNN then computes the feature map for that image. Then on the feature map runs a small-sized convolution network 3×3 to predict the bounding boxes of its relative category.

SSD Ingredients:

a. **Backbone** model usually used as a feature extractor by using a pre-trained image classification network. It is normally a network such as ResNet trained on ImageNet that eliminates the last fully connected (FC) classification layer. We have thus a deep neural network which can extract semantic meaning from the input frame while retaining, albeit at a lower resolution, and the spatial structure of the image.
b. **SSD head** model is just one or more layers of convolution applied to the backbone, and outputs are seen in the spatial location of the final layer activation as bounding boxes and object groups (Fig. 1).

3.2 Faster Region-Based Convolution Neural Networks (Faster R-CNN)

In R-CNN, we use selective search to extract the image region where this method extracts only 2000 regions from the images, and they are called region proposals. Object area proposal is rescaled to fixed image size, then applied to the CNN model which is pre-trained on ImageNet for feature extraction. SVM classifier predicts the object presence within each region proposal and also recognizes object classes. Disadvantages of this model are cannot implemented in real-time cases due to high training time consumption, and selective search method is a fixed algorithm, no learning occurs, and this leads to poor object region proposals generation.

Fast-RCNN, CNN model produces a fixed-length sequence, regardless of area size while performing object detection feature maps are calculated only once from the entire image, but still, the detection speed is limited.

In faster R-CNN, selective search is substituted by region proposal network (RPN). First, an image is taken into ConvNet, and feature maps are returned. From here, RPN applies on these maps of features, and we get the input image object proposals. It then resizes all proposals to the same size and transfers it to the FC convolution layer in order to classify the bounding boxes of the image.

3.3 Yolo

YOLO "You Look Only Once" is a neural network that can detect what is in the entire image and where the content is in a single pass. The regression-predicting bounding boxes around the objects are given with the class probabilities, and a number of objects can be identified at once.

a. YOLO v1
   YOLO inspired by a "GoogleNet" network. It contains 24 convolutional layers that act as feature extractors and two dense layers to predict. The neural network framework called Darknet, a network architecture created by YOLO's first author.
   The algorithm is based on splitting an image into a grid of cells. In addition to the class probabilities, their confidence scores are measured for each cell bounding box. The confidence is granted in terms of an the intersection with the union (IOU), which essentially calculates how often an object detected overlaps with the ground truth as a fraction of the total space (the union) which spans them together. In order to minimize the algorithm losses, the box positions position, their dimensions, the confidence scores for the forecast, and the groups predicted are taken into account.
b. YOLO v2
   Presents a few new features: Mainly anchor boxes (pre-defined box sets to allow the system to switch from predicting border boxes to predicting offsets) and more advanced features in order to predict smaller objects. YOLOv2 generalizes the image more effectively using a mechanism that spontaneously resizes images from time to time.
c. YOLO v3
   YOLO V3 for detection tasks that initially had a network of 53 layers trained on Imagenet with a fully convolutional layer of 106 underlying YOLO V3 architecture. That is why YOLO V3 is less sluggish than YOLO V2, but in all other respects, it performs better. Unlike using sliding window and region proposal-based techniques, YOLO provided with all the detail about the entire

image and the objects during training and the test time so it certainly encodes relevant data approximately classes as well as their appearance.

4 Optimization Hyperparameters

Tuning hyperparameters in real deep learning projects is the most essential way to implement a network that can predict a specific problem accurately. Common hyperparameters include the number of hidden layers, evaluation steps, batch size, the activation function, and how many times (epochs) training should be repeated.

4.1 Learning Rate

Model training will improve very slowly if your learning rate is kept low as it will make tiny adjustments to the weights in the network, and it could also lead to overfitting. When you set a high learning rate, it takes huge leaps bouncing around chaotically and missing the local optima making the training diverge. In other words, a large learning rate helps in regularizing the model to capture the true signal.

4.2 Batch Size

The batch size refers to the amount of network-disseminated training samples. During training, our aim is to achieve maximum performance by reducing the computational time required. The learning rate value does not affect the training time, but the batch size does which larger batch size enables using a large learning rate. However, with the reduced batch size, the final loss values will be lower.

4.3 Epochs

One cycle through the whole training data set refers to an epoch. The neural network usually takes more than a couple of epochs. Also, it is the number of times the algorithm runs over the entire dataset which one epoch means that the intern model parameters have been modified by every sample in the training data collection.

5 Experimental Setup

Our system is based on sequences from Shutterstock Web site for the surveillance cameras of the vehicles entry to the garage. We implemented the ALPR system for real-time applications based on the convolutional neural network architectures. A prototype will be presented to detect the license plate at the entry of Garage.

The efficiency of the algorithm for object detection depends upon several factors, dataset, data quality, algorithm parameters, i.e., training steps, weights, biases, and training rate. Experimental analyses were done to test the above algorithms based on the following methods.

1. Data Collection: We have taken training images from Shutterstock from the events of the different situations and collections on license plates in order to keep the various backgrounds in the images of the train.
2. Data Labeling: Using LabelImg, we annotated around 375 license plates then create a training folder with images and XML files for each image which contains the needed information about license plate objects, and the same applies to the test folder.
3. Data Preparation: For each algorithm, set the number of training steps, evaluation steps number, and model name, type of pipeline file, batch size, and initial learning rate. Also, modify pipeline.config file with adding training, validation data, and label map.
4. Model Training: Train the new models and finetune the new trained model using the pipeline.config file.
5. Test Models: We tested our algorithms over 10 video sequences, sequences with an average of 300 frames. Accuracy of the test depends on frames are correctly classified.

6 Results and Discussions

This section provides the results that have been obtained based on the convolutional neural network architectures as mentioned in the experimental setup section in order to recognize the vehicle license plate. Detail of findings that have been yielded through the implementation of this research will be discussed in this section (Figs. 2, 3 and 4).

Its performance will be analyzed in terms of accuracy and real response. Present work demonstrates the challenges of using video cameras to detect license plates from the video streams. The below table shows the used configuration and parameter of each algorithm (Table 1).

Fig. 2 Examples of license plate detection results from SSD

7 Conclusion and Future Work

In this work, we review the performance for automatic vehicle license plate detection and recognition based on the latest convolutional neural network architectures including Yolo V3, RCNN with ResNet backbone, RCNN with inception backbone, and their ensemble. This process consists of the following steps: image acquisition, license plate extraction, and recognition. We demonstrated through experiments on synthetic and real license plate data of vehicles at the entry of garage that the proposed system is not only highly accurate but is also efficient which the performance analyzed

Fig. 3 Examples of license plate detection results from faster-RCNN

Fig. 4 Examples of license plate detection results from YOLO

in terms of accuracy and real response. Our work demonstrates the challenges of using video cameras to detect license plates from the video streams. Also, it will share the early results of techniques to explore the hidden knowledge with real-time video stream systems to drive a real-time ALPR that is suitable for edge AI machines.

Table 1 Comparative evaluation based on used convolutional neural network architectures

Parameters	SSD	Faster-RCNN	YOLO
Accuracy	44%	91%	97%
Prediction time	appx (3.6 s)	appx (2.5 s)	appx (2 s)
Input image resolution	300 × 300	600 × 600	600 × 600
More data feeding	YES	NO	NO
Detect small objects	NO	YES	YES
Multiple objects detection	NO	YES	YES
Training steps	4000	6000	500
Learning rate	0.004	0.0002	0.001
Batch size	24	1	32
Framework	TensorFlow	TensorFlow	Keras

References

1. Shan Du, Ibrahim Mahmoud, Shehata Mohamed S, Badawy Wael M (2013) Automatic license plate recognition (ALPR): a state-of-the-art review. IEEE Trans Circuits Syst Video Tech 23(2):311–325
2. Han B-G, Lee JT, Lim K-T, Choi D-H (2020) License plate image generation using generative adversarial networks for end-to-end license plate character recognition from a small set of real images. Appl Sci 10:2780
3. Silvaa SM, Jungb CR (2020) Real-time license plate detection and recognition using deep convolutional neural networks. J Visual Commun Image Represent 71:102773. https://doi.org/10.1016/j.jvcir.2020.102773
4. Wang H, Hou J, Chen N (2019) A survey of vehicle re-identification based on deep learning. IEEE Access. 28(7):172443–69
5. Huang W (2010) Automatic vehicle license plate recognition system used in expressway toll collection. In: 2010 3rd international conference on computer science and information technology, vol. 6. IEEE
6. Saadouli G, Elburdani MI, Al-Qatouni RM, Kunhoth S Al-Maadeed S(2020) Automatic and secure electronic gate system using fusion of license plate, car make recognition and face detection. In: 2020 IEEE international conference on informatics, IoT, and enabling technologies (ICIoT). Doha, Qatar, pp 79–84. https://doi.org/10.1109/ICIoT48696.2020.9089615
7. Choudhury AR, Wael B, Ahmad R (2010) A real time vehicles license plate recognition system. In: Proceedings of the IEEE conference on advanced video and signal based surveillance
8. Lin W, Li Y, Yang X, Peng P, Xing J (2019) Multi-view learning for vehicle re-identification. In: 2019 IEEE international conference on multimedia and expo (ICME). Shanghai, China, pp 832–837. https://doi.org/10.1109/ICME.2019.00148
9. Lakshmanan L, Yash V, Raj G (2020) Deep learning based vehicle tracking system using license plate detection and recognition." arXiv preprint arXiv:2005.08641
10. Castro-Zunti DR, Yépez Y, Ko S-B (2020) License plate segmentation and recognition system using deep learning and Open VINO. IET Intell Trans Syst 14(2):119–126. https://doi.org/10.1049/iet-its.2019.0481, Print ISSN1751-956X, Online ISSN 1751-9578
11. Zhang C, Wang Q, Li X (2020) IQ-STAN: image quality guided spatio-temporal attention network for license plate recognition. In: ICASSP 2020–2020 IEEE international conference

on acoustics, speech and signal processing (ICASSP). Barcelona, Spain, pp 2268–2272. https://doi.org/10.1109/ICASSP40776.2020.9053966
12. Puarungroj W, Boonsirisumpun N (2018) Thai license plate recognition based on deep learning. Procedia Comput Sci 135:214–221. https://doi.org/10.1016/j.procs.2018.08.168
13. Weihong W, Jiaoyang T (2020) Research on license plate recognition algorithms based on deep learning in complex environment. IEEE Access 8:91661–91675. https://doi.org/10.1109/ACCESS.2020.2994287
14. Islam KT, Raj RG, Shamsul Islam SM, Wijewickrema S, Hossain MS, Razmovski T, O'Leary S (2020) A vision-based machine learning method for barrier access control using vehicle license plate authentication. Sensors 20:3578
15. Madhukar LS (2006) The recognition and detection system for theft vehicle by number plates. Trans 8(3)
16. Murthy CB, Hashmi MF, Bokde ND, Geem ZW (2020) Investigations of object detection in images/videos using various deep learning techniques and embedded platforms—a comprehensive review. Appl Sci 10:3280
17. Mohamed AT, Ezzat K, El Shal I, Badawy W (2021) On the application of hierarchical adaptive structured mesh "HASM®" codec for ultra large video format. In: ICSIE 2020: Proceedings of the 2020 9th international conference on software and information engineering (ICSIE), November, pp 135–139, https://doi.org/10.1145/3436829.3436870
18. Badawy W (2020) On scalable video codec for 4K and high definition video streaming – the hierarchical adaptive structure mesh approach "HASM-4k". In: 2020 IEEE 10th international conference on consumer electronics (ICCE-Berlin), Berlin, Germany, pp 1–5. https://doi.org/10.1109/ICCE-Berlin50680.2020.9352175
19. Wei Liu et al (2016) SSD: Single shot multibox detector. In: European conference on computer vision. Springer, pp 21–37

Chapter 22
Localization of Facial Images Manipulation in Digital Forensics via Convolutional Neural Networks

Ahmed A. Mawgoud, Amir Albusuny, Amr Abu-Talleb, and Benbella S. Tawfik

1 Introduction

The remarkable phenomenon, a troubling example of the societal threat posed by computer-generated spoofing images, is an important concern of digital forensic analysis. A spoof-video attack will make the Internet a target. Some of the tools available can be used for interpreting head and face movement in real time or for making visual images [1]. In addition, an attacker can also clone a person's voice (only a few minute are required to speak) and sync it with the visual portion for audiovisual spoofing, thanks to advances in voice synthesis and conversion [2]. Throughout the near future, such techniques will become widely accessible, so that everyone can generate profound information. In the visual domain, numerous countermeasures were introduced. Many of these were tested using only one or a few databases, including CGvsPhoto, Deepfakes and FaceForensics++ databases [3].

Cozzolino et al. [4] tackled many state-of-the-art spoofing detectors' transferability problems and developed an auto-encoding architecture to promote generalizations and easily adapt them to a new domain through easy finishing. For digital forensics, another significant issue is the location of compromised areas [5]. The shapes of the segmentation masks for the handling of facial images and video may display

A. A. Mawgoud (✉) · A. Albusuny
Information Technology Department, Faculty of Computers and Artificial Intelligence, Cairo University, Giza, Egypt
e-mail: aabdelmawgoud@pg.cu.edu

A. Abu-Talleb
Computer Science Department, Faculty of Computers Science, University of People, Pasadena, CA, USA
e-mail: aabdelmawgoud@pg.cu.edu.eg

B. S. Tawfik
Information System Department, Faculty of Computer Science, Suez Canal University, Ismailia, Egypt

A. E. Hassanien et al. (eds.), *Enabling Machine Learning Applications in Data Science*, Algorithms for Intelligent Systems, https://doi.org/10.1007/978-981-33-6129-4_22

Fig. 1 Early video frame (top left), changed video frame with the Face2Face technique, using the Deepfakes technique (bottom left, rectangular mask) and using the FaceSwap method (bottom right, multiple-faceted mask)

clues as shown in Fig. 1. The three most popular forensic segmentation methods are used: elimination, copy-moving and splicing. Such methods must process full-scale images, as with other segmentation tasks of images [6].

Rahmouni et al. [1] used a sliding window to process high resolution images which Nguyen et al. [7] and Rossler et al. [8] subsequently used. In spoofing pictures produced with the Face2Face process, this sliding window effectively handles parts. However, many overlapping windows must be marked by a spoofing method which takes a lot of calculating power. We have established a multi-task method for classification and segmentation of treated facial images simultaneously. Our auto-encoder consists of an encoder and a Y-form decoder and is semi-controlled. Classification is carried out by triggering the encoded features [9]. For segmentation, the output of the decoder is used, and the other output is used to restore the input data. These tasks (classification, segmentation and reconstruction) relay information, which improves the overall efficiency of the network [10].

This paper is structured as following. Section 2 includes the related work about previous projects for facial image manipulation detections, Sect. 3 describes the proposed solution of our approach followed by Sect. 4 which is the experiment itself through using two databases and finally Sect. 5 which is the summarization of the overall paper work.

2 Related Work

2.1 Manipulated Videos Generation

The creation of a digital photo realistic actor is a fantasy for those who work in computer graphics. An initial example is the Virtual Emily project, where the image of an actress and her actions to synthesize a numerical version were recorded by sophisticated tools [11]. During the time, the perpetrators could not access this technology, so a digital representation of a victim could not be produced. It changed after Thies et al. [12] performed a face reconstruction in real time in 2016. Following research, heads with basic specifications met by any average individual could be translated. The mobile Xpression App1 was later released with the same feature.

2.2 Manipulated Images Detection Phase

For the identification of manipulated images, multiple countermeasures were introduced. A common approach is to view a video as an image series and to insert the images. The noise-based solution is one of the best developed detectors by Fridrich et al. [13]. The enhanced version with the CNN demonstrated how easily automated feature extraction can be used for detection [14]. Take advantage of high-performance pretrained models among deep learning approaches to detection, fine-tuning and transfer. This is an efficient way of enhancing the efficiency of a CNN by using the CNN as pretrained function extractor. Some detection methods include the use of a restricted convolutionary layer, use a statistical pooling layer, use a two stream network, use a lightweight CNN network and use two layers at the bottom of one CNN. Cozzolino et al. [4] have developed a benchmark for the transferability of state-of-the-art detectors for use in unseen attack detection. We also introduced an auto-encoder-like architecture, which significantly improved adaptability. Li et al. [15] proposed using a time method to detect the blindness of the eye which has not been well replicated in fake footage. Our proposed approach offers segmentation maps of controlled areas in addition to conducting classification [16]. This information could be used to assess the authenticity of images and videos particularly if the classification task does not detect spoofed inputs.

2.3 Manipulated Regions Localization

There are two common approaches to the identification of distorted areas in images: the entire image section and binary classification using a sliding window on several occasions. The segmentation technique is also used for detecting attacks on elimination, copying and splicing. For forgery segmentation, semantic segmentation approaches may also be used [17]. The boxes which represent the limits of the manipulated regions should be returned instead of returning segmentation masks as a slightly different segmentation approach. For spoofing areas, the sliding fan method is used more and creates fake images or videos from real images by a computer. In this method, binary classificators are named at any location of the sliding window, to identify images as a spoof or a good faith. The length of the sliding window can be equal to (non-overlapped) the length of the window or less than (overlapped) the length [18]. The sliding window approach is more used to detect spot areas created by a computer to produce spoof pictures or videos. In this method, binary classifiers are named at any location of the sliding window, to identify images as a spoof or a good faith. The length of the sliding window can be equal to (non-overlapped) the length of the window or less than (overlapped) the length. Our method proposed takes the first approach but with one major difference: Instead of the entire image, only the facial areas are taken into account. It overcomes the problem of measurement expenditures in large numbers inputs.

3 Proposed Method

3.1 Overview

In comparison with other single-target approaches, the likelihood of an input being spoofed and the segmentation maps of manipulated areas are shown by our proposed technique in every context of the data, as shown in Fig. 2.

A collection of frames is used to handle video inputs. For this analysis, we focused on facial images so that the face parts are preprocessed [19]. In principle, the approach proposed can accommodate different input image sizes. However, to make it simple in class, before feeding them into your auto-encoder, we resize cropped images to 256–256 pixels.

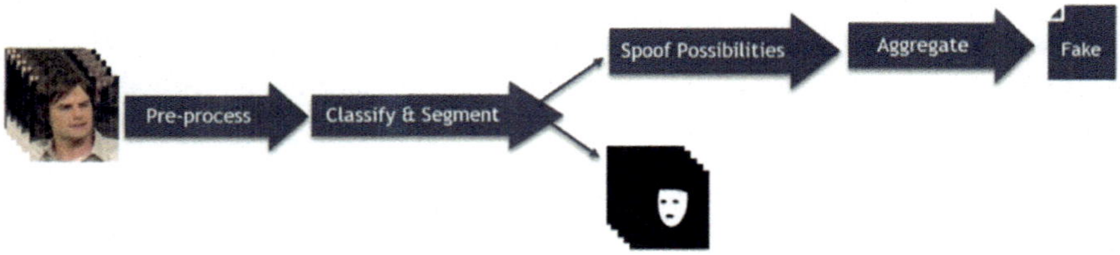

Fig. 2 Description of the formulated network

3.2 Y-shaped Auto-Encoder

The partitioning of latent characteristics and the y-figured nature of the decoder (motivated by work by Cozzolino et al. [4] enables the encoder to exchange useful information between classification, segmentation and reconstruction, enhancing overall efficiency through reduction in losses. Three specific forms of loss are activation loss D_{act}, segmentation loss D_{seg} and reconstruction loss D_{rec}. The accuracy of partitioning into the latent space is calculated by the activation of the two halves of the encoded characteristics given labels $y_i \in \{0, 1\}$, activation loss:

$$L_{\text{act}} = \frac{1}{N} \sum_i |a_{i,1} - y_i| + |a_{i,0} - (1 - y_i)| \tag{1}$$

where N is the sample number, $a_{i,0}$ and $a_{i,1}$ are the activation values and the L_1 standards for half latent characteristics, $h_{i,0}$ and $h_{i,1}$ (due to K the $\{h_{i,0}\,|h_{i,1}\}$ features):

$$a_{i,c} = \frac{1}{2k} \|h_{i,c}\|_{1,} \quad c \in \{0, 1\} \tag{2}$$

This ensures that, given an input xi of class c, the corresponding half of the latent features $h_{i,\,c}$ is activated $a_{i,c} > 0$. The other half, $h_{i,1-c}$, remains quested ($a_{i,1-c} = 0$). To force the two decoders, D_{seg} and D_{rec}, to learn the right decoding schemes, we set the off-class part to zero before feeding it to the decoders ($a_{i,1-c}\!: = 0$). We are using cross-entropy loss to determine the consistency between the divisional s_i mask and the ground-truth mask m_i corresponding to the input x_i: The loss is cross-entropy loss as the segmentation loss:

$$L_{\text{seg}} = \frac{1}{N} + \sum_i \|m_i \log(s_i) + (1 + m_i)\log(1 - s_i)\|_1. \tag{3}$$

The reconstruction losses calculate the contrast between the restored image and the original image using the L_2 distance ($\dot{x}_i = D_{\text{rec}}(h_{i,0}, h_{i,1})$). The reconstruction loss is for N samples

$$L_{\text{rec}} = \frac{1}{N_{2.}} \sum_i \|x_i - \dot{x}_i\| \tag{4}$$

The weighted average of the three losses is the overall loss:

$$L_{\text{rec}} = \gamma_{\text{act}} L_{\text{act}} + \gamma_{\text{seg}} L_{\text{seg}} + \gamma_{\text{rec}} L_{\text{rec}}. \tag{5}$$

We placed the three weights on equal footing (equal to 1). Unlike Cozzolino et al. [4], the classification task and the division task are equally important, and the restoration process plays a significant role in the division mission. The results from the various settings (described below) were experimentally compared.

3.3 Implementation

As seen in Fig. 3, the Y-shaped auto-encoder was introduced.

It is a fully integrated CNN, using three to three convolutional windows and three to three deconstructive windows with a stride of one interspersed (for the decoder). A batch layer and a linear rectified unit (ReLU) accompany each convolutional layer.

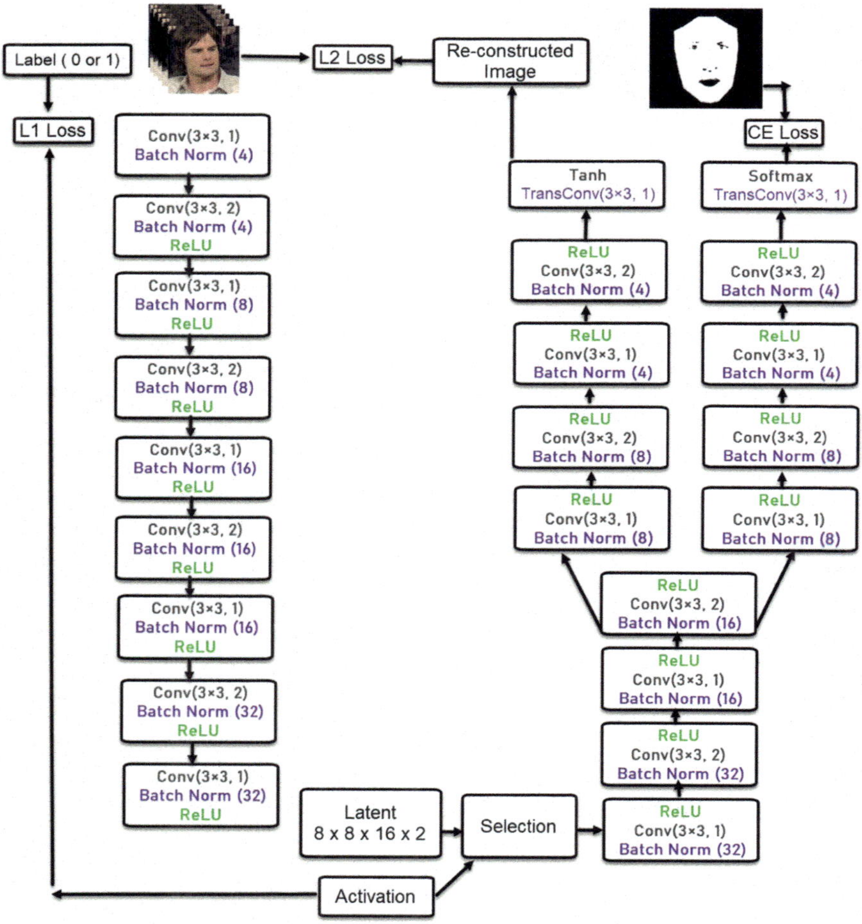

Fig. 3 Proposed Y-shaped auto-encoder to detect and segment manipulated facial images

Only the true half of latent characteristics ($h_{i,}$) will move through the selection block and zeros out in the other half ($h_{i,1-y}$). Thus, only the true half of latent functions is needed to decode by decoders D_{seg}; D_{rec}. The embedding dimension is 128, which has proven optimal. A softmax activation feature is used to generate segmentation maps for the segmentation branch D_{seg}. A hyperbolic tangent function (tanh) is used to form the output in the range for reconstruction branch D_{rec} $[-1;\ 1]$. For ease, we feed normalized images directly into the auto-encoder without turning them into residual images. Throughout further research, the advantages of the use of residual images throughout classification and segmentation tasks will be explored [20]. We trained the network after the work of Cozzolino et al. [4] with an ADAM optimizer of 0.001, a batch of 64, betas of 0.9 and 0.999 and an epsilon of 10−8. After that, we trained the network.

4 Experiments

4.1 Databases

Our proposed network has been tested using two databases:

Forensics Face and forensics Face++. The FaceForensics database contains 1004 true YouTube videos and their associated manipulated versions, divided into two sub-datasets:

- A dataset of 1004 counterfeit videos generated by the Face2Face system is repeated in the source to target; the source video (attacker) is different in each input pair for replay. Victim is different in the target pair.
- Self-reposition dataset with 1004 fake videos, generated again using the Face2Face method; the source and target videos are the same in each input pair for replication. Although this dataset does not matter from the viewpoint of the attacker, the source-to-target reenactment dataset poses a more demanding benchmark.

Every dataset was divided into 704 training videos, 150 validation videos and 150 test videos. The database also provided segmentation masks that suit the manipulated images. The H.264 codec2 is based on three compression rates, and compression light (quantization = 23) and heavy squeeze (quantization = 40) were used. The FaceForensics++ database is a modified FaceForensics database and comprises the Face2Face dataset, the FaceSwap3 dataset (graphic handling) and the Deepfakes4 dataset (deeplearning handling). There are 1,000 actual videos (1000 in each dataset) and 3000 manipulated videos. Every data package was divided into 720 training videos, 140 validation videos and 140 testing videos. For the same measurement values, the same three rates of compression based on the codec H.264 were used. We used only light-compressed videos for convenience (quantization = 23). The images from the Cozzolino et al. [4] videos have been extracted by using: 200 frames were

Table 1 Training and test datasets production

Name	Dataset source	Description	Manipulation approach	Videos number
Training	Source to target	All test usage	Face2Face	805 × 4
Test 1	Source to target	Seen attack matching	Face2Face	260 × 6
Test 2	Self-reenactment	Seen attack nonmatching	Face2Face	260 × 6
Test 3	Deepfakes	Unseen attack (deep learning)	Deepfake	240 × 4
Test 4	FaceSwap	Unseen attack (graphic based)	FaceSwap	240 × 4

used for the training of each training video, and 10 frames were used for both the evaluation and testing of each evaluation and test video. The rules for frame selection are not precise, so we have chosen the first (200 or 10) frames of each video and cut the facial areas. For the entire sample of ImageNet Large Scale Visual Recognition Challenge, we applied normalization to average (0:485; 0:456, 0:406) and standard deviation (0:229; 0:224; 0:225).

The datasets were constructed for training and research, as shown in Table 1. The Face2Face approach for generating manipulated videos has been used for test 1 and test 2 datasets. The pictures in test 2 were more difficult to identify than in test 1, because the source and objective videos for re-enacting were the same, and the reenactment pictures were of higher quality. Therefore, the match and the conditions of a clear attack are called Test 1 and Test 2. The Deepfake attack method was used in Test 3, while the FaceSwap attack method was used in Test 4, presented in the FaceForensics++ database. Both of these attack strategies have not been used to build the training set and were thus known as invisible attacks.

Table 2 Auto-encoder configuration

No.	Approach	Depth	Seg. Weight	Rec. Weight	Rec. Loss
1	*FT res*	Narrower	0.3	0.3	L_1
2	*FT*	Narrower	0.3	0.3	L_1
3	*Deeper FT*	Deeper	0.3	0.3	L_1
4	*Proposed old*	Deeper	0.3	0.3	L_1
5	*No recon*	Deeper	2	1	L_2
6	*Proposed*	Deeper	2	1	L_2

4.2 Training Y-Shaped Auto-Encoder

We built the configuration as shown in Table 2 to determine the contribution of each part in the Y-shaped auto-encoder. The methods of FT Res and FT are re-implementation with or without residual images by Cozzolino et al. [4]. We can also be interpreted without a segmentation branch as the *Y*-shaped auto-encoder. Lower FT is a lower, the same depth variant of FT as the proposed process. The old approach is the method proposed by using Cozzolino et al. [4] weighting settings. Researchers demonstrated that they can detect and identify image manipulations, outperforming human observers significantly. They use recent advances in deep learning, particularly with convolutional neural networks that can learn extremely powerful image features. Training a neural network allows them to tackle the detection issue. For this purpose, they collect a wide dataset of model-based methods for manipulation.

We have equipped the shallowest networks with 100 epochs, and the lower networks with 50 epochs because we need to converge longer than deeper networks. In each process, all the tests mentioned in this section were performed using the training stage of maximum accuracy for the classification task and an adequate segmentation loss (if available).

4.3 Dealing with Identified Attacks

Table 3 (Test 1) and Table 4 (Test 2), respectively, display the results for match and incompatibility conditions for seen attacks. The deeper networks (the last four) were much better defined than the lower networks.

The provided results in the proposed method in Cozzolino et al. [4] showed that that the detection rate in the manipulated images was not effective enough when it is compared with the previous contributions. The new methods used by the new settings for this segmentation task were higher than the old system, which used the old weighting settings, for the segmentation task.

When dealing with the mismatch condition for seen attacks, the efficiency of all methods was slightly degraded. The FT Res and the new methods proposed were best

Table 3 Test 1—image results

Method	Classification		Segmentation
	Acc (%)	EER (%)	Acc (%)
FT_Res	64.6	54.2	_
FT	62.31	52.88	_
Deeper_FT	64.45	48.11	_
Proposed_Old	**67.93**	47.31	95.34
No_Recon	65.97	46.97	**95.97**
Proposed_New	65.08	**45.05**	95.78

Table 4 Test 2—image results

Method	Classification		Segmentation Acc (%)
	Acc (%)	EER (%)	
FT_Res	64.6	54.2	_
FT	62.31	52.88	_
Deeper_FT	64.45	48.11	_
Proposed_Old	**67.93**	47.31	95.34
No_Recon	65.97	46.97	**95.97**
Proposed_New	65.08	**45.05**	95.78

adapted, as shown in their lower degradation scores. It illustrates the significance of using the residual (FT Res) images and the reconstruction (new weighting proposed approach for the Y-shaped auto-encoder). The reconstruction branch also helped to get the highest score for the segmentation task in the proposed new system.

4.4 Dealing with Un-identified Attacks

4.4.1 Evaluation Through Pretrained Model

All six approaches had slightly lower precision and higher EERs for invisible attacks, as shown in Table 5 (Test 3) and Table 6 (Test 4). In test 3, the shallower approaches, in particular FT Res, had greater adaptability. The more profound approaches, the tests of almost random grouping had a higher probability of overfitting [21]. Test 4 indicated in its best EERs that the decision thresholds were moved, although all the methods suffered from almost random classification accuracies. The findings of the segmentation were a fascinating finding. Although degraded, it still showed high segmentation accuracies, particularly in Test 4 in which FaceSwap was using a computer graphic method to copy the facial area from source to target. This knowledge for segmentation may also provide an important indicator in order to determine

Table 5 Test 3—image results

Method	Classification		Segmentation Acc (%)
	Acc (%)	EER (%)	
FT_Res	64.6	54.2	_
FT	62.31	52.88	_
Deeper_FT	64.45	48.11	_
Proposed_Old	**67.93**	47.31	95.34
No_Recon	65.97	46.97	**95.97**
Proposed_New	65.08	**45.05**	95.78

Table 6 Test 4 before fine-tuning—image results

Method	Classification		Segmentation Acc (%)
	Acc (%)	EER (%)	
FT_Res	64.6	54.2	_
FT	62.31	52.88	_
Deeper_FT	64.45	48.11	_
Proposed_Old	**67.93**	47.31	95.34
No_Recon	65.97	46.97	**95.97**
Proposed_New	65.08	**45.05**	95.78

the validity of the queried images while dealing with unexpected attacks [22] (Table 6).

4.4.2 Fine-Tuning Through Limited Data

For the finalizing of all methods, we used a FaceForensics + -FaceSwap validation package (a small set which is usually used to select hyperparameters during training which vary from the test set). To make sure the frequency data was small, for every video, we used only ten frames. The dataset was split into two parts: 100 videos from each training class and 40 for each assessment class. We trained them in 50 years and picked the best models based on their evaluation results.

Table 7 displays the results after the test 4 is done. Their classification and segmentation accuracies increased in relation to the small amount of data, respectively, by approximately 20 and 7%. The one exception was the old approach introduced—it did not enhance the segmentation accuracy.

The FT Res method was much better adapted than the FT system, which supports the argument of Cozzolino et al. [4] as demonstrated by the results of Table 7, the latest approach proposed had the most potential transferability against unknown attacks.

Table 7 Test 4 after fine-tuning—image results

Method	Classification		Segmentation Acc (%)
FT_Res	90.05 (37.65)	28.68 (**36.64**)	_
FT	81.91 (29.71)	36.67 (27.34)	_
Deeper_FT	93.11 (39.72)	28.4 (20.78)	_
Proposed_Old	89.68 (32.86)	31.81 (26.61)	95.41(1.27)
No_Recon	83.84 (39.18)	27.14 (29.14)	83.71(8.85)
Proposed_New	**94.82** (**30.75**)	**26.18** (29.08)	95.12(9.45)

5 Conclusion

The proposed neuronal network with a Y-shaped auto-encoder demonstrated its effectiveness, which is widely used by classifiers, for both classification and segmentation tasks without using a slack window. Sharing data among classification, segmentation and reconstruction tasks improved the overall performance of the network, in particular for malfunctions seen. However, by using only a few fine-tuning samples, the auto-encoder is easy to change for unseen attacks. Future research should specifically investigate how the residual images will influence the output of the auto-encoder, process high resolution images with no re-dimension and enhance its ability to deal with invisible attacks and enlarge it to the domain of the audiovisual.

References

1. Rahmouni N, Nozick V, Yamagishi J, Echizen I (2017) Distinguishing computer graphics from natural images using convolution neural networks. In: 2017 IEEE workshop on information forensics and security (WIFS). IEEE, pp 1–6
2. Mawgoud AA, Taha MH, Khalifa NE, Loey M (2020) Cyber security risks in MENA region: threats, challenges and countermeasures. In: International conference on advanced intelligent systems and informatics. Springer, Cham, pp 912–921
3. Gong D (2020) Deepfake forensics, an ai-synthesized detection with deep convolutional generative adversarial networks. Int J Adv Trends Comput Sci Eng 9:2861–2870. https://doi.org/10.30534/ijatcse/2020/58932020
4. Rossler A, Cozzolino D, Verdoliva L, Riess C, Thies J, Nießner M (2019) Faceforensics++: learning to detect manipulated facial images. In: Proceedings of the IEEE international conference on computer vision, pp 1–11
5. Kleinmann A, Wool A (2014) Accurate modeling of the siemens S7 SCADA protocol for intrusion detection and digital forensics. J Digit Foren Secur Law. https://doi.org/10.15394/jdfsl.2014.1169
6. Mawgoud A, Ali IA (2020) Statistical insights and fraud techniques for telecommunications sector in Egypt. In: international conference on innovative trends in communication and computer engineering (ITCE). IEEE, pp 143–150
7. Nguyen HH, Fang F, Yamagishi J, Echizen I (2019) Multi-task learning for detecting and segmenting manipulated facial images and videos. arXiv preprint arXiv:1906.06876
8. Rössler A, Cozzolino D, Verdoliva L, Riess C, Thies J, Nießner M (2018) Faceforensics: a large-scale video dataset for forgery detection in human faces. arXiv preprint arXiv:1803.09179
9. Mohan A, Meenakshi Sundaram V (2020) V3O2: hybrid deep learning model for hyperspectral image classification using vanilla-3D and octave-2D convolution. J Real-Time Image Process. https://doi.org/10.1007/s11554-020-00966-z
10. Zarghili A, Belghini N, Zahi A, Ezghari S (2017) Fuzzy similarity-based classification method for gender recognition using 3D facial images. Int J Biometrics 9:253. https://doi.org/10.1504/ijbm.2017.10009328
11. Saini K, Kaur S (2016) Forensic examination of computer-manipulated documents using image processing techniques. Egypt J Foren Sci 6:317–322. https://doi.org/10.1016/j.ejfs.2015.03.001
12. Thies J, Zollhofer M, Stamminger M, Theobalt C, Nießner M (2016) Face2face: real-time face capture and reenactment of rgb videos. In: Proceedings of the IEEE conference on computer vision and pattern recognition, pp 2387–2395

13. Mawgoud AA, Taha MHN, Khalifa NEM (2020) Security threats of social internet of things in the higher education environment. In: Toward Social Internet of Things (SIoT): enabling technologies, architectures and applications. Springer, Cham, pp 151–171
14. Fridrich J, Kodovsky J (2012) Rich models for steganalysis of digital images. IEEE Trans Inf Foren Secur 7(3):868–882
15. Bayar B, Stamm MC (2018) Constrained convolutional neural networks: a new approach towards general purpose image manipulation detection. IEEE Trans Inf Foren Secur 13(11):2691–2706
16. El Karadawy AI, Mawgoud AA, Rady HM (2020) An empirical analysis on load balancing and service broker techniques using cloud analyst simulator. In: 2020 international conference on innovative trends in communication and computer engineering (ITCE), Aswan, Egypt (2020), pp 27–32
17. Chu CC, Aggarwal JK (1993) The integration of image segmentation maps using region and edge information. IEEE Trans Pattern Anal Mach Intell 15(12):1241–1252
18. Wei Y, Feng J, Liang X, Cheng MM, Zhao Y, Yan S (2017) Object region mining with adversarial erasing: a simple classification to semantic segmentation approach. In: Proceedings of the IEEE conference on computer vision and pattern recognition, pp 1568–1576
19. Marshall, S.W., Xerox Corp, 2004. Method for sliding window image processing of associative operators. U.S. Patent 6,714,694.
20. Terzopoulos D, Waters K (1990) Analysis of facial images using physical and anatomical models. In: Proceedings 3rd international conference on computer vision, pp 727–728. IEEE Computer Society
21. Mawgoud AA (2020) A survey on ad-hoc cloud computing challenges. In: 2020 international conference on innovative trends in communication and computer engineering (ITCE), pp 14–19. IEEE; Heseltine T, Pears N, Austin J (2002), July. Evaluation of image preprocessing techniques for Eigen face-based face recognition. In: 2nd international conference on image and graphics, vol 4875. International Society for Optics and Photonics, pp 677–685.
22. Korshunova I, Shi W, Dambre J, Theis L (2017) Fast face-swap using convolutional neural networks. In: Proceedings of the IEEE international conference on computer vision, pp 3677–3685

Chapter 23
Fire Detection and Suppression Model Based on Fusion of Deep Learning and Ant Colony

Bassem Ezzat Abdel Samee and Sherine Khamis Mohamed

1 Introduction

Data appears that we are entering an age of greater, more forceful fires that murder more individuals, and do more harm to our environment, our framework, and our air. For case, a later consider appears that the State of California has experienced a fivefold increment in yearly burned zone since 1972. Usually reflected in wildfires getting greater and more damaging all over the world for case, in the greater USA, Australia, and South America. These patterns have been ascribed to numerous causes, such as hotter temperatures, dry spell, fuel collection (a few due to forceful fire concealment), dead vegetation, and expanded populace thickness close wildlands. These patterns are likely to proceed and maybe indeed quicken, likely driving to indeed more unsafe rapidly spreading fires within the future [1].

The specialists dependable for overseeing rapidly spreading fires must react to the expanding risk in a wide assortment of ways counting checking, concealment, departure, endorsed fire, and other implies of fills administration. In all cases, fire offices favor to be made mindful of all conceivable data approximately the fierce blaze some time recently beginning to bargain with it [2]. In a departure situation, each data around the region of the fire can spare lives and result in more secure and superior facilitated clearing. With respect to suppression, littler fires are clearly much less demanding to quench or divert but in colossal fierce blazes it is imperative to know the leading area to begin stifling the fire from. It is certainly more fetched viable for fire concealment organizations to allot an intemperate number of assets, keeping a fire at many section of land fire compared to the fetched of a weeks-long

B. E. A. Samee (✉) · S. K. Mohamed
Department of Information Technology, Institute of Graduate Studies and Research, University of Alexandria, 163 Horyya Road Elshatby, P.O:832, Alexandria 21526, Egypt
e-mail: igsr.bassem.ezzat@alexu.edu.eg

S. K. Mohamed
e-mail: sherinekhamis@alexu.edu.eg

A. E. Hassanien et al. (eds.), *Enabling Machine Learning Applications in Data Science*, Algorithms for Intelligent Systems, https://doi.org/10.1007/978-981-33-6129-4_23

fight against a 100,000 acre fire. Here comes the significance of utilizing diverse machine learning calculations to distinguish and smother [3].

Early inquire about on computer vision-based fire discovery and concealment frameworks were centered on the color of a fire inside the system of a rule-based framework, which is frequently touchy to natural conditions such as light and climate. So, encourage ponders included supplementary highlights to the color of a fire, counting region, surface, boundary, and movement of the suspected locale, with other sorts of decision-making calculations, such as Bayes classifier and multi-expert systems, in arrange to form a strong choice. All things considered, nearly all the inquire about tries to identify the fire and smoke in a single outline of Closed-Circuit Television (CCTV) or a constrained number of outlines in a short period [4]. In general, it isn't a simple assignment to investigate the inactive and energetic characteristics of differing fire and smoke to be misused in a vision framework, because it requires an expansive sum of domain knowledge. In the deep learning approach however, these investigation and exploitation forms can be supplanted by the preparing of a suitable neural network with a quate sum of data in arrange to dodge overfitting. This approach, in this manner, gets to be helpful once a dataset with numerous fire and smoke pictures or video clips has been built [5, 6].

There are a wide run of challenging choice and optimization issues within the zone of woodland fire administration, numerous of which would advantage from more responsive fire behavior models which may well be run rapidly and upgraded effectively based on unused information. For case, one straightforward choice issue is whether to permit a fire to burn or not, since burning may be a common shape of fuel lessening treatment. Answering this question requires an extraordinary bargain of costly reenactments to assess the arrangement alternatives. These recreations are built by a dynamic investigate community for woodland and wildland fire behavior modeling. Information are collected utilizing trials in genuine woodland conditions, controlled lab burning tests, physics-based fire modeling, and more. These handmade physics-based models' recreations have tall exactness but are costly to form and overhaul and computationally costly to utilize [7, 8].

Since Fire discovery and concealment field is considered an imperative field of machine learning, a model is recommended that employments Deep Learning to identify the fire and its range. Deep Learning was utilized as the numerous layers within the neural network make the method of extricating features to be more exact. Comes about are being more particular within the detection of the fire in [9]. Ant colony is utilized to choose the most excellent point to begin from within the concealment handle. Ant colony is utilized since of its capability in managing with dynamic data [10, 11].

That paper is structured as takes after: numerous of recent related works are put forward Sect. 2. The comprehensive description of the proposed model has been made in Sect. 3. In Sect. 4, a short overview is given for the results and description of the dataset is put forward. As a final point, the conclusion is annotated in Sect. 5.

2 Literature Review

Analysts have endeavored to utilize machine learning in combination with adj. symbolism in an arrangement of biological applications within the past. Disciple radar pictures and machine learning are utilized for the location of oil spills. Over the past two decades, fire discovery with the assistance of computer vision has ended up well known due to the progresses in gadgets, computers and sensors. Point by point audit on computer vision based fire discovery are detailed in. Among all the detailed works, as it were a modest bunch of them have combined on fire and smoke marks for fire detection.

Machine learning algorithms utilize a programmed inductive approach to recognize designs in information. Once learned, design connections are connected to other comparative information to produce expectations for data-driven classification and relapse issues. The work in [12] takes an assignment of administered lithology classification (topographical mapping) utilizing airborne pictures and multispectral obsequious information and compares the application of prevalent machine learning procedures to the same. A tenfold cross approval was utilized to choose the ideal parameters to be utilized in all the strategies. These chosen parameters were utilized to prepare the machine Learning classification models on the whole set of tests [12].

In 2015, the awesome thrive of hereditary programming, the analysts discus the application of a brilliantly framework based on hereditary programming for the forecast of burned regions in a wildfire circumstance. They moreover compare the hereditary programming strategy to state-of-the-art machine learning algorithms in fire detection and conclude that hereditary programming methods are way better. The major machine learning algorithms utilized are SVM with a polynomial part, irregular timberlands, spiral premise work organize, straight relapse, isotonic relapse, and neural systems. Experimentally set rule-based thresholding calculation is proposed in [13]. The creators utilized outline differencing, RGB-YCbCr color rules, corner point variety of candidate locale and variety in R color channel for fire location. The subsequent portion of the framework utilized histogram variety, wavelet vitality and correlation with background image for smoke detection.

Whereas rule-based algorithms presented are computationally reasonable, they are inclined to misclassification. On the other hand, utilizing machine learning algorithms such as support vector machine, neural network, Bayes classifier, fuzzy logic etc. can make the algorithms stronger beneath a wide set of scenarios. Researchers in [14] utilized a fuzzy logic classifier rather than experimentally set threshold to detect fire and smoke. The by and large algorithm employments a motion history picture and HSI color space to distinguish a candidate locale followed by flash examination within the candidate region. Moreover, after discovery, the persistently versatile cruel shift (CAMSHIFT) algorithm is connected to track the identified fire region. The creator detailed 87% and 77% fire and smoke location rate individually and 20 ms location time. Fuzzy logic classifier was moreover utilized in [15] to distinguish fire in YCbCr color space and rule-based thresholding in RGB to detect smoke.

Another machine learning based show displayed in [16] utilized Artificial Neural Network (ANN). They utilized two ANNs: one for fire and one for smoke and ideal mass transport-optical stream as highlights. Their report appeared promising comes about but the classifiers were prepared and tried with a restricted dataset. ANN and optical stream strategy for smoke acknowledgment are too connected in [17]. Their fire recognition strategy utilized an isolated algorithm where candidate region was segmented utilizing frame differencing and HSI color rules taken after by partitioning the frames into 8 × 8 pieces and analyzing the glint recurrence of each square. In any case, the utilize of transient data within the shape of optical stream makes these strategies computationally costly.

Although automated fire detection and suppression have been studied for few decades, there is still a room to make it extra efficient and practical in real applications. According to the aforementioned review, it can be found that past studies were primarily devoted to: (1) picture segmentation, (2) not addressing the best feature to increase the accuracy (3) doesn't take the dynamic parameters value in consideration during the suppression process. However, to the best of our knowledge, little devotion has been given to advising new technique to utilize the best feature in detecting and the best point to start the suppressing the fire.

3 Proposed Methodology

Nowadays, inaccessible control centers that are prepared with intelligent frameworks and run by master administrators are broadly utilized in about all major cities within the world to improve our way of life execution. In a fire detection framework, discovery can be made either by the human administrators who screen woods or by a committed subsystem that works on the premise of a foreordained algorithm. The proposed fire detection and suppression model uses satellite images and applies deep learning to detect the fire and its area. To diminish computational complexity, temporal features such as flickering examination, development, and movement etc. have been overlooked and as it were extricating color and surface data. At that point, the result of the deep learning (fire zone) is passed with other energetic parameters esteem to Dijkstra algorithm to choose the most limited way between fire points. Ant colony employments these ways to discover the most excellent point to begin concealment (suppression) from [18]. The full model description is in next paragraphs.

3.1 Preprocessing of Satellite Images

In the preprocessing phase, the model first starts making satellite images ready to be used in the neural network of the deep learning algorithm. Our purpose is to classify a sub-window x, of size 15 × 15 pixels extracted from an image, as a fire ($x \in V$) or as a non-fire ($x \in N$). To preprocess the images, Convolutional Neural Network is considered here. It is motivated from the works of Krizhevsky, Hinton and Sutskever [9]. The design of this organize is comparative to single layer neural arrange with a few extra layers. The essential design incorporates a convolutional layer comprising of predefined number of channels of measure 3 × 5 × 5, called the include maps, which learn highlights from the input image. Here neurons are as it were associated to a little region (locally associated) having size of the channels. This is often taken after by a pooling layer of settled measure 2 × 2. For each of the models we have utilized max-pooling. This max pooling layer down tests its input along its width and stature. This layer is taken after by a fully-connected layer where all the neurons considers each actuation within the past layer. Each layers of the show learns its weights and predispositions (obscure parameters) utilizing angle plummet in little mini-batches of preparing tests. The introductory values of weights and predispositions of all the highlight maps are relegated with arbitrary typical dispersion. There are a number of hyper-parameters like learning rate (η), decay of learning rate, cost regularization constant (ℷ), minibatch size, number of neurons in fully-connected layer which plays a vital role in achieving good performance [19].

3.2 *Fire Detection*

In our proposed model, a weighted Global Average Pooling (GAP) scheme is adjusted to extricate the spatial features. After immaterial bounding boxes are sifted out by thresholding their possess certainty score, the critical SRoFs and non-fire objects are chosen. The image which does not contain any particular little bounding box is treated as a non-fire protest whose bounding box covers the full picture with certainty score 1. Note that the critical SRoF or the non-fire protest has its own confidence score which can be utilized to require the weighted Hole. The spatial highlights are extricated from the final layer of CNN of the Faster R-CNN protest locator with d feature maps, where $d = 1024$. From each include outline fi, the scalar include esteem is decided as takes after:

$$v_i = \frac{1}{Z} \sum_{\substack{for\ each\ bounding\ box\ b\ of\ SRoFs \\ or\ non-fire\ objects}} w_b \cdot \sum_{(x,y)\in b} f_i(x, y), \tag{1}$$

where,

$$Z = \sum_{\substack{for\ each\ bounding\ box\ b\ of\ SRoFs \\ or\ non-fire\ objects}} w_b \cdot \sum_{(x,y)\in b} 1. \tag{2}$$

The vector $v = (v_1, v_2, \ldots, v_d)$ represents speaks to the amassed spatial include for SRoFs or non-fire objects recognized by Faster R-CNN in an picture or an outline of a video. In common, the conspicuous features among d can be found anticipating the bounding box of SRoFs or non-fire objects on the feature outline comparative to the lesson enactment outline. Since they are only spatial features which don't contain transient data, the include choice in our proposed model is exchanged to the taking after LSTM arrange of the transient accumulation in a short-term. After calculating the weighted regions in successive outlines, we take midpoints over a periodT_{ave}, then for everyT_{rep}, the consecutive average areas are given separately from or with the final fire decision to obtain a better understanding of the current dynamic behavior of the fire [20].

3.2.1 Long Short Term Memory (LSTM) for Detecting Fire

Long Short Term Memory (LSTM) organize comprises of two stages in which the number of memory cells in LSTM is decided tentatively. The brief transient features pooled through the LSTM arrange are utilized to create a short-term fire choice by two soft-max units, one for fire or the other for non-fire. The LSTM organize is independently prepared utilizing the weighted Crevice spatial features of CNN in bounding boxes. The fire judgment from LSTM is based on the temporally totaled spatial features within the SRoFs and the non-fire objects. Here, ready to consider extra temporal highlights related to the region of SRoFs. The numerous regions permit us to require the weighted sum of SRoFs, where the weights are given by the certainty score corresponding to the SRoF. In Eq. (2), Z can be treated as the weighted zone of objects in an outline. Be that as it may, we independently calculate the weighted ranges for fire and smoke objects to deliver a more exact elucidation [21].

3.3 Fire Points Paths

The most objective of this work is to supply instruments for supporting master clients in arranging, controlling and moderating errands for lightening the results of rapidly spreading fires in common situations. In arrange to reach this objective, we have designed a system for displaying a few of the foremost vital characteristics which are involved in a woodland. The most objective of this work is to supply instruments for helping master clients in arranging, controlling and relieving assignments for reducing the results of wildfires in common situations. In arrange to reach this objective, we have planned a system for displaying a few of the foremost critical characteristics which are included in a woodland wildfire.

Most of the rapidly spreading fires happens in forested zones. It is broadly known the expansive biodiversity and expansion of the different normal scenes. These wooden regions comprise of multiple species of trees, scrubland and other topographical components such as slopes, streams, valleys and knolls. A few characteristics are included within the regions in arrange to attain a reasonable demonstrate. A few of the forested zone highlights are particularly important, like vegetation volume, stickiness, temperature, height, slant and likelihood of fire proliferation. The advancement of the wind flow is additionally considered due to its impact within the fire wide spreading. For this reason, we have included important data for speaking to wind speeds and bearings.

We define a grid $G = \{0, 1, \ldots, n\} \times \{0, 1, \ldots, m\}$ where $n, m \in \mathbb{N}$. We define a *region of G*, as a tuple $r = (c, v, p, st, l, i, t, hum, wf)$ where $c \in G$ represents the coordinates of the region, v is the total volume of vegetation measured in m/ha^3, $p \in [0, 1]$ is the probability of fire propagation in this region, st indicates the state of the region, $l \in \mathbb{R}$ is the elevation of the land measured in meters, $i \in [0, 90]$ represents the inclination angle, $t \in \mathbb{R}$ is the average temperature measured in Celsius degrees, $hum \in [0, 100]$ represents the humidity level of the area and $wf \in [0, 360) \times [0, 12]$ indicates the direction measured in degrees and the wind speed, based on Beaufort scale, that affects the region.

The state of the region st takes values in {*Healthy, On Fire, Burned*}, where *Healthy* means that the region is clear of fire, *OnFire* indicates that the region is partially on fire and *Burned* represents that the fire burns down the region to the ground. Given a region $r = (c, v, p, st, l, i, t, hum, wf)$ we let pos(r) be equal to c. A *surface* modeled by a grid G is a tuple $= S_G(h, w, R, ld, st)$ where $h, w \in \mathbb{N}+$ correspond to the height and weight of a forest measured in meters, respectively, R is a set of regions of G, ld: $R \times R \rightarrow \mathbb{R} > \cup\{\infty\}$ is the *difficulty level of access function* and st: $R \rightarrow \{Clear, Low, Medium, High\}$ is the *wildfire function.* In order to ensure that the set of regions R completely covers the surface we need two conditions to be fulfilled:

- For all distinct $r, r' \in R$ we have pos(r) $\neq$ pos(r').
- For all $c \in G$ there exists $r \in R$ such that pos(r) $= c$.

Given$r_1, r_2 \in R$ such that pos(r_1) $= (x_1,y_1)$ and pos(r_2) $= (x_2,y_2)$ we say that r_1 and r_2 are *neighbours*, denoted by neigh(r_1,r_2), if and only if $|x_1 - x_2| \leq 1 \wedge |y_1 - y_2| \leq 1 \wedge (x_1 = x_2 \oplus y_1 = y_2))$. The function $ld(r_1,r_2)$ returns a value that represents the difficulty level to access between r_1 andr_2. The function returns ∞ for all the pairs (r_1,r_2) such that neigh(r_1, r_2), does not hold. We say that $\sigma = <r_1, \ldots, r_w>$ is a path of S_G if and only if for all $1 \leq i < w$, neigh(r_i,r_{i+1}) and for all $1 \leq i < j \leq w$ we have pos(r_i) $\neq$ pos(r_j). Abusing the notation, we will write $r \in \sigma$ if there exists $1 \leq i \leq w$ such that $r = r_i$. The function $st(r)$ returns a value that represents the intensity of a possible seat of fire in region r. If the region is burning the function returns either *Low, Medium* or *High* depending of the intensity of the fire. In other case, the value associated with the region is *Clear*.

3.4 Fire Suppression

After demonstrating surface, this surface speaks to a region where a few seats of fire have been identified. In this environment, the proposed demonstrate is expecting for determining the most excellent methodology to relieve and arranging the fire concealment. Once provided, the surface must be examined to set, up based on the highlights considered in our hypothetical demonstrate, the trouble level to get to between any two adjoining locales of the surface). The another step comprises within the calculation of the briefest ways between the diverse seats of fire. This step is performed by utilizing the briefest way Dijkstra algorithm [22]. The remove between two locales is measured by taking into consideration both the separate to the goal locale and the trouble level to reach it. Once the most limited ways have been calculated, the another step comprises in making a rearranged chart whose hubs compare to seats of fire. The edges of this graph contain the values gotten within the previous step. Once the rearranged chart has been built, within the taking after step the ACO algorithm is applied for dissecting and organizing seats of fire. It permits to choose a suitable technique to cover all the centers within the least sum of time with the least level of trouble for getting to them.

During the resolution of the problem, each ant leaves a pheromone trail on the connecting path among the points of fire. These pheromone paths will be reinforced if here exists a path between points of fire. Starting from the start point, the ants select the following step to be performed using the next equation:

$$p_{ij}^{k}(t) = \frac{[\tau_{ij}(t)]^{\alpha}[\eta_{ij}]^{\beta}}{\sum_{l \in N^k}\{[\tau_{ij}(t)]^{\alpha}[\eta_{ij}]^{\beta}\}}, \forall j \in N^k \tag{3}$$

where:

- τ_{ij}: The pheromone deposited by each ant on the path that join the points of fire i and j.
- $\propto$: The intensity control parameter.
- η_{ij}: The quality of the path from i to j. This factor is determined by $\eta_{ij}=st(j)/l$, where $st(j)$ is the intensity of the fire at j and l_{ij} is the cost of move from region i to region j.
- β: The visibility control parameter.
- N^k: Represent the set of focuses not yet visited by the ant k.

$$\tau_{ij} \leftarrow [(1-p)\tau_{ij} + \Delta\tau_{ij}^{\text{best}}]_{\tau_{\min}}^{\tau_{\max}} \tag{4}$$

$$\Delta\tau_{ij}^{\text{best}} = \begin{cases} 1/L_{\text{best}} & \text{if } (i, j) \in T_{\text{best}} \\ 0 & \text{i.o.c} \end{cases} \tag{5}$$

In order to maximize the accuracy of the algorithm, $\tau_{\max}$, $\tau_{\min}$ and L_{best} are initialized by following the next equations [23]:

$$\tau_{\max} = \frac{1}{\rho \cdot L_{gb}} \tag{8}$$

$$\tau_{\min} = \frac{1 - \sqrt[n]{p_{\text{best}}}}{\rho \cdot L_{gb}} \tag{9}$$

$$L_{\text{best}} = \frac{1}{d_{ib}} \tag{10}$$

where:

L_{gb} is the best tour from the beginning of the algorithm,
L_{ib} is the best tour in the current iteration,
P_{best} is a control parameter,
T_{best} represents the best tour.

Following, the information is provided to the fire fighters in a friendly way to ease the suppression of the wildfire.

4 Implementation and Experimental Results

4.1 Implementation

To analyze the potential benefits of the proposed system, we first evaluated their relevance and contribution to predictive accuracy; we proceed by comparing it to other status-of-the-art systems. The functions of the system are written in the suggested algorithm has been implemented in MATLAB (R2017b) simulator. MATLAB is a software development environment that proposes high-accomplishment numerical computation, data analysis, visualization capabilities and application development tools. It is an object-oriented programming language that may be able to be utilized to write code to run within the MATLAB surroundings, or in a wide variety of programs released by other companies which work with MATLAB. MATLAB has numerous boons: (1) it allows the testing algorithms immediately without recompilation. It permits typing something at the command line or executing a section in the editor and immediately demonstrates the results, and greatly facilitating algorithm development. (2) MATLAB's built-in graphing tools and GUI builder ensures that customizing the data and models assists the user interpret their data more easily for quicker decision making. (3) MATLAB's functionality may be able to be expanded by adding toolboxes [24], and it was run on a HP Pavilion—15-cs3008tx—Core™ i7 processor—8 GB DDR4-2666 SDRAM running windows 10.

4.2 Dataset

Inside the system of work conducted for the "Fire" Project concerning a visual estimation of the geometric characteristics of wildland fires, a database containing pictures and groupings of wildland fire pictures has been compiled in a dataset called "Corsican Fire Database". These pictures have been procured within the visual extend, within the close infrared run and beneath different conditions of situating, vision, weather, vegetation, separate to the fire and brightness. Each database picture is related with a set of data counting, for case, the diagram of the fire zone within the manually-extracted picture, the prevailing colour of the fire pixels, the rate of fire pixels secured with smoke, the nearness of clouds and the brightness of the environment. This database, named the Corsican Fire Database, right now contains 500 images obtained as it were within the visual extend, 100 picture sets obtained at the same time within the visual run and near infrared extend and 5 sets of picture sets obtained at the same time within the visual run and close infrared run. The Corsican Fire Database can be found at https://cfdb.univ-corse.fr/ and can be partially or fully downloaded.

4.3 Model Evaluation

The goodness-of-fit and prediction power of the forest fire susceptibility models were evaluated based on statistical measures such as overall success rate, positive predictive value, negative predictive value, specificity, sensitivity (Eqs. 11 and 12)

$$\text{Overall success rate} = \frac{\text{TP} + \text{NT}}{\text{TP} + \text{NT} + \text{FP} + \text{FN}}; \text{ Sensitivity } = \frac{\text{TP}}{\text{TP} + \text{FN}} \tag{11}$$

$$\text{Specificity} = \frac{\text{TN}}{\text{FP} + \text{TN}}; \text{ PPV} = \frac{\text{TP}}{\text{FP} + \text{TP}}, \text{ NPV} = \frac{\text{TN}}{\text{FN} + \text{TN}} \tag{12}$$

where TP (True Positive) and TN (True Negative) are samples in the training or validation datasets that area correctly classified. FP (False Positive) and FN (False Negative) samples in the training or validation datasets that are misclassified. Overall success rate is the number of forest fires and non-fire points that are correctly classified divided to the total points. Positive Predictive Value (PPV) is the probability of points that is classified to forest fires, whereas Negative Predictive Value (NPV) is the probability of points that is classified to non-fire. Sensitivity is the percentage of correct forest fires points whereas specificity is the percentage of correct non-fire points in the training or validation datasets [25].

Table 1 Performance of the three machine learning models (SVMC, RF, and MLP-Net) Compared with the suggested model

No.	Statistical index (%)	SVMC	RF	MLP-net	Suggested model
1	SENS	78.4	82.4	83.8	84.0
2	SPEC	76.4	74.1	83.9	84.7
3	ACC	77.4	77.6	83.8	87.2

4.4 Experimental Results

To validate the suggested model in fire detection, it was compared by some of the machine learning algorithms. In the first experiment, the suggested model was compared with Vector Support Machine (VSM) Model. The suggested model was l little bit better. That is because VSM doesn't work well with noisy datasets. Besides, in cases where the number of features for each data point exceeds the number of training data samples, the SVM will underperform. Deep learning can overcome these problems easily. The suggested model was compared with Random Forests (RF) in the second experiment. It gave better results as the RF can fit over noisy datasets but for data with different values, attributes with more values will have a greater impact on random forests, so the attribute weights generated by random forests on such data are not credible. The comparison between the suggested model and Multilayer perceptron neural network was done in the third experiment. Deep learning gave better results as the more the layers are the better the results. When there are more layers the features are dealt with in a better way. Table 1 shows the result acquired according to the accuracy and specify of each model.

To assess the model suggested for suppression, we have performed a few tests. For this, we have created a model based on the recommended system in Segment 3. This model has been actualized utilizing OMNeT++ 5.0. We have displayed two surfaces which have been arranged utilizing 5 seats of fire for each surface. In expansion, we have utilized values extended between 50 and 450 for the number of ants, and we have performed 50, 100, 200 and 400 cycles. Each locale in these surfaces cover a range of 30 m^2. The proposed calculation (see Sect. 3) has been utilized to examine these surfaces in order to discover the finest arrange to moderate the spreading of the rapidly spreading fire. The primary demonstrated surface comprises of 625 districts. This range is proportionate to a quarter of Central Stop. The length of the visit, when we consider 5 seats of fire, ranges from 22 to 49 units. Therefore, applying the calculation employing a diminished number of ants encompasses a coordinate effect on the quality of the arrangement. The higher is the number of ants the lower is the length of the arrangement. By the by, the constrained measurements of this surface don't permit an important decrease within the add up to length of the visit when expanding the number of ants and iterations.

The second modelled surface comprises of 2500 districts; whose expansion is identical to the region of Monte Carlo. In this case, the length of the visit, when 3 seats of fire are considered, ranges from 29 to 75 units. So also, to the past test, the application of the calculation employing a diminished number of ants does not deliver the way better arrangement. All things considered, in this case, the surface is longer than the past one and for this reason, there exist a pertinent diminishment within the add up to length of the visit when the number of ants and iterations are expanded. Additionally, it is worth noticing that the inconstancy of the whole length of the visit is more honed, due to the higher extension of this surface. In common, expanding the number of ants gives way better arrangements. This is often primarily due to the reality that the increment of the number of ants moreover increases the likelihood of finding the leading solution. Furthermore, when the number of ants used is reduced, the solution can be improved by increasing the number of iterations.

5 Conclusion

This paper proposes an automated Model is used in dealing with wildfire in forests. It detects fire and its area then it finds the best point to start suppressing the fire from. This is done by utilizing deep learning as it is known by its superiority over machine learning algorithms in dealing with features and noisy images. Ant colony is also used as it is known as the best metaheuristic search algorithm in dealing with dynamic parameters and it is known with minimizing total system losses. The objective function used was based on a reliability index, where Ant Colony algorithm was applied to solve discrete optimization problems. As a consequence, the acquired results were promising. To set a plan for future works, the may use some spatial analysis in detecting the fire and Shuffled frog leaping algorithm in the suppression step.

References

1. Alkhatib AA (2014) A review on forest fire detection techniques. Int J Distrib Sensor Netw 10(3):597368
2. Martell DL (2015) A review of recent forest and wildland fire management decision support systems research. Current Forest Rep 1(2):128–137
3. Subramanian SG, Crowley M (2017) Learning forest wildfire dynamics from satellite images using reinforcement learning. In: Conference on reinforcement learning and decision making
4. Cao Y, Wang M, Liu K (2017) Wildfire susceptibility assessment in southern China: a comparison of multiple methods. Int J Disaster Risk Sci 8(2):164–181
5. Hantson S, Pueyo S, Chuvieco E (2015) Global fire size distribution is driven by human impact and climate. Glob Ecol Biogeogr 24(1):77–86
6. Arnett JT, Coops NC, Daniels LD, Falls RW (2015) Detecting forest damage after a low-severity fire using remote sensing at multiple scales. Int J Appl Earth Obs Geoinf 35:239–246
7. Houtman RM, Montgomery CA, Gagnon AR, Calkin DE, Dietterich TG, McGregor S, Crowley M (2013) Allowing a wildfire to burn: estimating the effect on future fire suppression costs. Int J Wildland Fire 22(7):871–882
8. Finney MA, Cohen JD, McAllister SS, Jolly WM (2013) On the need for a theory of wildland fire spread. Int J Wildland Fire 22(1):25–36

9. Muhammad K, Ahmad J, Lv Z, Bellavista P, Yang P, Baik SW (2018) Efficient deep CNN-based fire detection and localization in video surveillance applications. IEEE Trans Syst Man Cybern Syst 49(7):1419–1434
10. Guan B, Zhao Y, Sun W (2018) Ant colony optimization with an automatic adjustment mechanism for detecting epistatic interactions. Comput Biol Chem 77:354–362
11. Khan S, Baig AR (2017) Ant colony optimization based hierarchical multi-label classification algorithm. Appl Soft Comput 55:462–479
12. Cracknell MJ, Reading AM (2014) Geological mapping using remote sensing data: a comparison of five machine learning algorithms, their response to variations in the spatial distribution of training data and the use of explicit spatial information. Comput Geosci 63:22–33
13. Castelli M, Vanneschi L, Popovič A (2015) Predicting burned areas of forest fires: an artificial intelligence approach. Fire Ecol 11(1):106–118
14. Ho CC (2009) Machine vision-based real-time early flame and smoke detection. Measure Sci Technol 20(4):045502
15. Çelik T, Özkaramanli H, Demirel H (2007) Fire and smoke detection without sensors: Image processing based approach. In: 2007 15th European signal processing conference. IEEE, pp 1794–1798
16. Kolesov I, Karasev P, Tannenbaum A, Haber E (2010) Fire and smoke detection in video with optimal mass transport based optical flow and neural networks. In: 2010 IEEE international conference on image processing. IEEE, pp 761–764
17. Yu C, Mei Z, Zhang X (2013) A real-time video fire flame and smoke detection algorithm. Proc Eng 62:891–898
18. Ayub U, Naveed H, Shahzad W (2020) PRRAT_AM—an advanced ant-miner to extract accurate and comprehensible classification rules. Appl Soft Comput 106326
19. Pal KK, Sudeep KS (2016) Preprocessing for image classification by convolutional neural networks. In: 2016 IEEE international conference on recent trends in electronics, information & communication technology (RTEICT). IEEE, pp 1778–1781
20. Kim B, Lee J (2019) A video-based fire detection using deep learning models. Appl Sci 9(14):2862
21. Zhou B, Khosla A, Lapedriza A, Oliva A, Torralba A (2016) Learning deep features for discriminative localization. In: Proceedings of the IEEE conference on computer vision and pattern recognition, pp 2921–2929
22. Cañizares PC, Núñez A, Merayo MG, Núñez M (2017) A hybrid ant colony-based system for assist the prevention and mitigation of wildfires in forests. In: 2017 2nd IEEE international conference on computational intelligence and applications (ICCIA). IEEE, pp 577–581
23. Stützle T, Hoos HH (2000) MAX-MIN ant system. Futur Gener Comput Syst 16(8):889–914
24. Katsikis V (ed) (2012) MATLAB: a fundamental tool for scientific computing and engineering applications, vol 3. BoD–Books on Demand
25. Bui DT, Bui QT, Nguyen QP, Pradhan B, Nampak H, Trinh PT (2017) A hybrid artificial intelligence approach using GIS-based neural-fuzzy inference system and particle swarm optimization for forest fire susceptibility modeling at a tropical area. Agric For Meteorol 233:32–44

Part III
Blockchain Technology, Social and Smart Networks Applications

Chapter 24
Sentiment Analysis for E-Learning Counting on Neuro-Fuzzy and Fuzzy Ontology Classification

Mohamed Sherine Khamis

1 Introduction

Nowadays, civilizations are living an inspiring transformation, maybe the utmost essential of the latest years, which, through the strong dispersal of the contemporary technologies, is deeply altering the nature of the interactions among countries, education, people, and cultures. This technological revolution is said that it clearly facilitated the procedure of globalization, information give-and-take, and learning [1]. In accordance with this scenario, a main role is played by the life-long learning, which lasts all along life and ambitions at improving people's realization both at personal and social level. Inside the learning society, to be unceasingly up-to-date is the crucial condition to live in and keep an eye on the modifications. A striking tool for the lifelong learning is equipped by the fusion of the innovative technologies tools in the formative procedure: the E-Learning. Near about twenty years, the 'E-Learning' occurrence has largely spread itself in the distance-learning scenery. Because of the habit of using Internet and its services, the user's assisting and tracking procedures might be easily integrated of the pedagogical and technological aspects for an energetic learning [2].

Inspirational and emotional aspects, among other aspects, look like to influence the schoolchild motivation and, in general, the outcome of the learning procedure [3]. As a consequence, in learning contexts, having the ability of identifying and managing enlightenment about the schoolchildren' emotions at a definite time might contribute to know their potential needs at that time. On one hand, adaptive E-Learning environments might make use of this information to satisfy those needs at runtime: They might deliver the user with recommendations about activities to confront or contents to communicate with, reformed to his/her emotional state at

M. Sherine Khamis (✉)
Department of Information Technology, Institute of Graduate Studies and Research, University of Alexandria, 163, Horyya Road Elshatby, 832, Alexandria 21526, Egypt
e-mail: sherinekhamis@alexu.edu.eg

A. E. Hassanien et al. (eds.), *Enabling Machine Learning Applications in Data Science*, Algorithms for Intelligent Systems, https://doi.org/10.1007/978-981-33-6129-4_24

that time [4]. Contrarily, enlightenment about the schoolchild emotions concerning a course might carry out as feedback for the teacher. This is specifically worthwhile for online courses, in which there is slight (or none) face-to-face interaction between schoolchildren and teaching staff, and as a consequence, there are fewer chances for teaching staff to get feedback from the schoolchildren. Generally, for a system to have the ability to take decisions counting on enlightenment about the users, it is necessary for it to get and accumulate enlightenment about them. One of the main traditional procedures to acquire enlightenment about users is composed of asking them to fill in questionnaires. In spite of that, the users might find this mission too time-consuming. Recently, non-intrusive techniques are preferred [5]. We also think that information for schoolchild models have to be acquired as inconspicuously as possible, yet without compromising the reliability of the model built.

The sentiment analysis (SA), which is commonly acknowledged as opinion mining or contextual mining, is employed in the natural language processing (NLP), computational linguistics, text analysis which helps to identify, systematically extract, and quantify the visceral information (subjective). The sentiment analysis actually works widely in the form of a customer's voice like reviews or responses on any material or item. Sentiment analysis includes data preprocessing, feature selection, and classification and then finds the polarity of data as shown in Fig. 1. Data preprocessing includes tokenization, stop word removal, stemming, lemmatization, etc. Tokenization is a mission of splitting a sequence of words into discrete words called tokens. Stop words are the words (is, am, are, in, to, etc.) which do not hold any opinion, so it is beneficial to remove them. Stemming is a mission of converting word's variant forms to its base form like helping to help [6].

When handling with users and sentiments, it is worthwhile to know the users' emotional state at a definite time (positive/neutral/negative), to deliver each of them with personalized assistance accordingly. Besides, it is also captivating to know whether this state corresponds to their "usual state" or, on the contrary, an obvious variance might have taken place. Behavior variances, as identified in the messages written by a user (when sentiment histories are available), might indicate modifications in the user's mood, and specific actions might be hypothetically desirable or recommended in such cases. With the main aim of extracting enlightenment about users' sentiments from the messages they write in Facebook and identifying modifications, we have developed a new and non-intrusive methodology for sentiment analysis in this social network [4].

It relies on a hybrid approach, conjoining lexical-based and machine learning algorithms. We have implemented this methodology in SentBuk, a Facebook application that retrieves the posts that are written by the operators and extracts enlightenment about their emotional state.

Sentiment analysis models are machine learning-based, lexicon-based, and hybrid methodology. In machine learning methodology, labeled dataset is used where the polarity of a sentence is already mentioned [7]. From that dataset, features are extracted and those features help to classify the polarity of the anonymous input sentence. Machine learning methods divided into supervised learning and unsupervised learning as mentioned in [8].

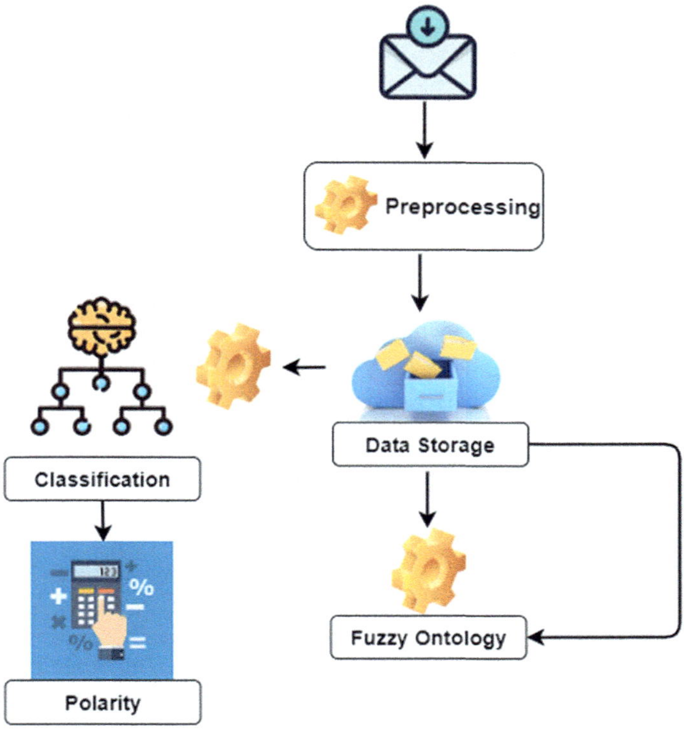

Fig. 1 Proposed sentiment analysis model

2 Literature Review

Sentiment analysis is looked at as the electronic scrutinize of opinions, sentiments, and emotions indicated in a written way [9]. In this regard, we take the simplified definition of sentiment as "a personal positive or negative feeling or opinion." An example of a sentence conveying a positive sentiment might be "I love it!," where "It is a terrible movie" conveys a negative one. A neutral sentiment does not express any feeling (e.g., "I am travelling to work"). A vast number of effort in this study area emphasis on classifying texts in accordance with their sentiment polarity, which might be positive, negative, or neutral [10]. That is why, it might be deliberated a text classification problem, since its objective composed of categorizing texts within classes by means of algorithmic methods. Existing techniques to inspirational computing and sentiment analysis fall into three main categories: knowledge-based techniques, statistical techniques, and hybrid techniques.

Knowledge-based techniques are common because of their accessibility and economy. Text is classified into effect categories on the basis of the presence of fairly unambiguous influence words, such as "happy," "sad," "afraid," and "bored." Popular causes of influence words or multiword expressions include the inspirational lexicon, linguistic annotation scheme, WordNet-Affect, Senti-WordNet, SenticNet, [11] and other probabilistic knowledge bases trained from linguistic corpora. Knowledge-based techniques have a major weakness: poor recognition of influence when linguistic rules are entailed. For example, although a knowledge base might acceptably classify the sentence "Friday was a contented day" as being contented, it is likely to fail on a sentence like "Friday wasn't a contented day at all." To this end, more sophisticated knowledge-based techniques exploit linguistics rules to distinguish how each specific knowledge base entry is utilized in text. The validity of knowledge-based techniques, besides, counts heavily on the depth and breadth of the used resources. Without a comprehensive knowledge base that involves human knowledge, in fact, it is not easy for a sentiment-mining system to seizure the semantics accompanying natural language or human behavior. Additional limitation of knowledge-based algorithms deceits in the typicality of their knowledge representation, which is habitually strictly defined and does not allow handling diverse concept nuances, because the inference of semantic and inspirational features accompanying concepts is bounded by the fixed, flat representation.

Statistical methods, such as assist vector machines and deep learning, have been famed with influence classification of texts, and researchers have utilized them on projects such as a movie post classifier and many others [12]. By feeding a machine learning algorithm, a large training corpus of inspirationally annotated texts, it is possible for the system to not only learn the inspirational valence of influence keywords (as in the keyword-spotting approach), but also to consider the significance of other random keywords (like lexical affinity) and frequencies of word co-occurrence. Yet, statistical methods are generally semantically weak—that is, lexical or co-occurrence elements in a statistical model have slight predictive value individually. As a result, statistical text classifiers work with acceptable accuracy only when given an appropriately large text input. Consequently, although these methods might be able to inspirationally classify a user's text on the page or paragraph level, they do not work well on smaller text units such as sentences or clauses.

Hybrid techniques to inspirational computing and sentiment analysis, finally, exploit both knowledge-based techniques and statistical methods to carry out duties such as emotion recognition and polarity identification from text or multimodal data. Sentic computing, [13] for example, investigates an ensemble of knowledge-driven linguistic patterns and statistical methods to infer polarity from text. Yunqing Xia and colleagues utilized SenticNet and a Bayesian model for contextual concept polarity disambiguation [14]. Mauro Dragoni and colleagues put forward a fuzzy framework that merges WordNet, ConceptNet, and SenticNet to extract key concepts from a sentence. Another system is iFeel. It permits manipulators generate their peculiar sentiment analysis framework by utilizing both SenticNet, SentiWordNet, and other sentiment analysis methodologies [15]. Jose Chenlo and David Losada utilized SenticNet to extract bag-of-concepts and polarity features for subjectivity identification and other sentiment analysis duties [15]. Jay Kuan-Chieh Chung and classmates utilized SenticNet concepts as stones and put forward a technique of unsystematic walk in ConceptNet to retrieve more concepts sideways with polarity scores. Other researches have put forward the joint utilization of knowledge bases and machine learning for Twitter sentiment analysis, short text message classification, and frame-based opinion mining [16].

Although schoolchildren' sentiment analysis has been studied for few decades, there is still a room to make it extra efficient and practical in real applications. According to the aforementioned review, it can be found that past studies were primarily devoted to: (1) words but not semantics, (2) collecting schoolchildren' opinions through questionnaires. Conversely, to the best of our knowledge, little devotion has been given to advising new technique that analyzes schoolchildren' sentiment through their posts or deal with the semantic of the sentences they write.

3 Proposed Methodology

This paper proposes a new model that analysis schoolchildren' sentiment through their posts, so it does not concentrate on syntax only. It focuses on automatic semantic evaluation. It also endeavors commonsense knowledge fuzzy ontologies, taxonomies, and as a consequence work on different domains. The main diagram of the suggested system is shown in Fig. 1. The following subsection goes through the detailed steps of the suggested model. The presented model is based on the extraction of enlightenment about users' positive/neutral/negative sentiments from the messages they write (the way in which we acquire the messages and handle the privacy matters will be described below).

3.1 Preprocessing of Posts

In the preprocessing phase, the system first goes through [17] a post and divides it into sentences. Then, it generates a duplicate of each sentence and does a number of preprocessing steps: tokenization, part-of-speech tagging, finding and labeling stop-words, punctuation marks, determiners, and prepositions, transformation to lower-case, and stemming. This step is important as the neuro-fuzzy network cannot deal with the post as a whole it needs a vector as an input.

3.2 Context Extraction from Facebook Posts

To begin with, POS tagging is carried out, and from this, it is seen that most of the context words are either nouns or phrases containing a noun. The opinion words are either adjectives or adverbs [18]. The Stanford Parser identifies the subject and its respective predicate and extracts the opinion words. The object is accompanying the subject or adjective verb, and the semantic feature of the context is analyzed. The post that has a single context and object is identified easily because the verbal phrase relates to the object. If more than one context is observed, then the polarity word has to be identified to map the proper context with the polarity of the opinion word. Mapping of the context with the opinion word and acquiring their relationship requires higher-level comparisons. To solve this problem, we use neuro-fuzzy classification, which is discussed in Sect. 3.2. The model uses SentiWordNet to identify the polarity of the post sentences [19].

3.3 Neuro-Fuzzy Classification

To prototypically model our fuzzy classifier for classifying positive and negative posts, we have defined five input variables to train the neural network. For finding out the value of input variables, we have calculated the statistical average polarity value of each of the four part-of-speeches and also for the emojis. A single post constitutes a single training unit. Input variable number 1 is defined by computing the average polarity of all nouns of specific post.

$$V_{\text{noun}} = \sum \frac{P(X_{\text{noun}})}{N_{\text{noun}}} \tag{1}$$

where $N_{\text{noun}} N_{\text{noun}}$ symbolizes the incidence of noun in a certain training unit, $X_{\text{noun}} X_{\text{noun}}$ is the polarity value for every token from the noun set, and $V_{\text{noun}} V_{\text{noun}}$ is the value of first input variable for a definite training unit. Applying the same idea, values of rest of the input variables for verb, adverb, and adjective are calculated. The average polarity of all adjectives of corresponding training unit calculation determines the second input variable.

$$V_{\text{adjective}} = \sum \frac{P\left(X_{\text{adjective}}\right)}{N_{\text{adjective}}} \tag{2}$$

The average polarity variable number 3 of corresponding training unit calculation determines the third input variable. After that, the average polarity variable number 4 of corresponding training unit calculation determines the fourth input variable.

$$V_{\text{verb}} = \sum \frac{P(X_{\text{verb}})}{N_{\text{verb}}} \tag{3}$$

$$V_{\text{adverb}} = \sum \frac{P(X_{\text{adverb}})}{N_{\text{adverb}}} \tag{4}$$

In any Facebook post, emoticons or emojis have indispensable role in deciding the polarity of this post. That is why, during data preprocessing stage, associated polarity of all the emojis of all training set data has been gotten, and the average polarity for of each training unit is considered as the fifth input variable of fuzzy neural network.

The neuro-fuzzy classification classifies the current post. A set of sixteen fuzzy rules is framed to extract the context from the post, and the posts are classified as they pass through the fuzzy rules. In this study, the context in the post sentence refers to the target word (noun/verb) and the relationship between the target word and the opinion word. Fuzzy rules are linguistic compositions casted as efficiently as possible to resolve tricky conditions that may not be recurrent. A fuzzy relation embodies the association amid one or more elements during rare happenings. The list of sixteen fuzzy rules framed here is publicized in Table 1. Table 1 encompasses keywords and notations, which are clarified in Table 2, so Table 1 must be read in conjunction with Table 2. Sequentially, for every post, the polarity value is calculated, thereby the polarity of the document is calculated. The document polarity refers to the polarity of the post, as every document contains the posts of the individual schoolchild [20, 21].

Table 1 Fuzzy rules for neuro-fuzzy classification

Rule 1: nphr	> < OAdj, TNOUN >
Rule 2: nphr	> < OB,TA2 > > + NOUN > < TA2,TA1 >
Rule 3: nphr	> < OB,HN > + clcomp > < OB,Wd > && dirobj > < Wd,TA2 > + NOUN > < A2,TA1 >
Rule 4: nphr	> < OB,HN > && dirobj > < OB,TA1 >
Rule 5: nphr	> < Wd,OB > && compadj > < Wd,TA1 >
Rule 6: nphr	> < Wd,OB > + compadj > < Wd,OB > && clmod > < TA2,Wd > && NOUN > < TA2,TA1 >
Rule 7: modadj	> < TA,Wd > + modadj > < Wd,OB >
Rule 8: modadj	> < TA,Wd > + Coand > < Wd,OB >
Rule 9: modadj	> < Wd,OB > + Coand > < Wd,TA >
Rule 10: nphr	> < OB,HN > + prwith > < > && NOUN > < TA2,TA1 >
Rule 11: nphr	> < TA, OB2 > && NOUN > < OB1,OB2 >
Rule 12: modadj	> < TA2,OB > + Coand > < TA2,TA4 > + NOUN > < TA4,TA3 > + Coand > < TA2,T5 > +
Rule 13: modadj	> < TNOUN,OAdj >
Rule 14: phralt	> < OAdj,TNOUN >
Rule 15: phralt	> < Wd, TNOUN > + nphr > < Wd,OAdj >
Rule 16: clcomp	> < Wd,OAdj > + nphr > < Wd,TNOUN >

Table 2 Description of the terms in Table 1

Term	Description
Nphr	It is a noun phrase
OAdj	It is the opinion of the adjective
TNOUN	It is the target noun/context
TA1, TA2	It is the target word. It is included to mine the posts with two target context
NOUN	Noun
HN	It is the head noun
Clcomp	It is an added complementary word to a noun which has an adjective
Wd	The first word encountered in the post
Dirobj	It is the direct object which is relating the verb
Compadj	It is a complementary word to the adjective
Clmod	A modifier for the noun phrase
Modadj	A word that modifies the meaning of a
Coand	It is a conjunction word 'and' which splits the posts
Prwith	It represents the 'with' preposition

3.4 Building Fuzzy Ontology

Fuzzy ontology is considered partial requirement to check any changes in the student's feelings. The first time the system is run, the ontology is built first, and then the post is analyzed through the neuro-fuzzy network. Every time after that, data from the data storage is added into the fuzzy ontology. If the fuzzy ontology is consistent, then student's feelings did not change. If the fuzzy ontology is not consistent, then student's feelings changed and needs to be classified through the neuro-fuzzy network again. The model starts building the base fuzzy ontology with a fuzzy ontology that encompasses the past posts of the schoolchild [22]. Fuzzy ontology is used as it solves the problem of the vagueness of the language. It is made up of a lattice of fuzzy ontologies that work as a group of basic logically-specified fundamentals (classes, relations, functions, instances). WordNet taxonomy is used to add synonyms (collected in synsets) and hypernyms to the fuzzy ontology.

4 SentBuk: An Application to Assist Sentiment Analysis in Facebook

For the purpose of implementing and test the methods described in the previous section, a Facebook application called SentBuk was created. The objectives of SentBuk are to access the text messages written by Facebook users, to collect all the data required, to use this information for sentiment analysis, and to deliver information regarding the analysis done [23].

5 Experimental Results

To analyze the potential benefits of the proposed system, we first evaluated their relevance and contribution to predictive accuracy; we proceed by comparing it to other state-of-the-art systems. Functions of the model are written in Python. Fuzzy ontology was built and checked with Protégé which is a free, open-source ontology editor and a knowledge management system. It was run on a HP Pavilion g6-1304tx–15.6″–2.5 GHz Intel Core i5-2450 M–16 GB RAM running windows 10. We select benchmark datasets issued by University of Charles, Faculty of Mathematics and Physics. It contains total 10.000 of posts. These posts are 2.587 positive, 5.174 neutral, 991 negative, and 248 bipolar.

In our first experimental, the suggested model was run on the dataset to evaluate its accuracy. Its accuracy was 94.6%. Table 3 shows the results that were acquired.

In the second experiment, the proposed model was compared by a model that utilizes two types of neural network. It used recurrent neural network and sequential conventional neural network. The proposed model proved that it was better in accuracy than the compared model. This is because these types of neural network deal with the words in a statistical way. It does not take into consideration the relationships and semantic of the words. Usage of fuzzy logic overtakes this drawback and leads to better results. Table 4 presents the results of the experiment.

The third experiment compares the proposed model with a model that uses ontology to analyze sentiments. Accuracy and specify were measured, and the proposed model gave better results. In this experiment, two datasets were used; the previously mentioned and Kaggle Facebook comments sentiment analysis dataset which contains 979 comment. 64% of the comments are positive, 28% are negative, and 8% neutral. Results are shown in Table 3. The proposed model gave better results because of using fuzzy ontology instead of the crisp ontology. Fuzzy ontology query

Table 3 Results from dataset

Positive	Neutral	Negative
2.447	4.894	937

Table 4 Comparison between proposed model and neural network models

Model	Accuracy (%)
RNN	82.4
SCNN	81.8
Proposed model	94.6

Table 5 Comparison of the proposed model with ontology-based system

Model	Accuracy (%)	Specify
Ontology-based system	93.38	0.614
Proposed model	94.6	0.844

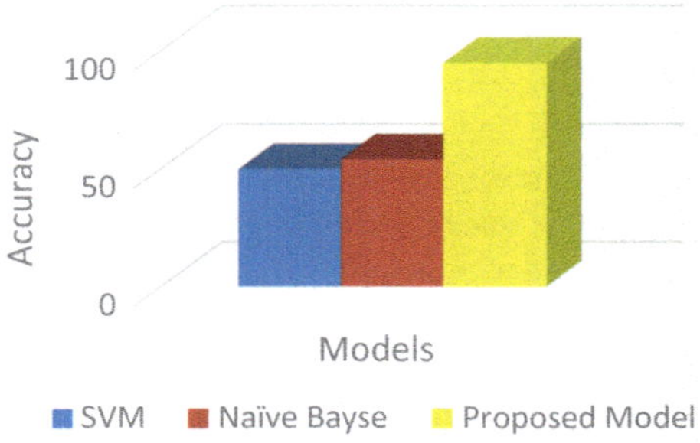

Fig. 2 Proposed model compared with machine learning-based models

depends on membership function rather than location in crisp ontology which makes the query more accurate in fuzzy ontology (Table 5).

Then, the proposed model was compared with the most popular machine learning techniques that are used in the field of sentiment analyses. These techniques are Naïve Bayes and support vector machine. Figure 2 shows the superiority of the proposed model. This superiority comes from the fact that these systems do not pay attention to the word order as these systems use statistical algorithms to extract the lemmas into a vector and then start dealing with these lemmas as numbers only. However in the suggested model, the usage of the fuzzy logic overtakes this draw back.

6 Sentiment Analysis and E-Learning

In the educational context, the capacity of knowing the schoolchild sentiment broadens the possibilities for E-Learning. That information is specifically worthwhile for adaptive E-Learning systems, which have the ability to guide each schoolchild throughout the learning procedure in conformity with his/her particular needs and preferences at each time. For the aim of doing so, they require to acquire and store enlightenment about each user in what is called the schoolchild model. We suggest to join in enlightenment about schoolchild sentiments in schoolchild models, thus this information might be utilized with adaptation purposes.

A dissimilar worthwhile exploitation of sentiment analysis in E-Learning embodies being able to categorize positive/negative emotions concerning an ongoing course and expending this information as feedback for the teaching staff or persons responsible for the course. The way of getting enlightenment about the schoolchild feelings concerning a specific course might be done in diverse ways. One of them composed of an approach as put forward in this study in combination with schoolchild selection and topic identification: (i) from all the messages accessible to SentBuk, only those written by the schoolchildren enrolled in the course are analyzed; (ii) from these messages, only the ones mentioning topics accompanying the course are considered (as well as the comments and likes associated to them). For the purpose of following this approach, a list of the schoolchildren' usernames in Facebook, as well as a dictionary of keywords associated to the course, need to be built. In such a way, sentiment analysis could pertain to the schoolchildren' sentiments concerning this specific course. This approach has the advantage that there is no reason to create specific Facebook groups or pages associated to the courses. In spite of that, it has the detriment that the schoolchildren have to give SentBuk authorization to access to their whole Facebook profile, which they might dislike due to privacy matters. Another drawback handles with the keyword dictionary creation and, more specifically, with its use to filter on the messages retrieved. Depending on the subject, the dictionary might contain either very technical words (e.g., Botanic or Chemistry), which is good for message filtering, or words utilized daily (e.g., Social Science or Humanistic). In the latter case, it might be much more complicated to determine whether the messages relate to the subjects taught.

7 Conclusion

The work described in this study demonstrates that it is feasible to extract enlightenment about the student's sentiments from the messages they post on Facebook with high accuracy. We have put forward a new methodology for sentiment analysis in Facebook. It assists, on one hand, to get enlightenment about the users' sentiment polarity (positive, neutral, or negative) in accordance with the messages they write, contrarily, to model the users' regular sentiment polarity and to identify significant emotional modifications. With the goal of making sure of the feasibility of our approach, we have created SentBuk, a Facebook application that retrieves the messages, comments and likes on the user's profiles, classifies the messages in accordance with their polarity, and builds/updates the user sentiment profile. The latter version of SentBuk focuses on "status messages," along with those comments and likes associated to these messages. Other types of messages are discarded since the large number of greetings that users write on others' walls gave rise to misleading results.

There are some advantages and difficulties through Facebook as a source of sentiment information. In Facebook, the discourse domain is unbounded. As a consequence, since the classifier created makes use of a lexicon-based approach, this approach might be worthwhile in other contexts in which sentiment analysis of English texts is required (e.g., marketing or politics). The work developed has proved to be good enough to deliver significant results and to be worthwhile in the context of E-Learning. In spite of that, some aspects could still be further evaluated or extended this can be handled in the future work. Other techniques such as fuzzy genetic algorithm can be used to classify student's posts.

References

1. Devi DN, Kumar CK, Prasad S (2016) A feature based approach for sentiment analysis by using support vector machine. In: Proceedings of the IEEE 6th international conference on advanced computing (IACC). India, pp 3–8
2. Pong-Inwong C, Rungworawut WS (2014) Teaching senti-lexicon for automated sentiment polarity definition in teaching evaluation. In: Proceedings of the 10th international conference on semantics, knowledge and grids. China, pp 84–91
3. Esparza GG, Díaz AP, Canul J, De-Luna CA, Ponce J (2016) Proposal of a Sentiment Analysis Model in Posts for improvement of the teaching-learning process in the classroom using a corpus of subjectivity. Int J Combinatorial Optimiz Problems Informat 7(2):22–34
4. Rajput Q, Haider S, Ghani S (2016) Lexicon-based sentiment analysis of teachers' evaluation. In: Applied computational intelligence and soft computing
5. Ravi K, Ravi V (2015) A survey on opinion mining and sentiment analysis: missions, approaches and applications. Knowl-Based Syst 89:14–46
6. Sharma D, Sabharwal M, Goyal V, Vij M (2020) Sentiment analysis techniques for social media data: a review. In: Proceedings of the 1st international conference on sustainable technologies for computational intelligence. India, pp 75–90
7. Pang B, Lee L (2008) Opinion mining and sentiment analysis. Found Trends Inf Retrieval 2(1–2):1–135
8. Kiprono KW, Abade EO (2016) Comparative Twitter sentiment analysis based on linear and probabilistic models. Int J Data Sci Technol 2(4):41–45
9. Wiebe J, Wilson T, Cardie C (2005) Annotating expressions of opinions and emotions in language. Language Res Eval 39(2–3):165–210
10. Taboada M, Brooke J, Tofiloski M, Voll K, Stede M (2011) Lexicon-based methods for sentiment analysis. Comput Linguistics 37(2):267–307
11. Tino H, Sebastian K, Andre B, Gunther H (2011) Sentiment detection with character n-grams. In: Proceedings of the 7th international conference on data mining (DMIN'11). USA
12. Cambria E, Olsher D, Rajagopal D (2014) SenticNet 3: a common and common-sense knowledge base for cognition-driven sentiment analysis. In: Proceedings of the 28th AAAI conference on artificial intelligence. Canada
13. Gonçalves P, Araújo M, Benevenuto F, Cha M (2013) Comparing and combining sentiment analysis methods. In: Proceedings of the 1st ACM conference on Online social networks. USA, pp 27–38
14. Cambria E (2016) Affective computing and sentiment analysis. IEEE Intell Syst 31(2):102–107
15. Recupero DR, Dragoni M, Presutti V (2015) ESWC 15 challenge on concept-level sentiment analysis. In: Semantic web evaluation challenges. Springer, Cham, pp 211–222
16. Thomas NT, Kumar A, Bijlani K (2015) Automatic answer assessment in LMS using latent semantic analysis. Procedia Comput Sci 58:257–264
17. Wang Y, Kim K, Lee B, Youn HY (2018) Word clustering based on POS feature for efficient twitter sentiment analysis. Hum-Centric Comput Informat Sci 8(1):17
18. Moshkin V, Yarushkina N, Andreev I (2019) The sentiment analysis of unstructured social network data using the extended ontology SentiWordNet. In: Proceedings of the 12th international conference on developments in esystems engineering (DeSE). Russia, pp 576–580
19. Katta P, Hegde NP (2019) A Hybrid Adaptive neuro-fuzzy interface and support vector machine based sentiment analysis on political twitter data. Int J Intell Eng Syst 12(1):165–173
20. Singh HR, Biswas SK, Purkayastha B (2019) A neuro-fuzzy classification system using dynamic clustering. In: Machine intelligence and signal analysis. Springer, Singapore, pp 157–170
21. Padmaja K, Hegde NP (2019) Twitter sentiment analysis using adaptive neuro-fuzzy inference system with genetic algorithm. In: Proceedings of the 3rd international conference on computing methodologies and communication (ICCMC). India, pp 498–503

22. Cobos R, Jurado F, Blázquez-Herranz A (2019) A content analysis system that supports sentiment analysis for subjectivity and polarity detection in online courses. IEEE Revista Iberoamericana De Tecnologías Del Aprendizaje 14(4):177–187
23. Kaur W, Balakrishnan V, Singh B (2020) Improving teaching and learning experience in engineering education using sentiment analysis techniques. In: Proceedings of the IOP conference series: materials science and engineering, vol. 834(1). Malaysia, p 012026

Chapter 25
Abnormal Behavior Forecasting in Smart Homes Using Hierarchical Hidden Markov Models

Bassem E. Abdel-Samee

1 Introduction

The amount of grown-up people and handicapped people who are in need for peripheral aid in their daily life is speedily increasing as uncovered by the latest statistics on worldwide population as testified in [1]. Those statistics show an intensification in the population group aged 65 or over. That is leading to deliver intensification to a series of complications in caring for older people and people with disabilities. Other than that, the European welfare model is regarded as not sufficient to satisfy the needs of the accumulative population, and cumulative the number of care delivers is not a realistic solution. It has been recognized that the usage of contemporary technologies as some illustrated intelligent surroundings can aid those people. Intelligent surroundings can enhance the lifestyle of grown-up people, protecting their privacy, and letting them live in their respective homes in place of care homes or hospitals for longer. By virtue of this, medical care costs for each human being are expected to be reduced [2].

Automated recognition of human activity is a crucial part and parcel of intelligent computer vision systems. Throughout the former years, numerous researches came to a result of simplifying automatic human action/activity recognition in videos. For human activity inspection, a precise recognition of the atomic activity in a video stream is the core constituent of the system, and also the most crucial, as it has an impact the accomplishment considerable. Although numerous recognition researches have been done in uncontrolled surroundings by putting into consideration real-world scenarios, recognizing human activity in a monitoring video remains challenging

B. E. Abdel-Samee (✉)
Department of Information Technology, Institute of Graduate Studies and Research, University of Alexandria, 163, Horyya Road Elshatby, 832, Alexandria 21526, Egypt
e-mail: igsr.bassem.ezzat@alexu.edu.eg

A. E. Hassanien et al. (eds.), *Enabling Machine Learning Applications in Data Science*, Algorithms for Intelligent Systems, https://doi.org/10.1007/978-981-33-6129-4_25

owing to the incredible intra-class variation in activity caused by divergent visual appearances, motion variations, temporal variability, etc. Inspection and apprehension of human activities, gestures, and facial expressions collectively are responsible for significant cues for behavior detection [2].

Automated recognition of human activity systems usually faces numerous challenges that are as follows: First is the challenge of in elevation volume of data. With the expansion of the scalability of the network, heterogeneous physical and virtual devices are relatable to the surroundings, which are continuously generating huge volume of data. It not only brings challenges in storing such data. Second challenge is that the surroundings are not controlled. There has been usually an emphasis that all the accessible assets are interconnected; despite that, it results in vastly changeable and uncontrolled network. As an illustration, the dynamically discovered new services, the device failure as a consequence of energy usage or other unforetold reasons, and so on. Third, human behavior detection has gained an increasing attention, and considerable efforts had been made. Despite that, it is quite in its prelim stages in providing proper datasets [3].

Abnormal human activity recognition (AbHAR) techniques may have the capability of being broadly written off as into two-dimensional AbHAR systems and three-dimensional AbHAR systems on the prime sort of input fed to the system. A two-dimensional AbHAR model is presented with 2D silhouettes for sole human being AbHAR in the bulk of mentioned techniques. Whereas, three-dimensional AbHAR systems are delivered with deepness silhouettes and skeleton structures of the human being for AbHAR as shown in [4].

As a result, in that section, we encompassed sole human being AbHAR techniques supporting elderly health care issues and AAL. Multiple human beings AbHAR is a generalized case of crowd activity analysis, which are utilized interchangeably in that review. Occlusions and doubts in jam-packed scenes with intricate activities and scene semantics are numerous of the contests got to have to be addressed for crowd activity analysis. It is methodically inspected in the section of transportation and mutual security. Then again, the level of complication is extraordinary to designate the entire crowd activity with the assist of a feature. There have been two techniques to investigate human activity (1) spot group activity patterns—abnormal activity localization (2) symbolize an individual's behavior, i.e., trajectory-based technique and shape dynamics-based technique [5].

Since abnormal human activity detection is considered a nourishing branch in the region of computer vision, a model has been advocated that uses hierarchical hidden Markov model to detect the abnormal behavior of elder people during the course of their everyday living activity. Hierarchical hidden Markov models are utilized as it is a statistical technique that operates well with a diminutive dataset or unsatisfactory training data. What is more, the statuses of the Hierarchical Hidden Markov Models surrender to chain of observation images instead of sole reconnaissance images as is the case for the standard Hidden Markov Models statuses [6, 7]. That paper is organized as takes after various of later related works are put forward Sect. 2. The comprehensive description of the proposed model has been made in Sect. 3. In Sect. 4, the results and discussions on the dataset are looked at. As a final point, the conclusion is annotated in Sect. 5.

2 Literature Review

Anomaly detection in ADL has received tremendous attention over the years with divergent computational methodologies applied for the detection of various forms of anomalies. Hoque et al. [8] proposed a system called "Holmes" for detecting ADL anomalies took advantage of density-based spatial clustering of applications by means of noise (DBSCAN). In the same way, in [9], the number of sensor occasions, duration and time of an activity in an automatic control of home appliances is taken out and clustered with DBSCAN. Instances with unusual duration or irregular occasions are classified as anomalous. Jakkula et al. [10] used a method of detecting temporal relation amidst activities which can be classified as anomalous. Self-organizing map (SOM) was used to detect artificially induced anomalies in occupant behavior relating to room occupancy [11, 12].

Divergent variants of recurrent neural network (RNN) as an illustration vanilla RNN, long short-term memory (LSTM), and gated recurrent unit (GRU) have been employed in [5] to learn a human usual behavioral pattern and find a deviation from the learned pattern. Lotfi et al. [13] used echo status network (ESN) for the detection of anomalies in ADL from the unprocessed binary sensor data. Hidden Markov model (HMM) is trained to pick up the activity series over a period of time and chain that do not adapt to what is being learned as an anomaly are also classified, then a fuzzy rule-based system (FRBS) concludes if the detected sequence is an actual anomaly [14].

In [15], a combination of convolutional neural network (CNN) and LSTM is utilized to detect simulated anomalies in ADL data. Their technique is to generate synthetic anomalies mimicking the behavior of early dementia sufferers as an illustration disturbed sleep, repeated activities in an unknown order, etc. The core dataset serves as training data for the normal class while the synthesized anomalous data serves as training data for the anomalous class. The data is then transferred into a CNN with the aim of learning the encoding while LSTM is utilized to learn the activity chain of the behavioral routine. While that method demonstrates an encouraging result, it has one drawback that is the researcher is not able to possibly generate synthetic data for each and every sole category of anomaly. Due to this, anomalous occurrences that are not generated may not be identified by the model.

According to the aforementioned review, it has been found that past studies were primarily devoted to: (1) utilization of diverse low-key technological devices that are readily accessible, which made them depend mostly on the use of monitored call centers not the actual behavior (2) classification rather than detection of behavior. Despite that, to the best of our knowledge, little devotion has been given to advising new technique to detect abnormal behavior.

3 Proposed Methodology

That paper proposes a new model to tackle two issues: (a) modeling and learning intricate behaviors from a human behavior in automatic control of home appliances and (b) identifying the behavior from new activities if it is normal or abnormal. The core diagram of the suggested model is demonstrated in Fig. 1. The following subsection describes in detail the steps of the model.

3.1 Hierarchical Hidden Markov Algorithm

A discrete hierarchical hidden Markov algorithm (HHMM) is formally defined by a 3-tuple $\langle \varsigma, \gamma, \theta \rangle$: a topological structure ζ, a surveillance alphabet y, and a family of parameters θ. The topology ζ specifies the number of levels D, the status space at each level, and the parent-children relationship amidst levels d and $d + 1$. The statuses at the lowest level (level D) are called production statuses. The statuses at a higher-level $d < D$ are termed abstract statuses. Only production statuses emit surveillance. Given a topological specification ζ and the surveillance space y, the family of parameters θ is defined as follows [16].

Suppose that $B_{\mathcal{Y}|\mathcal{P}}$ is the probability of noticing $\mathcal{Y} \in y$ given that the production status is $\mathcal{P}$. Respectively, every abstract status $\mathcal{P}^*$ at level d and the family of its children ch($\mathcal{P}^*$), we denote $\pi^{d,\mathcal{P}^*}$ as the preliminary distribution over ch($\mathcal{P}^*$), $A_{i,j}^{d,\mathcal{P}^*}$ as the transition probability from child i to child j ($i, j \in$ ch($\mathcal{P}^*$)), and $A_{i,\text{end}}^{d,\mathcal{P}^*}$ as the probability that $\mathcal{P}^*$ terminates given its present child is i. The family of parameters θ is $\left\{B_{\mathcal{Y}|\mathcal{P}}, \pi^{d,\mathcal{P}^*}, A_{i,j}^{d,\mathcal{P}^*}, A_{i,\text{end}}^{d,\mathcal{P}^*} | \forall (\mathcal{Y}, \mathcal{P}, d, \mathcal{P}^*, i, j)\right\}$.

An abstract status $\mathcal{P}^*$ at level d may be able to be executed as follows. First, $\mathcal{P}^*$ selects a status i at the lesser level $d + 1$ from the preliminary distribution $\pi^{d,\mathcal{P}^*}$. At that point, i is executed while waiting for it terminates. At that time, $\mathcal{P}^*$ may be able to terminate with probability $A_{i,\text{end}}^{d,\mathcal{P}^*}$. If $\mathcal{P}^*$ does not terminate, it continues to select a status j for execution from the distribution $A_{i,j}^{d,\mathcal{P}^*}$. The loop continues until $\mathcal{P}^*$ terminates. The implementation of $\mathcal{P}^*$ is similar to the implementation of an abstract policy π^* in the AHMM [16], apart from that π^* selects a lesser level policy π base only on the status at the bottom level, not on the policy π' selected in the former footstep. Nevertheless, the theoretical status may be a special case of the theoretical arrangement within the AHMEM [17] (an extension of the AHMM).

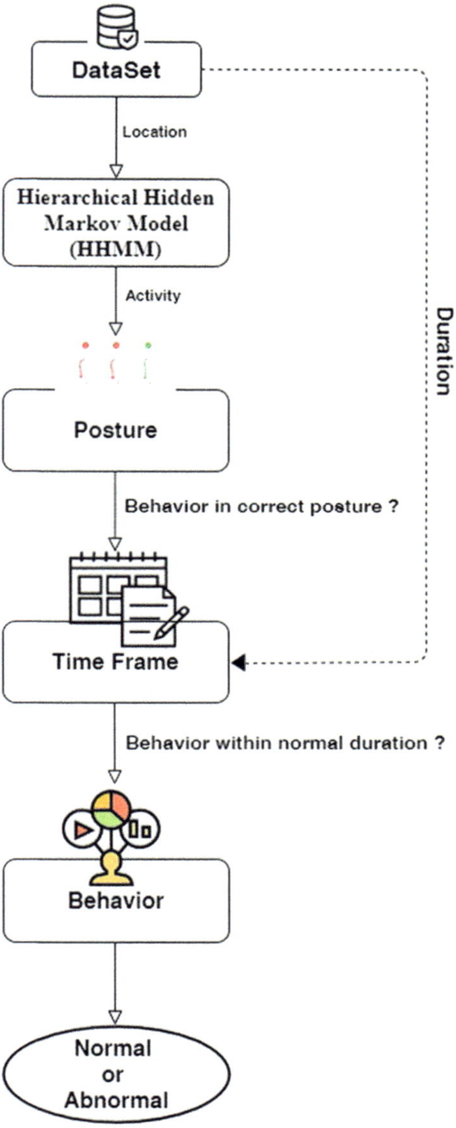

Fig. 1 The proposed abnormal behavior detection model

A representation of the HHMM as a DBN is delivered in [18], which defines a joint probability distribution (JPD) over the family of all variables $\{\mathcal{X}_t^d, e_t^d, \mathcal{Y}_t | \forall(t, d)\}$ where $\mathcal{X}_t^d$ is the status at level d and time t, e_t^d represents if $\mathcal{X}_t^d$ terminates or not, and $\mathcal{Y}_t$ is the surveillance at time t.

3.2 Learning Parameters in the HHMM

We need to learn the family of parameters θ of the HHMM from a surveillance sequence $\mathcal{O}$. A proposed technique counting on the EM algorithm and the asymmetric inside outside (AIO) algorithm to estimate θ. For that technique, the family of hidden variables is:

$$\mathcal{H} = \{\mathcal{X}_t^d, e_t^d | t = 1, \ldots, d = 1, \ldots D\},$$

where t is the length of the surveillance sequence. The family of observed variables is $\mathcal{O} = \{\mathcal{Y}_1, \ldots, \mathcal{Y}_T\}$. Assume that τ is the sufficient statistic for θ. The EM algorithm re-estimates θ by first calculating the expected sufficient statistic (ESS) $\overline{\tau} = \mathrm{E}_{\mathcal{H}|\mathcal{O}_\tau}$. Then, the result is normalized to obtain the new value for θ. The ESS for the parameter $A_{i,j}^{d,\mathcal{P}^*}$, as an illustration, is $\frac{\Sigma_{t=1}^{t=T-1} \xi_t^{d,\mathcal{P}^*}(i,j)}{\Pr(\mathcal{O})}$, where $\xi_t^{d,\mathcal{P}^*}(i,j) = \Pr\left(\mathcal{X}_t^d = \mathcal{P}^*, \mathcal{X}_t^{d+1} = i, \mathcal{X}_{t+1}^{d+1} = j, e_t^d = F, e_t^{d+1} = T, \mathcal{O}\right)$. The ESS for $A_{i,j}^{d,\mathcal{P}^*}$ can be computed by the AIO algorithm. The ESS for the other parameters of θ is computed in a similar manner. The intricate for the AIO algorithm is considered to be cubic in the length of the surveillance sequence, nonetheless linear in the number of statuses of the model [18].

3.3 Exact and Approximate Inference for the HHMM

The AIO algorithm may be able to be utilized directly to derive an exact filtering algorithm as follows. At time t, $\xi_{t-1}^{d,\mathcal{P}^*}(i.j)$ is computed by the AIO algorithm. Summing $\xi_{t-1}^{d,\mathcal{P}^*}(i,j)$ over $\mathcal{P}^*, i, e_{t-1}^d$ and e_{t-1}^{d+1}, we obtain the probability $\Pr\left(\mathcal{X}_t^{d+1} | \mathcal{Y}_1, \ldots, \mathcal{Y}_T\right)$. Note that, at the next time, $\xi_t^{d,\mathcal{P}^*}(i,j)$ may be able to be derived from $\xi_{t-1}^{d,\mathcal{P}^*}(i,j)$ by stretching that probability one more time slice. Thus, the intricate of the exact filtering algorithm is $O(T^2)$. That algorithm may be able to be well suited for short-term recognition, but it may not be realistic for a real-time recognition mission when the surveillance length T grows. Alternatively, the RBPF has been successfully deployed in the AHMM [19] and may be able to be readily adapted for the HHMM. Denote: $\mathcal{X}_t^{1:D} \triangleq \{\mathcal{X}_t^1, \ldots, \mathcal{X}_t^D\}$, $e_t^{1:D} \triangleq \{e_t^1, \ldots, e_t^D\}$, $e_{1:t}^{1:D} \triangleq \{e_1^{1:D}, \ldots, e_t^{1:D}\}$, $\mathcal{Y}_{1:t} \triangleq \{\mathcal{Y}_1, \ldots, \mathcal{Y}_t\}$ [19].

Algorithm 1

Begin

For $t = 1, 2, \ldots$

/ sampling step */*

For each sample $i = 1, 2, \ldots, N$

Absorb yt into $C_t^{(i)}$

Canonicalize $C_t^{(i)}$ update weight $C_t^{(i)} = B_t\ (yt)$

Sampling $e_t^{1:D\ (i)}$ $from\ C_t^{(i)}$ and Pr $(e_t^{1:D}|x_t^{1:D})$

/ re-sampling step */*

Normalize the weight $\widetilde{w}^{(i)} = \frac{w^{(i)}}{\sum_{i=1}^{N} w^{(i)}}$

Re-sample the sample set according to $\widetilde{w}^{(i)}$

/ exact step */*

For each sample $i = 1, 2, \ldots\ldots, N$

Compute $C_{t+1}^{(i)}$ from $C_t^{(i)}$ and $e_t^{1:D\ (i)}$

/ Estimation step */*

Compute Pr $(x_{t+1}^d | y1{:}t) \approx \frac{1}{N} \sum_{i=1}^{N} C_{t+1}^{(i)}\ x_{t+1}^d$

The conception of the RBPF is that the filtering distribution at time *t*—that is, $Pr\left(\mathcal{X}_t^{1:D}, e_t^{1:D} | \mathcal{Y}_{1:t}\right)$—is approximated via a two-step procedure: (1) sampling the Rao-Blackwellised (*RB*) variable $e_t^{1:D}$ from the present *RB* belief status $B_t = \Pr\left(\mathcal{X}_t^{1:D}, e_t^{1:D}, \mathcal{Y}_t \middle| e_{1:t-1}^{1:D}\right)$ and (2) updating the *RB* belief status B_t exploiting strict inference. At, respectively, time step t, the model keeps up a family of *N* samples, each involving the distribution of one slice network $C_t \triangleq \Pr\left(\mathcal{X}_t^{1:D} \middle| e_{1:t-1}^{1:D}\right)$. The RB belief status B_t may be able to be acquired directly from C_t by adding in the network representing the conditional distribution of $\mathcal{Y}_t$ and $e_t^{1:D}$.

A new sample $e_t^{1:D}$ may be able to be acquired from the canonical form of C_t (after absorbing $\mathcal{Y}_t$). At the next time slice, C_{t+1} is built by projecting C_t over one slice counting on the sampled value of $e_t^{1:D}$. The intricate of the RBPF algorithm for the HHMM is $O\ (N\ D)$, where *N* is the number of samples and *D* is the depth of the model. The complete algorithm is demonstrated in Algorithm 1 [19].

4 Implementation and Experimental Results

4.1 Implementation

To analyze the potential benefits of the proposed system, we first evaluated their relevance and contribution to predictive accuracy; we proceed by comparing it to other status-of-the- art systems. The functions of the system are written in the suggested algorithm has been implemented in MATLAB (R2017b) simulator.

4.2 Dataset

CASIA activity recognition dataset is employed. It is a benchmark dataset which contains a collection of video chain of human activities captured by divergent cameras placed in outdoor locations at divergent viewing angles. It entails two human beings' interactions done by two human subjects. Five most common interactions among them are 1. Fight: two subjects fighting with each other, 2. Overtake: one subject overtaking another one, 3. Rob: one subject robing another one, 4. Follow: one subject following another one till the end 5. Meet and part: two subjects meeting each other and then departing.

4.3 Evaluation Criteria

To evaluate the accomplishment of the hierarchal Markov models, we must calculate the foretelling error. The error function indicates how the foretelling of our network is close to the target values and, as a result, what adjustment should be applied to the weight and bias in the learning algorithm in each iteration. As a result, we evaluate the root mean square error (RMSE) given by:

$$\text{RMSE} = \sqrt{\frac{1}{N}\sum_{i=1}^{N}(Y_i - X_i)^2}$$

where N is the number of learning surveillances, Y_i represents the foretold data, and X_i represents the real data of the ith surveillance.

4.4 Results

To ensure a secure life at home, we must know all the routine activities of the occupant so as to be able to remotely recognize (without disturbing) unusual activities. In our work, we focus on to monitoring activities of everyday living of the grown-up people to inspect any presence of the unusual behavior. Activities of everyday living ADLs are a basic self-care mission. They comprise grooming, unary function, washing, dressing, and eating.

First, the test datasets from the participants were used to compare the capability of AML–ADL to identify deviations from normal routines with a benchmark method put forward by Jos et al. (2015). That benchmark method uses Gaussian mixture models (GMM) to detect deviations from the usage patterns of electrical devices. For each appliance, the probability of the appliance usage during intervals of the day is modelled. For each day of the monitoring stage, the benchmark method calculates the probability union amidst all appliances to score the occupant's routine (activity level). The scores of activity level obtained from that technique range from 0 to 1 where scores near 1 indicate a normal performance of ADLs and scores near 0 indicate the opposite. Figure 2 demonstrates the daily scores of activity level obtained from proposed method and the benchmark method for the testing datasets from Gary and Debby.

In the second experiment, the behavior breakfast was foretold in place of the correct lunch and snacking behaviors. That is because the behaviors breakfast, lunch, and snacking are all located in the kitchen. As a result, the foretelling accuracy was 96.2% exploiting the hierarchical hidden Markov models. After the hierarchical hidden Markov models results were acquired, the posture of the per son was checked. There were no warnings when the posture was checked on any of the days given for testing because there was no abnormal behavior in the testing dataset. Finally, time rule was applied to the algorithm to check the duration. That was compared by the results from exploiting probabilistic analysis to detect abnormal behavior in everyday life. The foretelling accuracy was found by computing the percent—age of the real or actual sequence status that agreed with the foretold sequence. Hence, the formula:

Fig. 2 Proposed model versus benchmark method

$$\text{Accuracy} = \frac{\sum \text{actual state}}{\sum \text{predicted state}}$$

The results of the warning show that the algorithm may be able to detect if the user has spent either too much or too little time performing a behavior. Most of the warnings are for the behavior breakfast, because the duration of that behavior varied a great deal. One reason for that is that the behavior breakfast was foretold in place of the real behavior snack or lunch, which usually takes less time (snack) or more time (lunch). Other warnings resulted in the behavior grooming and leaving. Hence, the algorithm detected that the human being spent less time or more time than usual engaging in those behaviors and issued warnings. Figure 3 demonstrates the routine of numerous activities and each elapsed activity by week during a month in the goal to know approximately the routine of each activity (Table 1).

During the third experiment, the proposed algorithm was compared with other models that utilized semi-hidden Markov algorithm and neural networks. The comparison criteria were accuracy rate, previously detection and correct detection of the proposed algorithm excelled in accuracy, previously detection and correct detection which is demonstrated in Table 2. That excellence comes from the capability

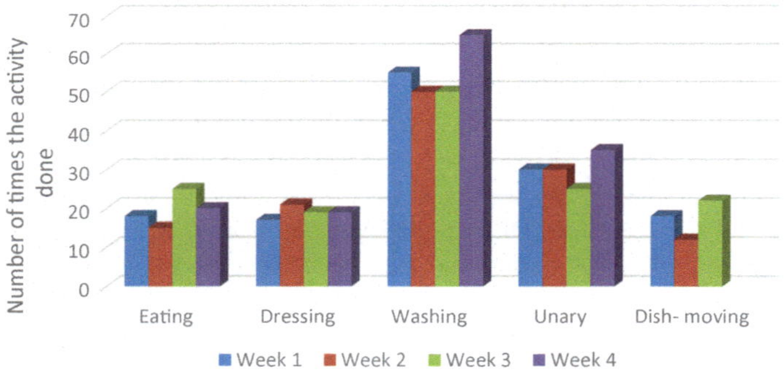

Fig. 3 Routine of numerous elapsed activities by week during a month

Table 1 Warnings in behaviors when duration is checked

Day	Foretold behavior	Proposed model time	Probabilistic analysis time
1	Grooming	**00:00:07**	00:00:09
2	Sleeping	**00:00:07**	00:00:11
3	Leaving	**00:12:52**	00:13:00
4	Breakfast	**00:23:08**	00:22:49
5	Grooming	**00:00:05**	00:00:08
6	Leaving	**00:10:32**	00:11:59
7	Grooming	**00:00:03**	00:00:06

Table 2 Comparison amidst the proposed model and semi-hidden Markov model and neural networks models

	Proposed model	Semi- Hidden Markov model	Neural Networks model
Accuracy rate	**96.2**	80.15	72.36
Previously detection	**80.9**	78.48%	79.17
Correct duration	**74.35**	68.89	70.74

of hierarchical hidden Markov models to work well with a diminutive dataset or unsatisfactory training data.

5 Conclusion

The efficient monitoring of activities of everyday living regarding monitored human beings in automatic control of home appliances and the detection of the unusual scenarios represent on the crucial issues that have been investigated in the recent years. The identification of any unusual behavior or situation is very critic. In that paper, we proposed a foretelling unusual behavior algorithm exploiting hierarchical hidden Markov model. The proposed algorithm is counting on the routine of everyday life, the routine duration of activities, and the number of routines of elapsed activity. From the proposed algorithm, we may be able to effectively recognize the unusual activities and then evaluate the behavior status of our elderly. The purpose was to increase the certainty of the foretelling. The delivered results counting on the root mean square error confirm that our proposed algorithm gives better results if compared to other models proposed previously in the literature. It also showed consistency in the foretelling. Nevertheless, a limitation was that the dataset utilized in the deliver study is a combination of activity and behavior.

Our future work aims at multiple occupancy and prediction of abnormal behavior. For the work put forward in that paper, only a limited number of discrete sensors were used. Despite that, more research is required when a combination of discrete sensors (occupancy, door entry point,…) and continuous sensors (temperature, humidity,…) are used.

References

1. Riboni D, Bettini C, Civitarese G, Janjua ZH, Helaoui R (2015) Fine-grained recognition of abnormal behaviors for early detection of mild cognitive impairment. In: 2015 IEEE international conference on pervasive computing and communications (PerCom). IEEE, pp 149–154

2. Lundström J, Järpe E, Verikas A (2016) Detecting and exploring deviating behaviour of automatic control of home appliances residents. Expert Syst Appl 55:429–440
3. Nigam S, Singh R, Misra AK (2019) A review of computational approaches for human behavior detection. Arch Comput Techniq Eng 26(4):831–863
4. Dhiman C, Vishwakarma DK (2019) A review of state-of-the-art techniques for abnormal human activity recognition. Eng Appl Artif Intell 77:21–45
5. Arifoglu D, Bouchachia A (2017) Activity recognition and abnormal behaviour detection with recurrent neural networks. Procedia Comput Sci 110:86–93
6. Garcia-Constantino M, Konios A, Nugent C (2018) Modelling activities of daily living with petri nets. In: 2018 IEEE international conference on pervasive computing and communications workshops (PerCom Workshops). IEEE, pp 866–871
7. Garcia-Constantino M, Konios A, Ekerete I, Christopoulos SR, Shewell C, Nugent C, Morrison G (2019) Probabilistic analysis of abnormal behavior detection in activities of daily living. In: 2019 IEEE international conference on pervasive computing and communications workshops (PerCom Workshops). IEEE, pp 461–466
8. Hoque E, Dickerson RF, Preum SM, Hanson M, Barth A, Stankovic JA (2015) Holmes: a comprehensive anomaly detection system for daily in-home activities. In: 2015 international conference on distributed computing in sensor systems. IEEE, pp 40–51
9. Fahad LG, Rajarajan M (2015) Anomalies detection in smart-home activities. In: 2015 IEEE 14th international conference on machine learning and applications (ICMLA). IEEE, pp 419–422
10. Jakkula V, Cook DJ, Crandall AS (2007) Temporal pattern discovery for anomaly detection in a automatic control of home appliances.
11. Novák M, Biňas M, Jakab F (2012) Unobtrusive anomaly detection in presence of elderly in a smart-home surroundings. In: 2012 ELEKTRO. IEEE, pp 341–344
12. Novák M, Jakab F, Lain L (2013) Anomaly detection in user daily patterns in smart-home surroundings. J Sel Areas Health Inform 3(6):1–11
13. Lotfi A, Langensiepen C, Mahmoud SM, Akhlaghinia MJ (2012) Automatic control of home appliances for the elderly dementia sufferers: identification and prediction of abnormal behaviour. J. Ambient Intell Humanized Comput 3(3):205–218
14. Forkan ARM, Khalil I, Tari Z, Foufou S, Bouras A (2015) A context-aware approach for long-term behavioural change detection and abnormality prediction in ambient assisted living. Pattern Recogn 48(3):628–641
15. Arifoglu D, Bouchachia A (2019) Detection of abnormal behaviour for dementia sufferers using convolutional neural networks. Artif Intell Med 94:88–95
16. Jakkula V, Cook D (2011) Detecting anomalous sensor events in automatic control of home appliances data for enhancing the living experience. In: Workshops at the twenty-fifth AAAI conference on artificial intelligence
17. Yahaya SW, Langensiepen C, Lotfi A (2018) Anomaly detection in activities of daily living using one-class support vector machine. In: UK workshop on computational intelligence). Springer, Cham, pp 362–371
18. Dreiseitl S, Osl M, Scheibböck C, Binder M (2010) Outlier detection with one-class SVMs: an application to melanoma prognosis. In: AMIA annual symposium proceedings, vol. 2010. American Medical Informatics Association, p 172
19. Theissler A (2017) Multi-class novelty detection in diagnostic trouble codes from repair shops. In: 2017 IEEE 15th international conference on industrial Informatics (INDIN). IEEE, pp 1043–1049

Chapter 26
The Classification Model Sentiment Analysis of the Sudanese Dialect Used Into the Internet Service in Sudan

Islam Saif Eldin Mukhtar Heamida and A. L. Samani Abd Elmutalib Ahmed

1 Introduction

Opinion or sentiment analysis is a procedure used to determine whether the text set is positive, negative [1], or neutral as described by Paul Ekman (1992). In text analysis [2], natural language processing (NLP) and machine learning (ML) techniques are combined to assign feelings to subjects, groups, or entities within a phrase [3, 4]. Arabic is a very complicated language, as it is not sensitive to letters, does not have capital letters and has some variables in the form of spelling or typing. Usually, one word contains more than one suffix. Commonly, the word has different meanings, such as the word Ramadan, which can be a person's name or a month [5]. Arabic is divided into three types: Standard Arabic, modern Standard Arabic, and dialect Arabic. The Arabic language used in social media is usually a mixture of modern Standard Arabic and one or more Arabic colloquial dialects. Current opinion analysis studies focus on English and other languages but a few focus on Arabic [6]. Nafissa Yusupova, Diana Bogdanov, Maxim Boyko applied the analysis of emotions to the text in Russian based on the methods of machine learning, as well as Ghada al-Waked, Taha Osman, Thomas Hughes Roberts studied the analysis of emotions when dealing with the Saudi dialect, which does not conform to the official grammatical structure of modern classical Arabic. People use dialects according to their different Arab nationalities, such as Al-Shami, Iraq, and Egypt. We can also note the use of social media dialects that became more popular after the Arab spring revolutions (2013, Salem and Mortada). As a result, social media is rich in opinions [7]. Global data has recently revealed that the Arabic language ranks fourth in the world in terms of Internet usage and the number of Internet users in the Arabic language by the end of 2014, reaching more than 135 million users (Habash 2014) [8].

I. Saif Eldin Mukhtar Heamida (✉) · A. L. Samani Abd Elmutalib Ahmed
Faculty of Computer Science and Information Technology, Al Neelain University, Khartoum, Sudan

A. E. Hassanien et al. (eds.), *Enabling Machine Learning Applications in Data Science*, Algorithms for Intelligent Systems, https://doi.org/10.1007/978-981-33-6129-4_26

The Sudanese are definitely part of the Arab world. Official statistics indicate that the number of social media users in Sudan has reached 15 million people. However, independent bodies estimated at about 17 million and confirmed that Facebook is the most prevalent, followed by WhatsApp and then Twitter, Sudanese youth used these sites, in particular Twitter and Facebook, to show, clarify and address their issues by conveying a real image without restrictions or barriers. This open platform, which has allowed young people to reflect their opinions and make changes to our societies, has increased the needs of stakeholders and researchers to apply their opinions to Arabic texts in order to study those opinions, particularly in the field of services provided by companies to the public. The aim of this study was to analyze the opinions of the Internet service in Sudan, which was conducted on 1050 Facebook comments in Arabic using the modern, colloquial, and colloquial Sudanese language on the Internet. In the absence of a dictionary for the analysis of opinions on the Sudanese colloquial dialect, we have created a lexicon containing 1000 words categorized into positive words and negative.

2 Related Word

A lexicon-based sentiment analysis tool for general Arabic text used in daily chat and conversation and in social media has been suggested. SVM, Naïve Bayes, and maximum entropy works are used in this study. However, they have shown that SVM has achieved higher accuracy. Their tool must rely in part on human-based texts to overcome the problems that arise from the use of generic, non-standard Arabic texts. Al-Subaihin et al [9].

Ghadah Alwakid, Taha Osman, Thomas Hughes—Roberts [10], challenges in sentiment analysis for Arab social networks, apply emotional analysis when dealing with the Saudi dialect, which does not conform to the official grammar structure of modern classical Arabic (MCA). Targeting the use of unemployment in Saudi Arabia, this paper examines the key challenges faced by researchers in the analysis of emotions in the informal Arabic language and documents how these challenges can initially be addressed through pre-language processing of raw text and supervised machine learning (ML) to determine the polarity of emotions. The initial empiric evaluation produced satisfactory results for the classification of emotions, while demonstrating the usefulness of using ML to identify dominant features in the determination of the polarity of emotions, which, in turn, could be used to build a knowledge base for the problem that could further improve the accuracy of the classification of emotions.

Sentiment Analysis on Social Media [11]. Federico Neri, Carlo Alibrandi, studied sentiment analysis on more than 1000 Facebook posts about newsletters, comparing the sentiment for Rai—the Italian public broadcasting service—toward the emerging and more dynamic private company La7. This study maps the results of the study with observations made by the Osservatorio di Pavia, an Italian research institute specializing in theoretical and empirical media analysis, engaged in the analysis of political communication in the mass media. This study also takes into consideration the data provided by Auditel on newscast audiences, the correlation between the analysis of social media, Facebook in particular, and the quantifiable data available to the public domain.

Shubham Gupta, Diksha Saxen, Joy Joseph, Prof. Richa Mehra [12]. They provided an overview of the different techniques used in text mining and sentiment analysis for all tasks. It also categorized sentiment analysis at sentence level and sentiment analysis at document level. Sentiment analysis on Twitter using machine learning. The study used a knowledge base approach, and a machine learning approach to analyze the feelings of the text. Twitter posts about electronic products such as mobile phones, laptops, etc., were analyzed using a machine learning approach. By conducting an analysis of emotions in a particular area, it is possible to determine the effect of domain information in the classification of emotions. Enter a new direction to categorize Tweets as positive and negative and draw people's opinions about products applied was sentiment analysis of Hollywood movies on Twitter [13]. The authors Hodeghatta, Umesh, analyzed the tweets of six Hollywood movies and understood the sentiments, emotions, and opinions expressed by the people from nine different locations, across four different countries. The model is rained using Naive Bayes and MaxEnt machine learning methods, using Python and natural language tool libraries. They have taken into account the time stamp, user name, geolocation and the actual tweet message. These were tested for both unigram and bigram. The MaxEnt unigram classifier provided the best accuracy of 84%.

Apply emotion analysis to text in Russian based on machine learning approaches [14], description of the researchers Nafissa Yussupova, Diana Bogdanova, Maxim Boyko, the problem of classifying emotions in text messages in Russian using machine learning methods—Naive Bayes translator and support vector. One of the advantages of the Russian language is the use of a variety of decimal places depending on deviation, times, and grammatical gender. Another common problem in classifying emotions for different languages is that different words can have the same meaning (synonyms) and thus give equal emotional value. Thus, it was their task, how the reduction affects the accuracy of the classification of emotions (or other, with or without endings), and how the results are compared to the Russian and English languages. To evaluate the effect of synonyms, use a method when words of the same meaning are grouped into a single term. They used lemmatization libraries and synonyms to solve these problems. The results showed that the use of lemmatization of texts in Russian improves the accuracy of the classification of emotions. On the contrary, the classification of emotions for texts in English without the use of lemmatization gives a better result. The results also showed that the use of synonyms in the form had a positive effect on accuracy.

3 Tools and Techniques

- **NLP with Python**
  In this study, the researcher uses a natural language processing kit with Python. Natural language processing tools with Python NLTK is one of the leading platforms for working with human language data and Python, and the NLTK module is used to process natural languages. NLTK is literally the abbreviation stand of the natural language toolkit [15].
- **Classification Techniques**
  The researcher used two types of classification methods: SVM and Naïve Bayes (NB).
- **Support Vector Machines Classification Approach**
  Support vector machine (SVM) is one of the common classification technology methods that analyzes data and identifies patterns (SVM 2015) (Sayad 2010). Vector technology has recently gained importance in the field of machine learning as it was devised to solve learning problems Patterns and classification by determining the hyper-plane level of the data to be separated. Actually, the main goal of the vector technology is to find the best separator level for the data to be separated and classified into two halves. Consequently, the best separation of data is carried out by the level separator with the greatest distance between the nearest statement and the training data, and the data that is closest to the separation plane is called the support vector. Furthermore, the distance between the margin and the nearest margin is called the margin. Also, the vector can divide the given data into two classes by either a linear classifier or a nonlinear classifier. In fact, the nonlinear compiler comes from some classification issues that do not have a simple separator level to be used as a separation criterion. Finally, the nonlinear classifier is created using the concept of the kernel [16, 17].
- **Naïve Bayes Classification Approach**
  In the field of machine learning, NB is a collection of probability works based on the application of Bayes' theory with strong (assuming) assumptions of independence between features (Bayesian network models). Since the 1960s, Naive Bayes has been widely studied. It was introduced in the text retrieval community in the early 1960s and continues to be a popular method of classifying texts, distinguishing documents as belonging to one category or another (such as spam, mathematics, politics, etc.) with word frequencies as advantages of appropriate preprocessing. In this field, it is competitive with more advanced methods, including vector support machines [18].

4 The Proposed Classification Model

The researcher developed a model of learning classification using polarity features based on the polarity dictionary created by the researcher as shown in the Fig. 1 below in a number of steps.

Data Label

This is the first step, and the data was classified as positive, negative, and neutral by experts specialized in the Sudanese dialect, 538 positive, and 512 negative.

These data were available on social media platforms, but it may be a challenge for some communication sites to provide short-coded software interfaces (APIs) that allow those who want to obtain data from their platform to obtain them easily for example twitter site.

Data Preprocessing

The second stage involves cleaning up the data. Remove duplicate words or characters, tokenization as well as stop-words, and normalize, useless data such as URLs, tags, English characters and numbers, e.g., (@ and %), have been deleted when cleaning data. As far as the removal of duplicate letters and words is concerned, repeated letters or words have been deleted for the purpose of emphasis, for example, "كويييسة". Or repeat the word several times like "الشبكة سيئة سيئة". This is common practice in Arabic comments and other social media, where one message is repeated several times.

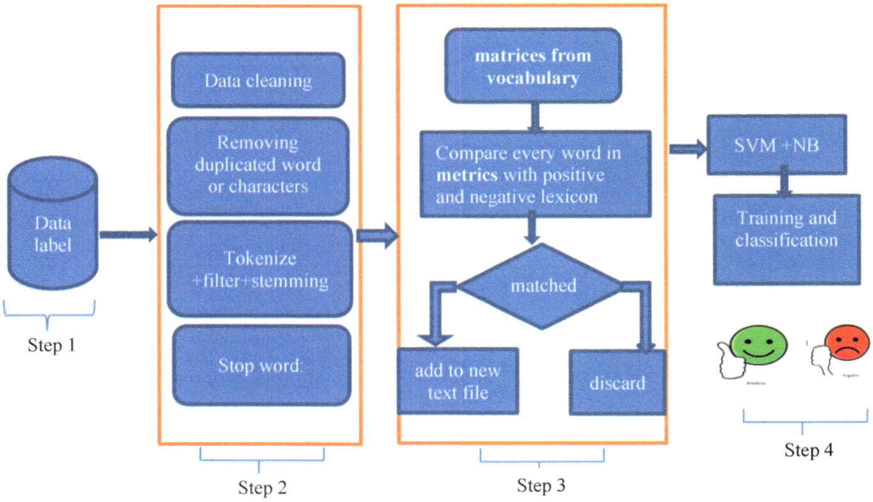

Fig. 1 Learning classification model using polarity features

In the normalization process, the standardization of Arabic letters, we have eliminating diacritics and nonlinear characters as well, replacing letter Hamza (ء), replacing (أ)and (إ)with (ا), also the letter (ي)with (ى)and replacing (ة)with (ه).

The text used in social media, in particular the comments, presents many challenges compared to the officially organized text. Analyzing text in Arabic is a very difficult task due to spelling inconsistencies, the use of associated words, and the lack of capital letters such as those in English that would be used to define features, an issue that requires normalization. In tokenization, words are separated from the comments by symbols. These symbols can be words, even a letter. Each alphanumeric string between two white spaces is extracted.

In Remove, the stop words, remove prepositions and some phrases and words that are not needed, removing any word that does not have any meaning (preposition) من ، في ، علي ، الي and some words that are entered into the Sudanese vernacular such as بس, مره, اخري.

Stemming

In the third step, delete any extension marks (prefixes added to the beginning of a word, additions added to the middle of a word, or / and suffixes added to the end of a word) of words to reduce. These words into root wards are under the assumption that words with the same stem are semantic related. For example, converting "الزيادة" and "زادت" and "زايدنها" to "زاد" as stem word.

The stem is a method for the unification of distinct forms of words. This can help to reduce the size of the vocabulary and therefore the results (Table 1).

Matrix from Vocabulary

In this fourth step, a matrix can be mathematically created that can be represented as shown in Fig. 1. It contains a set of rows "data information" representing file/document stem comments and a set of columns representing "header information" all vocabulary/terms in the comments file after processing operations. The last

Table 1 Example of preprocessing a comment

Preprocessing steps	Comments after preprocessing
The original comment	@ sudani!!سودانى في الشبكه زفتة زفتة وزبالة وتكلفتها غاااااالية
Data cleaning	سودانى في الشبكة زفتة زفتة وزبالة وتكلفتها غاااااالية
Normalization	سوداني في الشبكة زفته زفته وزباله وتكلفتها غاااااليه
Removing duplicated word or characters	سوداني في الشبكة زفتة وزبالة وتكلفتها غاليه
Tokenization	سوداني ،في ، الشبكة ، زفتة ، وزبالة ،وتكلفتها، غاليه
Stop remove word	سوداني الشبكة زفتة وزبالة وتكلفتها غاليه
Stemming	سوداني الشبكة زفت زبالة تكلفة غالى

columns are either class or label. They are positive, negative, or neutral, as pre-classified manually, determining the value of each intersection entry in the matrix by writing the categorized word after matching the words with positive and negative words in the polar lexicon.

$$\mathrm{X} = \begin{bmatrix} x_{00}x_{01} & \cdots & x_{0n} \\ \vdots & \ddots & \vdots \\ x_{m0}x_{m1} & \cdots & x_{mn} \end{bmatrix}$$

where

$$n = \text{size of vocabular}$$

$$m = \text{number of comments}$$

and

$$\{x : [x_{mn} = 1],\ \text{for positive word}\}$$
$$\{x : [x_{mn} = 1],\ for\ Negative\ word\}.$$

Sentiment Classification
In the last step, the data is divided into a data set for training and testing. (Dividing the data into two sections: 70% for training and 30% for testing). The training data set is used to construct a classification model based on the SVM and NB classifiers. Based on their polarity, the data was classified into positive and negative categories. The test data set is used to predict the polarity of comments.

5 Experimentation

Data Collection
In this work, the researcher focuses on collecting data from Facebook, which is one of the most popular social media. For this purpose, the data was collected manually because Facebook policy does not make the data available to researchers, and it focused on the collection of comments written in the Sudanese colloquial. Therefore, these data were collected manually in this study. The collected data is the subscribers' opinions of the Internet service who use Sudani Telecom Company from the Sudani Telecom Company page and a number of Sudanese active groups on the site.

As the data collected contained unrelated data from addresses and people's names, as a result, the researcher cleaned it, as it only kept the text comments in the preprocessing phase. It also includes numbers, signs, symbols, and spelling errors that have been addressed and programmatically cleaned during the data processing phase. The data collected has different sizes and categories used to train and test for the model.

Tools

Work environment CPU/Intel Core Pentium i5 processor, 8 GB RAM, 64-bit operating system, Windows 10 system was appropriate. In addition, we used the Anaconda platform to apply our data by implementing the classification model learning steps using the polarity glossary of our approach, and MS Excel 2016 to set up a data file, convert it to CSV files, and pass the result to Jupyter as a compiler tool.

6 Result and Discussion

Four different measures—precision, recall, accuracy, and F-measurement for both SVM and NB classifiers—were used to assess the validity of the test comment ratings as positive or negative. The confusion matrix is shown in Table 3[18].

	PREDICTED VALUES	
	NEGATIVES	POSITIVES
NEGATIVES	TRUE NEGATIVES (TN)	FALSE NEGATIVES (FP)
POSITIVES	FALSE POSITIVES (FN)	TRUE POSITIVES (TP)

In the above table, there are four parameters: true positive (TP), false positive(FP), false negative (FN), and true negative (TN).

TP: The number of comments correctly categorized as positive.
TN: The number of comments correctly classified as negative.
FP: The number of comments that have been classified as positive but negative.
FN: The number of comments categorized as negative but positive. Therefore, the formula for measures is as follows:

Precision = TP/(TP+FP)
Recall = TP/(TP+FN)
Accuracy = (TP+TN)/(TP+FP+FN+TN)
F-measure = 2* (Precision*Recall)/ (Precision*Recall)

The Tables show the results of each classifier with data set (Tables 2, 3 4 and 5):

From Table 3, we notice that support vector machine achieved good results for recall which equals to 83%.

In Table 5, Naïve Bayes achieved good result for precision equals to 83.5% and recall equals to 74%.

Table 2 True positive and true negative for the support vector machine

Total = 252	Positive	Negative
Positive	TP **152**	FN **12**
Negative	FP **23**	TN **65**

Table 3 Class precision, recall, accuracy, and F-measure for the support vector machine

Classifier	Precision	Recall	Accuracy	F-Score
SVM	85.6%	83.2%	86.5%	84.5%

Table 4 True positive and true negative for the Naïve Bayes

Total = 252	Positive	Negative
Predicted positive	TP 158	FN 6
Predicted negative	FP 42	TN 46

Table 5 Class precision, recall, accuracy, and F-measure for the Naïve Bayes

Classifier	Precision	Recall	Accuracy	F-Score
Naïve Bayes	83.5%	74%	80%	76.5%

7 Conclusion

This paper examined the sentiment analysis of the Internet service in Sudan for a Sudanese telecommunications company for Arabic comments written in the Sudanese Colloquial. Approximately 1048 comments concerning with the Internet service were analyzed. A new lexicon of Sudanese Colloquial was built, consisting of 1000 words. The data is divided into training and testing groups. The SVM and Naïve Bayes were applied to reveal the polarity of the comments given to the training group. Test group results showed that SVM achieved the best accuracy, which is equal to 86.5%, while Naïve Bayes achieved the best accuracy is equal to 80%.

This study is of great value in determining the positive and negative opinions of Internet service companies so that they can evaluate their services and what their customers are satisfied with. So that, it decides which measures can be taken to maintain and improve the performance of its products and services. In addition, this lexicon, which was designed for the Sudanese Colloquial, can be used to analyze sentiment order in order to reveal negative and positive comments on the services sector and on a number of other areas, such as political, economic, marketing, and other fields. In light of the above results, it is recommended that broadening the plurality of lexicons will be of great value in many areas as well as creating and designing libraries and tools that help deal with Sudanese vernacular, in general, and sarcastic expressions and comments in particular.

References

1. Pang B, Lee L (2009) Opinion mining and sentiment analysis. Comput Linguist 35(2):311–312
2. Al-Hasan AF A sentiment lexicon for the palestinian dialect. The Islamic University, Gaza, Building
3. https://monkeylearn.com/sentiment-analysis/
4. Balahur A, Mohammad S, Hoste V, Klinger R (2018) Proceedings of the 9th workshop on computational approaches to subjectivity, sentiment and social media analysis. In: Proceedings of the 9th workshop on computational approaches to subjectivity, sentiment and social media analysis.
5. Heamida ISAM, Ahmed ESAE Applying sentiment analysis on Arabic comments in Sudanese dialect
6. Hussein DMEDM (2018) A survey on sentiment analysis challenges. J King Saud Univ-Eng Sci 30(4):330–338
7. Liu B (2012) Sentiment analysis and opinion mining. Synth Lect Hum Language Technol 5(1):1–167
8. https://www.sarayanews.com/article/292487
9. Al-Subaihin AA, Al-Khalifa HS, Al-Salman AS (2011) A proposed sentiment analysis tool for modern Arabic using human-based computing. In: Proceedings of the 13th international conference on information integration and web-based applications and services, pp 543–546
10. Alwakid G, Osman T, Hughes-Roberts T (2017) Challenges in sentiment analysis for Arabic social networks. Procedia Comput Sci 117:89–100
11. Neri F, Aliprandi C, Capeci F, Cuadros M (2012) Sentiment Analysis on social media. ASONAM 12:919–926
12. Saxena D, Gupta S, Joseph J, Mehra R (2019) Sentiment analysis. Int J Eng Sci Mathe 8(3):46–51
13. Hodeghatta UR (2013) Sentiment analysis of Hollywood movies on Twitter. In: 2013 IEEE/ACM international conference on advances in social networks analysis and mining (ASONAM 2013). IEEE, pp 1401–1404
14. Yussupova N, Bogdanova D, Boyko M (2012) Applying of sentiment analysis for texts in Russian based on machine learning approach. In: proceedings of second international conference on advances in information mining and management, p. 14
15. Perkins J (2010) Python text processing with NLTK 2.0 cookbook. Packt Publishing Ltd.
16. James G, Witten D, Hastie T, Tibshirani R (2013) An introduction to statistical learning, vol. 112, p. 18. Springer, New York
17. Kubat M (2017) An introduction to machine learning. Springer International Publishing AG.
18. https://medium.com/sifium/machine-learning-types-of-classification-9497bd4f2e14

Chapter 27
Implementing Blockchain in the Airline Sector

Samah Abuayied, Fatimah Alajlan, and Alshymaa Alghamdi

1 Introduction

Recently, rapid developments in technologies have given rise to a growing trend in intelligent transportation systems (ITS) [1], which facilities many transportation modes [2]. However, intelligent transportation is a critical mile-stone, and One of its significant issues is the rigorous security risks within untrusted environments [3]. Moreover, integrating blockchain in this area has proven a high impact, since it's enabled high-security measures in data management [1]. On the other hand, blockchain, as a distributed ledger technology, has started in 2008 and guaranteed to improve the meaning of the digital transaction. It has gained more significant interest within a short period and based on IBM [4] blockchain defined as an immutably shared ledger for putting up transactions historically. The concept of blockchain has the potential to create an efficient business network that is cost-effective for trading. However, blockchain is not a single technology even though it is described as innovative and still needs more researching efforts to achieve the maturity level. Additionally, new scientific research [5] in the blockchain had a focus on the main challenges on it, such as usability, privacy, security, and system resource consumption. However, the airline sector, as one of the transportation modes, has faced a significantly increased rate in different types of issues. However, the strength of blockchain technology appears to combine with the airline sector to allow secure data sharing and processing among all parts. The main features of blockchain that make it useful in enhancement airline transportation methods are its ability to store any information, thus events, object transactions, and electronic transactions. Moreover, blockchain could offer legal support for flights and passenger information management and airline traffics. Generally, blockchain serves to resolve issues related to transparency, security, trust, and control [6]. The main aim of this paper is to review

S. Abuayied (✉) · F. Alajlan · A. Alghamdi
Computer Science and Information System College, Umm AL-Qura-University, Mecca, Kingdom of Saudi Arabia

A. E. Hassanien et al. (eds.), *Enabling Machine Learning Applications in Data Science*, Algorithms for Intelligent Systems, https://doi.org/10.1007/978-981-33-6129-4_27

the current investigations in integration blockchain with the airline sector as a way of enhancing the aviation industry. The paper has structured as following: Sect. 2 consists of a brief description of blockchain. Section 3 introduces the main airline sector issues. Section 4 presents blockchain novel studies in airline sector, which divided into 4 subsections. The last section concludes the whole paper.

2 A Background on Blockchain

A blockchain is a distributed ledger. In other words, blockchain is a distributed database with specific properties [7]. However, blockchain is an element in the idea of Bitcoin, which is a combination of peer to peer networks, software, and protocols. Additionally, the Bitcoin network consists of multiple elements:

- Digital keys: refer to public and private keys wheres, the private key used for signing the Bitcoin transaction, and the public key used for verifying the transaction to prove the ownership. Elliptic curve cryptography (ECC) used to generate the keys in the Bitcoin network. Additionally, the public key generated form it's corresponding private key.
- Addresses: a Bitcoin address consists of 26 to 35 characters created by hashing public key with its private key. However, the Bitcoin address can be presented as a QR code for readability and Facilitates the address distribution process.
- transaction: it's the primary process in the Bitcoin network. It merely consists of sending a Bitcoin a specific Bitcoin address.
- blockchain: include all the Bitcoin transaction. The block in blockchain identify by hash and linked to its previous block in the chain.
- Miners: it's the first node in the blockchain, which calculates the Proof of Work (POW). However, POW is a combination of complex computation when creating a new block, and all the blockchain nodes compete to solve the POW to became the miner node.
- Wallets: It's an electronic wallet that allows blockchain users to transfer and manage the bitcoin.

In the blockchain, information stored in a distributed manner where different databases in different computers in distributed places in the world which, store the same duplicated ledger. These databases structured as blocks and each block referred to as a node in the blockchain. Additionally, each block identifies by hash and linked to its previous block Fig. 1 shows the concept of blockchain.

Besides, blockchain considered a new field, which needs more experiments to solve the gaps. Additionally, these gaps had addressed by some research. [7, 8] as following:

- The idea of POW made blockchain costly and resource-consuming because it needs complex computation.
- The reduction of regulation could cause risks to the users.

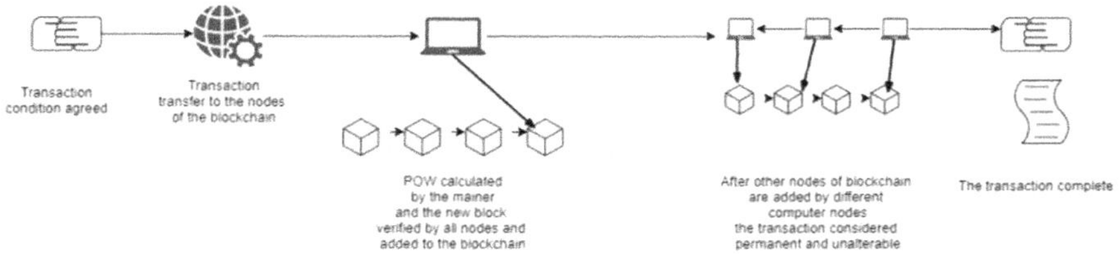

Fig. 1 The blockchain concept

- Blockchain is a complex system that could be difficult to understand by users.

According to [9, 10], several hurdles associated with the use of blockchain technology in transportation.

3 Airline Sector Issues

The growth in airline transportation demands leads to an increase in the need for integration of robustness technologies with airline systems. Currently, the radar systems which are ground-based and Mode-C transponders are relied mainly on upon and used by the Air Traffic Control (ATC) for tracking air-craft. However, it suffers several risks, including operational, physical, electronic warfare, and cyber-attack risks and those associated with radars that were secondary-surveillance divested [11]. On the other hand, the common issues in airlines have classified into two categories air side area and land-side area issues, and they affect each other. However, airline issues considered as complex issues, because it depends on a massive amount of variety of services and resources, that consist of multiple stochastic relations. Additionally, Theses issues affect airport managers, employees, and passengers' satisfaction with the offered services. Research in the same area [12] has addressed some issues related to passenger Check-in and security processes at Naples Capodichino Airport. The study introduced a new model using the surrogate Method to optimize management costs and the passengers' unification. Moreover, one of the most used technology in the transportation system is an intelligent transportation system (ITS) which aim to integrate communication technologies into transportation infrastructure to improve efficiency and safety. However, ITS has addressed multiple security privacy issues. Research in the same area [13] had presented ITS problems and challenges in more detail. Additionally, EUROCONTROL, a European Organization interested in Safety Air Navigation, recently launched an air traffic management (ATM) 2020 [14], which is a research program for enhancing airline system information management, where flight and passenger's information can suffer from different kinds of security attacks, such as eavesdropping and jamming attacks. Additionally, all these issues seem to be solved by blockchain as a secure ledger, so integrating blockchain achieves security and system efficiency goals in the airline transportation system. On the other hand, a blockchain radar-based systems framework that enables the ADS-B systems proposed by Reisman [15], which proved security and privacy goals for in managing air traffic. Moreover, [10, 16] assumed that the application of the blockchain in transport and logistics expected to have an impact on the marketing and consumption of goods. Additionally, the blockchain application in logistics expected to provide an advantage in terms of mobility, transparency, and decentralization. Moreover, the security threat son airline traffic management system had divided into two different categories as it assumed in the paper [14]:

- Data level security threats: consist of multiple unreported routes modification issues and unauthorized users' actions.
- Service-level security threats consists of disruption availability of the system services resulted by malicious actions or unintentional actions

4 Blockchain Novel Studies in Airline Sector

This section presents a comparative review of current studies in introducing blockchain for enhancing and facilitating airline transportation systems. However, depending on the fields of the reviewed studies, we have divided this section into four main subsections: flight information management, aircraft, and airport infrastructure maintenance, passengers information management, and Air traffic control.

4.1 Flight Information Management

The study [14] has proposed a blockchain-based flight planning model, which shown in Fig. 2. However, the model consists of multiple components, whereas the primary aeronautical environment assumed as a 3D coordinate in respect of time.

$$x(t), y(t), z(t) \tag{1}$$

The first component in the flights planning model [14] the nodes defining by a unique ID, and classified into three different categories as following:

- User node: denote as U and consist of the set of system users.
- Approver node: denote as A and consist of the set of airline navigation service providers.
- External node: denote as E and consist of the set of ATM network sources.

Each node in the model can identify by triple include unique ID and 3D coordinate variables.

$$(\mathrm{ID}, xa, ya, za) \tag{2}$$

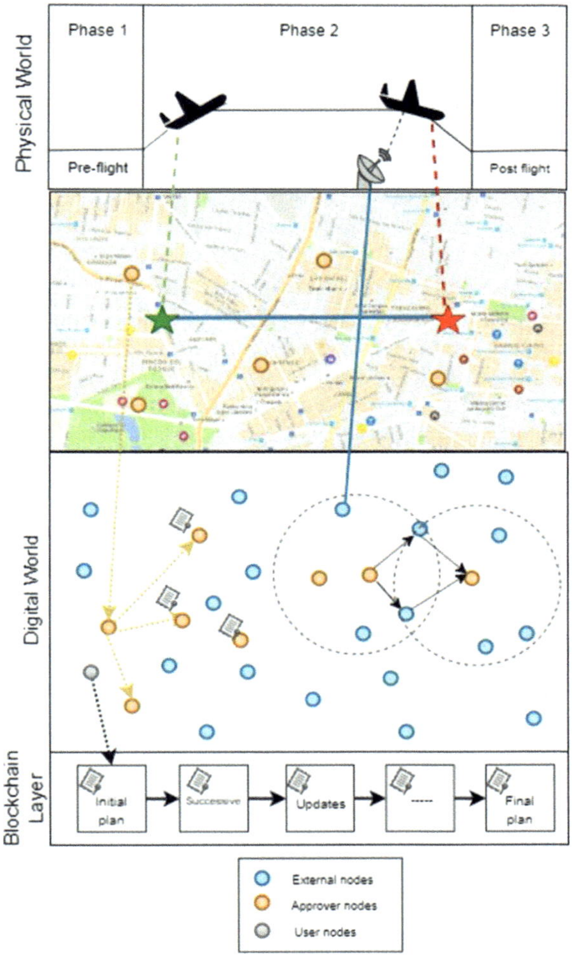

Fig. 2 Blockchain based flight planning system

Additionally, the model also includes two different layers, which are physical and logical layers. The model workflow [14] includes multiple steps starting by initializing the flight information followed by external source information collection, analyzing users' information, acceptance flight plan, and record the changes.

4.2 *Aircraft and Airport Infrastructure Maintenance*

Recently, blockchain had played a proper role in aircraft and maintenance; a clear example of this is the logbook framework proposed by Aleshi et al. [17]. However, the aircraft logbook had addressed some defects, such as theft or loss of the physical logbook. However, keeping aircraft logbooks accurate and safe became a significant concern to many aircraft owners because aircraft will become un-airworthy without these records. To illustrate this, research in the same area [17] has proposed a blockchain-based aircraft logbook to achieve data integrity and safety. The secure aircraft maintenance records (SAMR) has built using. Linux software called Hyperledger and Python. Additionally, to gain the idea behind blockchain distributive ledger, the study had used the algorithm Proof of Elapsed Time to display the global system state accurately. Furthermore, integrating blockchain with supply chain management (SCM) and operation management (OM) can enhance the airport's infrastructure. To illustrate this, a study [18] on the Italian airport infrastructures had reviewed and analyzed the election of blockchain integration with SCM. Moreover, the Italian air-port had adopted Airport Collaborative Decision Making (ACDM) technology, which is one of the leading blockchain applications in the airport. However, ACDM is an advanced platform able to optimize different airport operations and improve flight predictability and gate management. As a result, the study [18] assumed that blockchain integration with SCM needs more investigation to understand the effects of blockchain on SCM in the airport industry. A typical platform as Hyper-ledger fabric, which is a blockchain framework explicitly designed to deal with applications related to commercial enterprise [19].

4.3 Passengers Information Management

Blockchain is emerging as a reliable technology that can use in the management of passenger's information during the flights. The technology offers opportunities that allow secure biometric authentication of passenger's identities during their journey, thus, helping to eliminate the possibility of requiring multiple travel documents for the verification of travelers [20]. Also, it enables passengers not to disclose their data in the process. However, the SITA Lab used blockchain to introduce the idea of virtual passports that can use on a single wearable device or mobile equipment to minimize the cost, complexity, and liability of document checks, while the passenger is on transit [21, 22]. Moreover, this innovation will allow passengers to have a valid token on their mobile phones containing biometric and some other personalized data [21]. Con-sequently, in any place of the world, the passenger only needs to scan their devices to obtain their identity and determine whether or not they are authorized to travel. Additionally, the main advantage behind this technology is allowing to store passengers' biometric details in a distributive ledger [22]. On the other hand, by implementing this technology, Passengers deserve better in terms of data privacy, encryption, securing, tamper-proof, which are the most critical issues that must take into consideration when processing passenger information [22]. Besides, once the passenger's information has been entered into the system, changing it is not easy or even not possible. Therefore, the use of blockchain in passenger management eliminates any need for any third-party ownership. However, users identified with their unique ID, which then encrypted in the system [21]. Moreover, The ID considered as an encryption key only shared between two communication lines or encrypted entities [22].At any point, if it's necessary, the key can be changed to ease communication and offer more protection user identity. Additionally, the client and the server are the only parties that can share the key and the ID. Client interactions with other nodes within the same network happen through the server [22].

4.4 Air Traffic Control

Due to the increasing rate of airline traffic, there is a high demand for en hancing the infrastructure of airline systems. However, many research had integrated blockchain in this area to solve some issues in controlling the air- line traffic. to illustrate this, the study in [23] had proposed a Hyper-ledger Fabric system that depicts a unique number of features in design innovations, which aim to distinguish it from other technology systems, making it more convenient for air traffic control [23]. However, Hyper-ledger features include membership permission, private channels, and decentralized implementations among subsystems. Additionally, the permission process [24] uses the proof of work protocol and defines some management features, such as identifiable participants, transaction confirmation, latency, authentication, and high transactions. However, the system required a registration step before allowing user access. On the other hand, the aircraft operators are, therefore, provided for a standardized PKI infrastructure and acquire PKI supplied synchronous key. On the other hand, the research [25] has proposed a System Wide In- formation Management (SWIM) registry-based blockchain for the Brazilian airline traffic management system. However, the SWIM registry consists of multiple layers, each layer responsible for a specific service provided by the system, in which the central architecture includes three entities: the service providers, consumers, and regulatory. Moreover, the proposed model in [25] has integrated blockchain with SWIM in the Brazilian airline traffic management system in which each block presents a specific service. To illustrate this, Fig. 3 shows two kinds of blockchain integration into the SWIM. However, in part A form Fig. 3, each stakeholder allow to read or insert information, where in part B from the same Figure shows the way of distributing the services over the airline system using the idea of blockchain. According to the perspective natural of blockchain, any data or services had stored in the SWIM-based blockchain model will not be allowed to delete. Additionally, Table 1 summarize the blockchain novel studies in airline sector section

5 Conclusion

The main aim of this paper was to present a detailed literature review of the current investigations in introducing blockchain, which is an emerging technology in the field of aviation, and the airport sector. This technology will facilitate its use in

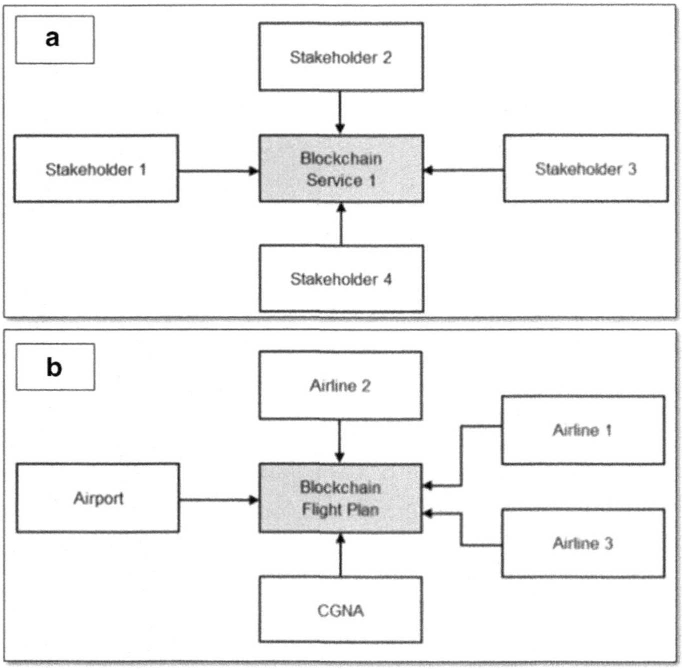

Fig. 3 A blockchain as a stakeholder's services, **b** blockchain for flight plan

Table 1 Summarizing blockchain investigations in airline sector

Paper	Category	Proposed model
[14]	Flight management	Blockchain flight planning model
[26]	Air traffic control	Blockchain air traffic flow management
[17]	Aircraft maintenance	Blockchain aircraft logbook
[23]	Air traffic control	Blockchain Hyper-ledger system
[25]	Air traffic control	SWIM registry with blockchain
[21]	Passenger information management	Managing passenger's information by blockchain

the field of aviation in all its forms. Many people believed that the technology of the block chain could not replace the traditional system and would be considered a weakness for airports, but it is otherwise has increased confidence and enhance its use at airports. However, in this research we focused on the scientific research that connected and integrated blockchain in the airline as a public transportation mode. Moreover, blockchain theologies are in its early stage of development, especially in when introducing it in the aviation sector, and there are a limited number of blockchain systems applied in this area, which encourage many scientific researcher to spend more effort and time in investigate this research area.

Acknowledgements The research for this paper was supported by Umm Al-Qura University. We gratefully acknowledge the support and generosity of the University, without this study could not have been completed.

References

1. Hîrtan L-A, Dobre C (2018) Blockchain privacy-preservation in intelligent transportation systems. In: 2018 IEEE International conference on computational science and engineering (CSE)
2. Merrefield C et al. (2018) What blockchains could mean for government and transportation operations, John A. Volpe national transportation systems Center (US)
3. Namiot D, Pokusaev O, Kupriyanovsky V, Akimov A (2017) Blockchain applications for transport industry. Int J Open Info Technol 5
4. Shirani A (2018) Blockchain for global maritime logistics issues in information systems 19
5. Kamath R (2018) Food traceability on blockchain: Walmart's pork and mango pilots with IBM. J British Blockchain Assoc 3712
6. McPhee C, Ljutic A (2017) Blockchain technology innovation management Review 7
7. Astarita V, Giofré VP, Mirabelli G, Solina V (2020) A review of blockchain-based systems in transportation. IEEE Internet of Inform 11
8. Bai CA, Cordeiro J, Sarkis J et al (2020) Blockchain technology: business, strategy, the environment, and sustainability. Bus Strat Environ 29
9. Carter C, Koh L (2018) Blockchain disruption in transport: are you decentralised yet?
10. Dobrovnik M, Herold DM, Fu¨rst E, Kummer S (2018) Blockchain for and in logistics: what to adopt and where to start. Logistics 2(18): Multidisciplinary Digital Publishing Institute
11. Akmeemana C (2017) Blockchain takes off, Dagstuhl Reports
12. Adacher L, Flamini M (2020) Optimizing airport land side operations check-In, passengers' migration, and security control processes. J Adv Transport
13. Hahn DA, Munir A, Behzadan V (2019) Security and privacy issues in intelligent transportation systems: classification and challenges. IEEE Intell Transp Syst 55
14. Clementi MD, Kaafar M, Larrieu N, Asghar H, Lochin E (2019) When air traffic management meets blockchain technology: a blockchain-based concept for securing the sharing of flight data
15. Reisman RJ (2019) Air traffic management blockchain infrastructure for security, authentication and privacy
16. Andrej S (2018) Blockchain digital transformation and the law: what can we learn from the recent deals?. CBS Maritime Law Seminar Series on March 22
17. Aleshi A, Seker R, Babiceanu RF (2020) Blockchain model for enhancing aircraft maintenance records security. IEEE
18. Di Vaio A, Varriale L (2020) Blockchain technology in supply chain manage ment for sustainable performance: evidence from the airport industry. Int J Info Manage 52
19. Ranjan S, Negi A, Jain H, Pal B, Agrawal H (2019) Network system design using hyperledger fabric: permissioned blockchain framework. In: 2019 Twelfth international conference on contemporary computing (IC3), IEEE pp 1–6
20. AÍvarez-Díaz N, Herrera-Joancomartí J, Caballero-Gil P, Smart contracts based on blockchain for logistics management. In: Proceedings of the 1st international conference on internet of things and machine learning

21. Schiller, Jerzy E, Niya SR, Timo S, Burkhard S (2019) Blockchain: a distributed solution to automotive security and privacy. IEEE 55
22. Butt TA, Iqbal R, Salah K, Aloqaily M, Jararweh Y (2019) Privacy management in social internet of vehicles: review, challenges and blockchain based solutions. IEEE Access 7
23. Lu Y (2018) Blockchain and the related issues: a review of current research topics. J Manage Analy IEEE 5
24. Valenta M, Sandner P (2017) Comparison of ethereum, hyperledger fabric and corda
25. Bonomo IS, Barbosa IR, Monteiro L, Bassetto C, DeBarros BA, Borges VRP, Weigang L (2018) Development of swim registry for air traffic management with the blockchain support. In: 2018 21st International conference on intelligent transportation systems (ITSC)
26. Duong T, Todi KK, Chaudhary U, Truong H-L (2019) Decentralizing Air traffic Flow management with blockchain-based reinforcement learning. In: 2019 IEEE 17th International conference on industrial informatics (IN-DIN)

Chapter 28
Vehicular Networks Applications Based on Blockchain Framework

Mena Safwat, Ali Elgammal, Wael Badawy, and Marianne A. Azer

1 Introduction

Automotive research and development sector works on vehicles to be connected and interacted directly with each other and with the road infrastructure. Vechile to everything connectivity introduces substantial benefits in terms of safety, real-time information sharing and traffic efficiency. Today, our vehicles are already equipped with numerous number of sensors to collect the information about the vehicle and its surroundings [1].

Vehicular ad hoc network (VANET) is a special type of mobile ad hoc network. VANETs contain some fixed infrastructures and sensors, where each vehicle acts as mobile node that can carry and relay data. In road vehicles, exchange messages with each other and with roadside units (RSUs); they communicate directly without intermediate node(s) or indirectly over the intermediate node(s). Each vehicle communicates with the neighboring RSUs to inform them about its own information including speed, location and heading. Also, vehicle gets traffic conditions/status of a specific road [2].

M. Safwat (✉) · A. Elgammal
Valeo, Giza, Egypt
e-mail: me.safwat@nu.edu.eg

A. Elgammal
e-mail: a.elgammal@nu.edu.eg

M. Safwat · A. Elgammal · W. Badawy · M. A. Azer
School of Information Technology and Computer Science, Nile University, Giza, Egypt
e-mail: wbadawy@nu.edu.eg

M. A. Azer
e-mail: mazer@nu.edu.eg

M. A. Azer
National Telecommunication Institute, Giza, Egypt

A. E. Hassanien et al. (eds.), *Enabling Machine Learning Applications in Data Science*, Algorithms for Intelligent Systems, https://doi.org/10.1007/978-981-33-6129-4_28

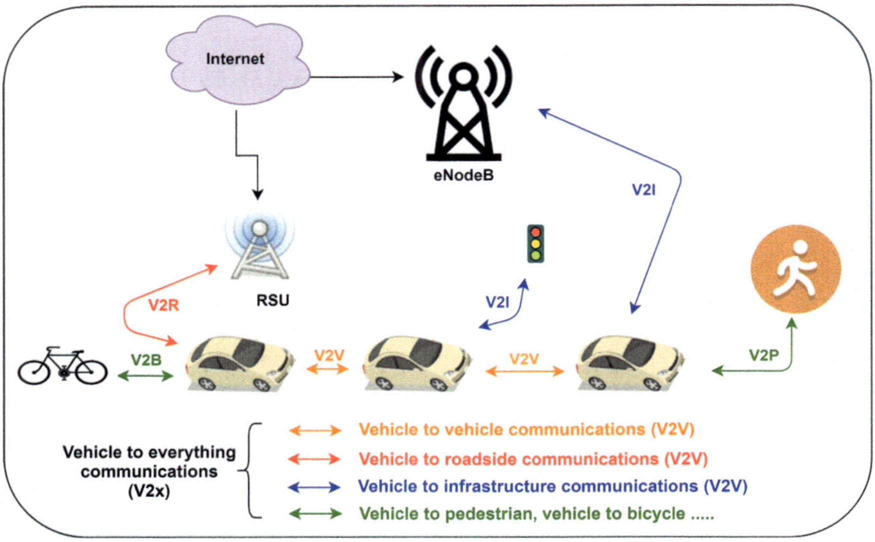

Fig. 1 VANET landscape

Mobility is one of the VANETs special characteristics, as it is very challenging to establish and maintain end-to-end connections in a vehicle ad hoc network (VANET) as a result of vehicle high speed and varied distance between vehicles. Vehicular communication networks include specific approaches, such as Vehicle-to-Everything (V2X), Vehicle-to-Vehicle (V2V), Vehicle-to-Pedestrian (V2P), Vehicle-to-Device (V2D), Vehicle-to-Home (V2H), Vehicle-to-Grid (V2G) and Vehicle-to-Infrastructure (V2I) as shown in Fig. 1 [3].

The second technology discussed in our paper is the Blockchain. A Blockchain could be defined as a shared database among its nodes and allows them to transact their valuable assets in a public and anonymous setup without the reliance on an intermediary or central authority. A Blockchain from implementation aspect could be described as a composition of a decentralized consensus mechanism, a distributed database and cryptographic algorithms. More specifically, transactional data are stored in a potentially infinite sequence of cryptographically interconnected data blocks.

In addition to decentralization, there are many advantages of Blockchain-based systems, such as the provision of a transparent, complete and intrinsically valid historical transaction log and the absence of a central point of (potential) failure. These characteristics facilitate cost-efficient microtransactions, save times, reduce the risks and increase the trust of writing contracts. The first mainstream used application for Blockchain system is Bitcoin cryptocurrency [4]. In this article, authors survey the proposed deployment architectures of Blockchain in VANETs networks and then the applications and the challenges to be mitigated after integrating Blockchain in vehicular network.

The remainder of this paper is organized as follows. Section 2 illustrates Blockchain architecture and example for the deployment in vehicular networks background required information. Section 3 demonstrates and lists the applications of Blockchain in VANETs. Finally, conclusions and future work are present in Sect. 4.

2 Blockchain Background Information

The key concepts necessary to better understand the issues covered by this survey are presented in this section. The background information focuses on the Blockchain technologies and the proposed Blockchain architectures in VANETs.

2.1 Overview of Blockchain

The Blockchain is actually one of the distributed ledger systems with the most research. It is built on a network called peer to peer. This network is used for maintaining a consistent database among all the Blockchain nodes. An ordered list of chained blocks containing transactions is stored in the Blockchain ledger "DataBase": currency, data, certificate, etc. All nodes of the network have the same copy of Blockchain ledger, and without the consent of most network members (51%), this ledger cannot be updated [5].

A consensus in the Blockchain is a mechanism tolerant to faults, which is used to achieve the requisite agreement about a single network state. It is a set of rules that decide upon the contributions of the Blockchain's various participating nodes to solve the Byzantine general problem or double spending problem. The first consensus algorithm was the Proof-of-Work (PoW) algorithm developed by the developers of Bitcoin [4]. This algorithm presents certain limitations (energy efficiency, 51% attack). Other algorithms were developed to deal with this: Proof-of-Stake (PoS), Proof-of-Authority (POA), etc.

The basic features of the Blockchain are listed as follows [6]:

(1) Immutability: When recording and confirming a piece of information, then it cannot be modified or deleted in the Blockchain networking. Neither can information be added arbitrarily.

(2) Privacy and anonymity: The Blockchain affords users anonymity. A user can anonymously join the network. The information about the user cannot be known by other users. It means the personal information is confidential, safe and private.

(3) Distributed and trustless environment: Any node that can be connected to the Blockchain, without a trusted central authority, can synchronize and verify all the Blockchain's content in a distributed way. It provides security and prevents a single point of failure.

(4) Reliable and accurate data: The data in the Blockchain are accurate, reliable, consistent, and timely mainly due to the decentralized network. Blockchain network can stand malicious attacks and hasn't a single fault point.
(5) Faster transactions: Setting up a Blockchain is very simple, and the transactions are confirmed very quickly. Processing the transactions or events requires only a few seconds to a few minutes.
(6) Transparency: It is fully transparent, since it stores details of each single transaction or event that takes place within a Blockchain network. The transactions can be interpreted transparently by anyone in the network.

Because of these features, the Blockchain may be an approach to dealing with the limitations of vehicle networks.

2.2 *Blockchain Scheme in VANET*

Shrestha [5] has proposed a new type of Blockchain to use the definition of an immutable public distributed database for security purposes dissemination of messages in the VANET. However, the essence of the new proposal is different from the Bitcoin, as we are concerned with event notifications, not transactions of cryptocurrency. Worldwide, there are millions of cars, so if every country handles a Blockchain separately, then there will be less scalability things relative to the global Blockchain.

The components of the proposed Blockchain study described in Fig. 2 and identified as follows:

(1) Vehicles: There are two types of nodes in the vehicle, that are, normal node and full node. The normal nodes help in the generation of messages during accidents, and the transmission and verification of messages received. And other nodes are full nodes and have a high level of confidence and a powerful computing capacity responsible for mining the blocks.
(2) Roadside Unit (RSU): The RSUs are responsible for authentication and are used for V2I communication and for supplying vehicles within its contact

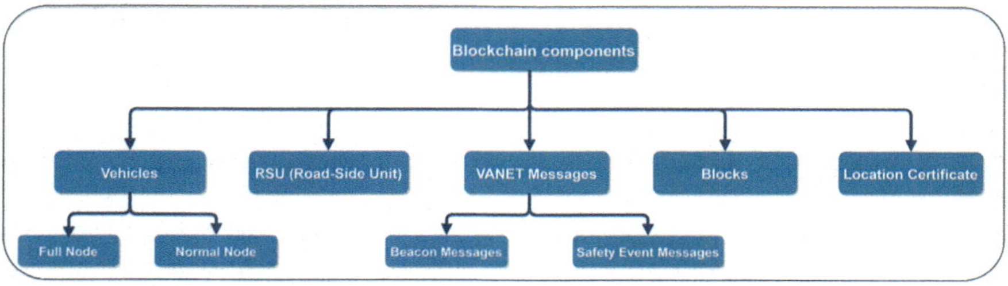

Fig. 2 Proposed blockchain components

range with a location certificate. The legitimate RSU generates a genesis block based on the local events that happen.

(3) VANET messages: The VANET contains basically two types of messages. They are safety event messages and beacon messages. The safety event messages are broadcasted when there are critical road events such as traffic accidents and road hazards. Beacon signals are periodically transmitted to warn adjacent vehicles about the driving status and locations of vehicles in order to gain mutual information for traffic management among other vehicular nodes on the route.

(4) Blocks: A block is a block header and a block body. The block header contains the previous block's hash, difficulty target, nonce, Merkle root and timestamp. The block body consists of a list of safety event messages that behave as transactions in the block body. The block body includes a list of security event messages which serve as block body transactions.

(5) Location certificate: A proof of location (PoL)-based location certificate is used to provide proof of a vehicle's location at a given time. Each vehicle needs the PoL to check that the vehicle is near the location of the incident [7]. For an event message that assists in the Blockchain, the PoL is used as a location proof too. The RSU serves as a validator to give the vehicles within its communication range a location certificate. The GPS is not usable as proof of a vehicle's location because it can be spoofed easily [8]. The PoL is secure because without the valid RSU signature, the vehicles can not create a fake location certificate.

The Blockchain scheme suggested for secure message propagation is shown in Fig. 3. The Blockchain is downloaded and updated for all vehicles in the network. In this proposal, the Blockchain functions as a centralized public ledger, which along with incident notifications, stores the full history of the confidence rates of vehicles in the database. The vehicle that witnesses an event, such as a crash, must broadcast the event message of the incident to nearby vehicles in the Blockchain network with many parameters. When a new event message is received by other vehicles, they first verify that it is in the same area, based on the LC contained in the event message [9]. Then, the neighboring vehicles check other event message parameters. To prevent spamming, denial-of-service and other nuisance attacks against the system, each vehicle independently checks each event message before propagating it further.

The mining vehicle uses policy on message verification to know the trustworthiness message as shown below: (1) Check the prior trust level of the sender vehicle from the main Blockchain. (2) Verify PoL based on certificate of location. (3) Assess if the knowledge is firsthand. (4) Time stamp check.

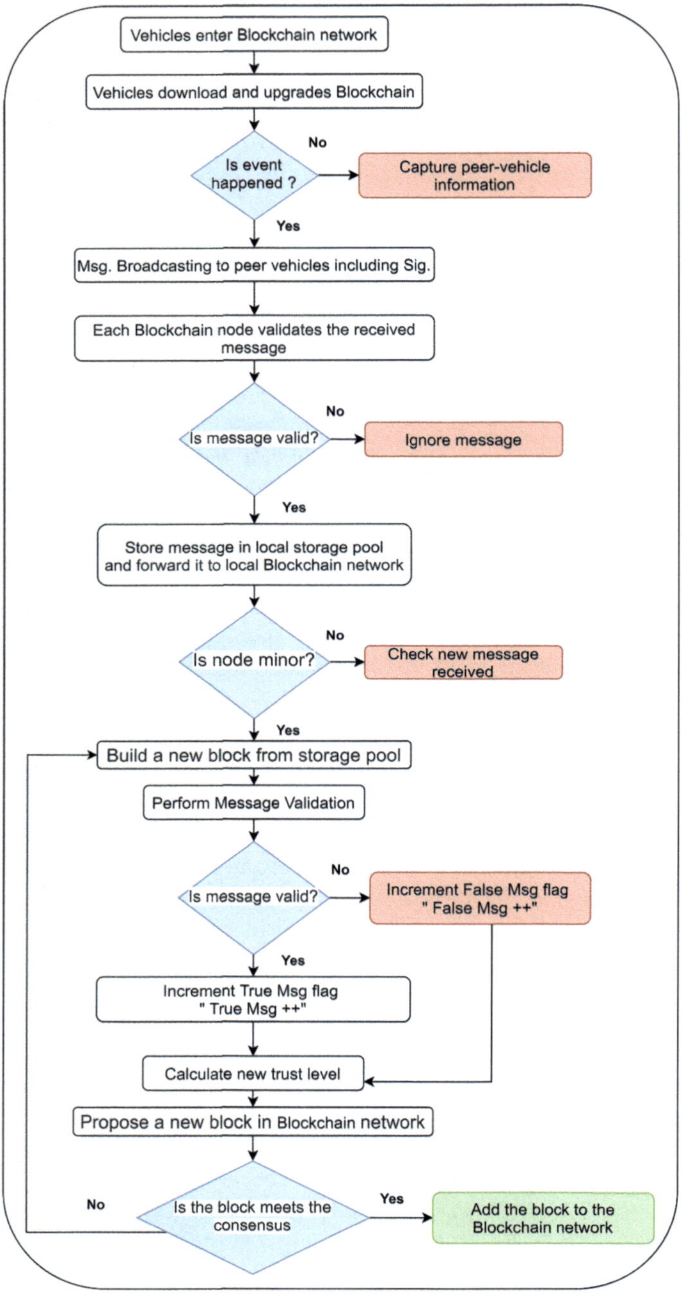

Fig. 3 Proposed blockchain schema in VANETs

3 Blockchain Platform Application in VANETS

Many ICN researchers have started to explore the application technology solutions of Blockchain networks. We will briefly present and review existing vehicular applications over Blockchain in this section.

3.1 *Electronic Payment Schemes*

Deng [10] proposes two electronic payment systems "V-R transaction and V-Rs transaction" for VANETs. All transactions are performed automatically via smart contracts, based on Blockchain technology. Only RSUs engage in the consensus process, and vehicles can access data through RSU, which guarantees the rapid synchronization of data deposited in the Blockchain across all individuals.

Transaction V-R refers to execution of transaction with one car and one RSU, the vehicle being the payer unit and the RSU is the payee unit. A park toll collection program is illustrated as a "V-R transaction" example. When the vehicle is able to reach the car park, it gets WAVE service advertisement (WSA) on the parking application page that requires the service channel (SCH), to be used by RSU on the control channel (CCH). The vehicle then switches to the designated SCH and communicates with RSU to submit a request for parking and obtains the corresponding facility. The vehicle sends out requests for departure to RSU after exiting the car park. Upon completion of the contract, the vehicle charges to RSU and collects the receipt as shown in Fig. 4 for transaction initiation procedures and Fig. 5 for transaction confirmation procedures.

Transaction V-Rs refers to the transaction being carried out between one vehicle and multiple RSUs. By using an electronic toll collection (ETC) as V-Rs transaction example, the driver gets the ETC service from the RSU1 at the CCH as they get ready to reach the highway. The vehicle then switches to SCH and interacts with RSU1 to submit requests for entry. When the vehicle leaves the highway, the vehicle likewise communicates with the RSU2 and sends a request for departure. While the vehicle charges and receives the receipt from its ETC account.

3.2 *Flashing Over the Air (FOTA)*

In the automotive industry, vehicle software is growing exceptionally. Today, vehicles software becomes a significant part. In case of the software update is required and high warranty is mandated, Original Equipment Manufacture (OEM) request a recall for all same model vehicles to the service center. FOTA comes to the commercial automotive industry to decrease the software update time and cost. Also, FOTA will increase the OEM credibility with customer.

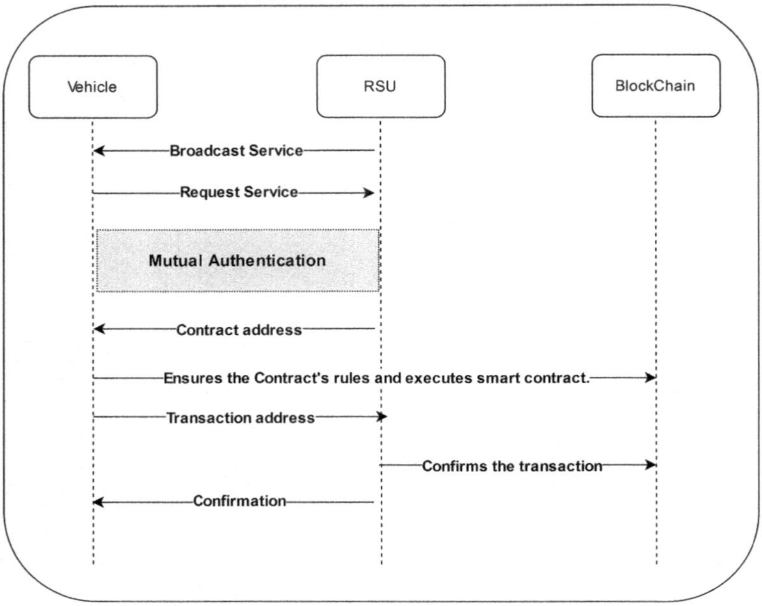

Fig. 4 Transaction initiation procedures

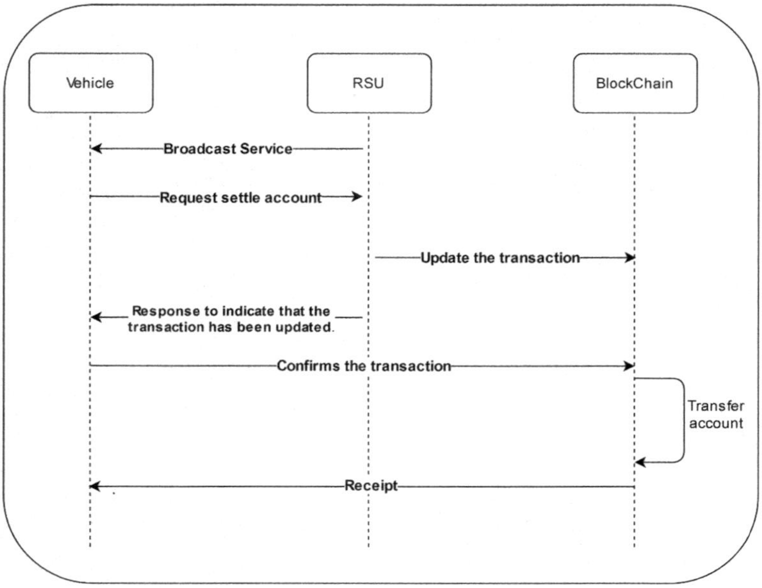

Fig. 5 Transaction confirmation procedures

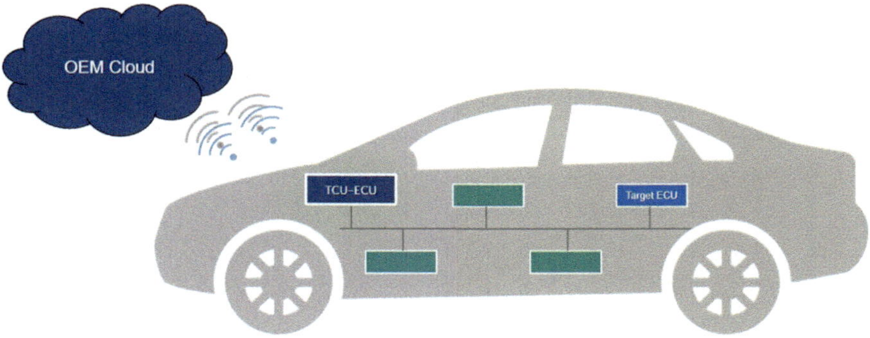

Fig. 6 Flashing over the air

Since automotive industry FOTA updates have a high potential and considered impact, the research community has highly attracted to this aspect. Researchers propose numerous security vulnerabilities and issues. This redirect research is to overcome these vulnerabilities and create a trust FOTA updates for vehicles.

Flashing over the air (FOTA) updating approach is an appropriate choice to update the ECUs software over wireless network. FOTA does not require any human interface as the software occurs while the vehicle is running and remotely.

OEM server announces an updated software version is released to all the concerned vehicle models, and every OEM uses cloud storage to store new software releases so that the app can be downloaded by its customers. Once the TCU "the only unit inside the vehicle which has an IP and connected to the network" requests the newer software release. Then, TCU forwards the received software data to the targeted ECU which is connected thought the internal communication bus as illustrated in Fig. 6.

Steger [11] proposes a proof-of-concept for the vehicle software update over the air implementation based on Blockchain.

The entire method of software upgrading is shown in Fig. 7 and listed below.

Vehicle creates a genesis transaction as an initial transaction required to participate in the Blockchain network. The software provider, who may be a particular OEM department or a manufacturer supplying the embedded applications to the ECU, releases a new software version. The software provider forwards update transaction request to OEM publisher for sharing a new software version, and then, software provider sends a new software version to OEM publisher. The OEM publisher signs the software binary file using his private key upon receiving binary file. The OEM publisher divides the software binary file to small blocks to match the Blockchain block size which is 1 megabit per block, and then, the OEM publisher sends the software data blocks to all the concerned vehicle models. Finally, all concerned vehicle models verify the transaction with the OEM publisher [12].

Blockchain adds to FOTA mandatory required specification 1—decentralized consensus and confirmation for data blocks with the participation of all network

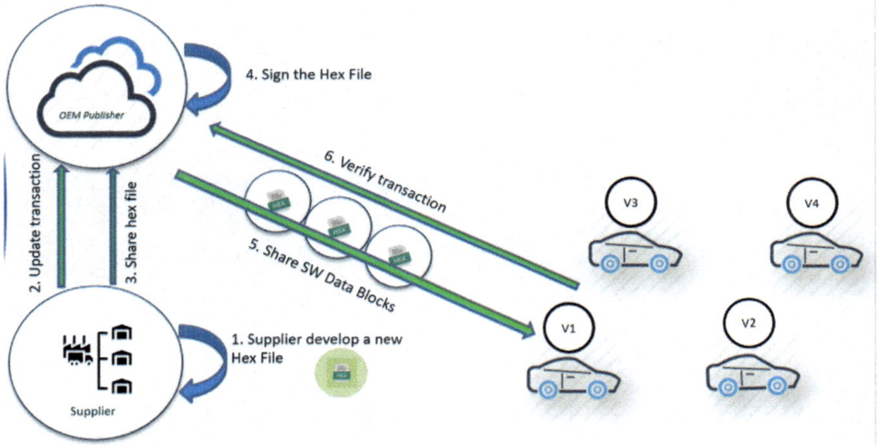

Fig. 7 Flashing over the air based on blockchain

node 2—privacy with utilized manner of public key (PK) exchanging 3—security from consensus algorithm and appending the blocks to the linked chain of blocks.

3.3 *Car-Sharing Services*

Car-sharing providers are rapidly increasing. The provision of such highly distributed services requires a secure and reliable connection between smart vehicles, car-sharing service providers and service users. A reliable contact channel is necessary to safely share data including vehicle location, auto unlock keys and user payment information.

Blockchain's decentralized nature is tailored to these highly distributed services, which include providing the car's location to a user, handling the interconnection between the user and the car (i.e., unlocking and using the car) and billing/payment upon use of the ride-sharing service.

In addition, Blockchain safely interconnects the individuals involved while preserving the user's privacy (e.g., no connection between the user's actual identity and a certain route driven) and the vehicle from unauthorized access (i.e., only registered and approved users are able to find, unlock and use a car) [13].

3.4 *Electric and Smart Charging Vehicle Services*

Electric vehicles (EVs) become an essential demand to reducing greenhouse CO_2 gas emissions and in facilitating the electric mobility by using the high penetration of renewable energy sources. A connection between EVs and everything else such as owner smart home and mobile devices could be utilized within several sophisticated services. Some of these services could be listed as (1) Tracking and personalizing the vehicle owner travel/movement rate, and by modulating these dates, we could get the needed information and accordingly guarantee that vehicle if fully charged before the owner trip. (2) Ensure the cheapest and most efficient charging cycle and avoid the high loaded periods.

These services acquire a communication between roadside units (RSU), electric supply stations, IoT devices and vehicles. Also, all aforementioned nodes have a shared data for tracking. Blockchain is proposed to handle the security, shared electric charging data integrity and immutability in a permissioned ledger. Accordingly, smart vehicles electric services could be enriched [13].

3.5 *Forensics Applications of Connected Vehicles*

Forensics of vehicles has become a vital aspect of the construction and operating life cycle of a vehicle. Involved stakeholders include inspectors from insurance companies and law enforcement departments involved in prosecuting accidents and accident events. In recent years, insurance providers and companies which provide their employees with vehicles for business-related activities have also used the forensic capability [14].

Capabilities such as gathering data inside and outside vehicles may have a significant effect on automotive forensics, which helps to determine the causes behind the incidents. With the development of self-driving cars, that are vulnerable to failures and cyber attacks, this domain will become even more essential [15].

The investigation into the accident in the current VANETs architecture lacks some elements which are completely required for the effective resolving the dispute. These can be identified as follows:

- The collected evidence does not provide an detailed vehicle history due to insufficient memory.
- There is no procedure for all stakeholders to integrate records, including other vehicles, road conditions, suppliers and services centers.

Block for forensic platform, a Blockchain-based automotive forensics network gathers components of the vehicles and associated business under the same umbrella [16]. In particular, the proposed system:

- Provides a lightweight privacy-aware Blockchain by capturing all involved parties such as drivers, service centers, manufacturers and police agencies without needing a trustworthy third party in the case of an accident.
- Establishes a system for vehicle forensics analysis that includes all the details required for a comprehensive and detailed vehicle forensics solution

4 Conclusions and Future Work

In automotive industry, vehicles software is increasing with the control of the performance of critical autonomous driving systems such as acceleration, auto-parking braking and steering. This software must perform the functionality and meet a hard-real-time tasks with a very tight timing constraints and handle spurious data in a safe manner.

Software also validates extensively all sensors input data to ensure the reliability and robustness. Once the vehicles start the communication between each other and with every surrounding nodes, this open and public network will introduce numerous security challenges. Researchers proposed already exist platform (Blockchain) to overcome many of these challenges through building the distributed ledger and consensus and approval from all nodes to be added to linked chain of data blocks.

In this paper, we surveyed the introduced vehicular ad hoc networks (VANETs) applications based on Blockchain technology for smart connect vehicles.

For the future work, we summarize the directions for future research:

- **Mobility**: Frequent vehicle mobility increases the packet and the overhead processing arising from the handing over process. New mobility approaches to reduce the overhead can be implemented.
- **New FOTA platform**: We plan to propose an implementation for flashing over the air (FOTA) platform based on Blockchain. This platform provides enhanced flashing time including the secure, private and immutable features based on Blockchain technology considering the mobility factor.
- **Caching data**: Every connected vehicle must access data from a cloud (e.g., software update), which includes overhead packets and overlay delays. The implementation of OEMs caching will reduce these overheads.
- **Key management**: May vehicle holds several keys to connect with RSU or other vehicles, keys could be changed over the vehicle lifetime. Managing keys raises a new threat to research.

Acknowledgements The authors would like to acknowledge Valeo Interbranch Automotive Software, Egypt, and Nile university research center.

References

1. Massaro E et al (2016) The car as an ambient sensing platform [point of view]. Proc IEEE 105(1):3–7
2. Othmane LB et al (2015) A survey of security and privacy in connected vehicles. In: Wireless sensor and mobile ad-hoc networks. Springer, Berlin, pp 217–247
3. Amadeo M, Campolo C, Molinaro A (2016) Informationcentric networking for connected vehicles: a survey and future perspectives. IEEE Commun Mag 54(2):98–104
4. Nakamoto S (2019) Bitcoin: A peer-to-peer electronic cash system. Tech. rep. Manubot

5. Shrestha R et al (2020) A new type of blockchain for secure message exchange in VANET. Digital Commun Networks 6(2):177–186
6. Mendiboure L, Chalouf MA, Krief F (2020) Survey on blockchain-based applications in internet of vehicles. Comput Electr Eng 84:106646
7. Dasu T, Kanza Y, Srivastava D (2018) Unchain your blockchain. In: Proceedings of the symposium on foundations and applications of blockchain, vol 1, pp 16–23
8. Tippenhauer NO et al (2011) On the requirements for successful GPS spoofing attacks. In: Proceedings of the 18th ACM conference on computer and communications security, pp 75–86
9. Shrestha R, Bajracharya R, Nam SY (2018) Blockchain based message dissemination in VANET. In: 2018 IEEE 3rd international conference on computing, communication and security (ICCCS). IEEE, New York, pp 161–166
10. Deng X, Gao T (2020) Electronic payment schemes based on blockchain in VANETs. IEEE Access 8:38296–38303
11. Steger M et al (2018) Secure wireless automotive software updates using blockchains: A proof of concept. In: Advanced microsystems for automotive applications 2017. Springer, Berlin, pp 137–149
12. Lee B, Lee J-H (2017) Blockchain-based secure firmware update for embedded devices in an Internet of Things environment. J Supercomput 73(3):1152–1167
13. Dorri A et al (2017) Blockchain: a distributed solution to automotive security and privacy. In: IEEE communications magazine, vol 55(12), pp 119–125
14. Mansor H et al (2016) Log your car: the non-invasive vehicle forensics. In: 2016 IEEE Trustcom/BigDataSE/ISPA. IEEE, New York, pp 974–982
15. Baig ZA et al (2017) Future challenges for smart cities: cyber-security and digital forensics. Digit Investig 22:3–13
16. Cebe M et al (2018) Block4forensic: an integrated lightweight blockchain framework for forensics applications of connected vehicles. IEEE Commun Mag 56(10):50–57

Author Index

A. E. Hassanien et al. (eds.), *Enabling Machine Learning Applications in Data Science*, Algorithms for Intelligent Systems, https://doi.org/10.1007/978-981-33-6129-4